The National Parks of Indonesia

Jatna Supriatna • Chris Margules

The National Parks of Indonesia

Foreword by
Prof. Dr. Ir. Siti Nurbaya, M.Sc.
Minister of Environment and Forestry,
The Republic of Indonesia

Supported By

Jatna Supriatna
Department of Biology
University of Indonesia
Depok, Indonesia

Chris Margules
Centre for Tropical Environmental and Sustainability Science
James Cook University
Cairns, QLD, Australia

Jointly published with Yayasan Pustaka Obor Indonesia
ISBN of the Co-Publisher's edition: 978-623-321-175-8
The edition is not for sale in Indonesia. Customers from Indonesia please order the print book from OBOR Books.

ISBN 978-3-031-76637-4   ISBN 978-3-031-76638-1 (eBook)
https://doi.org/10.1007/978-3-031-76638-1

© The Editor(s) (if applicable) and The Author(s), under exclusive license to Springer Nature Switzerland AG 2022, 2025

This work is subject to copyright. All rights are solely and exclusively licensed by the Publisher, whether the whole or part of the material is concerned, specifically the rights of reprinting, reuse of illustrations, recitation, broadcasting, reproduction on microfilms or in any other physical way, and transmission or information storage and retrieval, electronic adaptation, computer software, or by similar or dissimilar methodology now known or hereafter developed.
The use of general descriptive names, registered names, trademarks, service marks, etc. in this publication does not imply, even in the absence of a specific statement, that such names are exempt from the relevant protective laws and regulations and therefore free for general use.
The publishers, the authors, and the editors are safe to assume that the advice and information in this book are believed to be true and accurate at the date of publication. Neither the publishers nor the authors or the editors give a warranty, express or implied, with respect to the material contained herein or for any errors or omissions that may have been made. The publishers remain neutral with regard to jurisdictional claims in published maps and institutional affiliations.

Cover illustration: Book Cover created already by OBOR

This Springer imprint is published by the registered company Springer Nature Switzerland AG
The registered company address is: Gewerbestrasse 11, 6330 Cham, Switzerland

If disposing of this product, please recycle the paper.

TABLE OF CONTENTS

Foreword, *Prof. Dr.Ir. Siti Nurbaya, M.Sc*		viii
Preface, *Jatna Supriatna*		xi
Introduction		xiii
Acknowledgements		xxviii
Profile of Jatna Supriatna and Chris Margules		xxix
Chapter 1.	Sumatra	1
	1. Gunung Leuser National Park	12
	2. Kerinci Seblat National Park	20
	3. Siberut Nasional Park	30
	4. Batang Gadis National Park	36
	5. Bukit Tiga Puluh National Park	42
	6. Tesso Nilo National Park	47
	7. Berbak National Park	52
	8. Bukit Dua Belas National Park	57
	9. Sembilang National Park	60
	10.. Bukit Barisan Selatan National Park	66
	11. Way Kambas National Park	72
	12. Zamrud National Park	78
	13. Gunung Maras National Park	84
Chapter 2.	Java and Bali	87
	14. Kepulauan Seribu Marine National Park	93
	15. Ujung Kulon National Park	100
	16. Gunung Gede Pangrango National Park	109
	17. Gunung Halimun Salak National Park	118
	18. Karimun Jawa Islands National Park	125

	19. Gunung Ciremai National Park	130
	20. Gunung Merapi National Park	134
	21. Gunung Merbabu National Park	139
	22. Gunung Bromo Semeru Tengger National Park	145
	23. Alas Purwo National Park	152
	24. Meru Betiri National Park	158
	25. Baluran National Park	164
	26. Bali Barat National Park	171
Chapter 3.	Kalimantan	178
	27. Gunung Palung National Park	182
	28. Bukit Baka Bukit Raya National Park	188
	29. Danau Sentarum National Park	193
	30. Betung Kerihun National Park	198
	31. Tanjung Puting National Park	203
	32. Sebangau National Park	209
	33. Kutai National Park	214
	34. Kayan Mentarang National Park	222
Chapter 4.	Nusa Tenggara	227
	35. Gunung Rinjani National Park	231
	36. Komodo National Park	238
	37. Kelimutu National Park	245
	38. Laiwangi-Wanggameti National Park	249
	39. Manupeu-Tanah Daru National Park	253
	40. Gunung Tambora National Park	257
	41. Moyo Satonda National Park	264
Chapter 5.	Sulawesi	269
	42. Bunaken National Park	273
	43. Bogani Nani Wartabone National Park	280
	44. Lore Lindu National Park	286
	45. Taka Bonerate National Park	292
	46. Rawa Aopa Watumohai National Park	297
	47. Wakatobi Islands National Park	303

	48. Bantimurung Bulusaraung National Park	308
	49. Togean Island National Park	313
	50. Gandang Dewata National Park	318
Chapter 6.	Maluku (Molucca)	321
	51. Manusela National Park	324
	52. Aketajawe-Lolobata National Park	331
Chapter 7.	Papua	336
	53. Cenderawasih Bay Marine National Park	341
	54. Lorentz National Park	346
	55. Wasur National Park	353
References		358
Index		366

**MINISTER OF ENVIRONMENT AND FORESTRY
THE REPUBLIC OF INDONESIA**

FOREWORD

Indonesia harbors many of the most biologically rich and diverse ecosystems on Earth. It spans over 17,500 islands strung out along the equator, including some of the largest in the world. Indonesia is a megadiverse country with at least 700 mammal species, 280 of which are endemic, at least 411 species of reptile, 150 are endemic, at least 270 amphibians, 100 are endemic, and at least 1788 species, of which 471 are endemic. There are also 30-35,000 plant species in Indonesia. Our waters hold the richest marine biodiversity in the world, including more than 80,000 km^2 of coral reefs, 14 % of the Earth's coral reefs containing more than two-thirds of the world's hard coral species.

As part of its commitment to the Convention on Biological Diversity, Indonesia has protected more than 27 m ha (22.147 m ha of its terrestrial area and 5.0 m ha of our marine area) which includes 55 State or National parks. The 55 National (State) Parks currently declared span the length and breadth of the country. National (State) Parks comprise a multitude of scenic wonders —mountains, lakes, rivers, deserts and sea-coast; abundant and diverse wildlife; and profusions of wild flowers; display the ancient imprints of volcanoes; provide for many of Indonesia's charismatic species. As a home for flagship species such as Sumateran orangutan, tiger, rhinos and elephants, as well as Tapanuli orangutan, Borneo orangutan and Javan rhino, Indonesia is second to no other country in the world. National parks are a primary contributor to that protection. They also provide habitat for most of Indonesia's 61 primate species, 38 of which are endemic, and contribute significantly to the protection of all other species. Our marine national parks cover large tracts of coral reefs and seagrass meadows. Yet few are known beyond their immediate areas. This book introduces you of the wildest and most scenic of the parks, as well as ones that stand out for their cultural or historical significance.

Indonesia's National Parks also protect our awe-inspiring landscapes and seascapes, from coral reefs and endless white sandy beaches to vast tracts of mangroves to lowland and montane rainforests to alpine valleys an equatorial glacier in Papua.

Protecting biodiversity and landscapes and seascapes is a significant role of National Parks. Still, they also provide many other benefits, such as helping to achieve the UN's sustainable development goals, securing vital ecosystem services, supporting local livelihoods, and supporting humans and nature alike to adapt to the impacts of climate change.

Without National Parks, achieving UN Development Goals 14 Life below Water and 15 Life on Land would be challenging. National parks also support many other goals such as Goal 3, Good Health and Well-Being, Goal 6, Clean Water and Sanitation, Goal 12, Responsible Consumption and Production, and Goal 13, Climate Action.

Particular ecosystem service is the protection of catchments areas, watershed protection as water resources for fresh drinking water for people and nearby cities. More over the discovery of beneficial new compounds and the commercial development of products derived from biodiversity as we called bio-prospecting. The Indonesian domestic herbal medicines ("jamu") industry is strong, and there is an increasing demand in the developed world for natural products. This represents a potential source of revenue for local communities, whose knowledge of traditional medicine can point to suitable species. The recent and very populer is forest healing and eco-tourism for leissure-time, which is also gaining both domestic and international contributing increasingly to our foreign exchange revenue. Local and adat community should benefit from the parks they live beside.

The most considerable contribution from the National Park is the value of carbon either in the forest and coastal area. Indonesia has introduced new rules on carbon pricing to set up a market mechanism to help achieve the country's greenhouse gases (GHG) emission reduction target by 2030. H.E. President Joko Widodo signed the "Economic Value of Carbon" Regulation No 98/2021 ahead of the COP26 in Glasgow in 2021. Moreover the operation of GHG emission reduction form Forest and Land Use is guided by "Indonesia FoLU netsink 2030", which is legally binded by Ministerial Decree No 168/Menlhk/PKTL/PLA.1/2022 issued in February 2022 for plan of operation of FoLU and by Ministerial Regulation No 21 year 2022, for Technical Procedures on Carbon Pricing, issued in October 2022, prior to COP 27 in Sharm el Sheikh in 2022.

This publication will enhance awareness on the values of national parks and provide necessary information about the parks, hence encouraging more people to visit and enjoy our National (State) Parks. I do hope this book will also increase our knowledge as well as promoting the importance of other existing conservation areas in Indonesia.

Minister of Environment and Forestry
Government of Indonesia,

Prof. Dr.Ir. Siti Nurbaya, M.Sc

PREFACE

I never tire of the beauty of the natural landscapes of Indonesia, the mountains, jungles, gardens, seas and the people who inhabit them. Thousands of islands, both small and large lying between the two big oceans, the Pacific and the Indian, are like pearls on the equator. Looking at the map, those islands appear as the green of the forest against the blue of the oceans. Many scientists since the early 17th century have come to explore the biodiversity and landscapes of Indonesia and their legacy can be seen in the many scientific articles that they have produced.

However, my greatest admiration is reserved for Sir Alfred Russell Wallace, the co- discoverer, with Charles Darwin, of the theory of evolution. He captured my attention in junior high school and stayed with me throughout my doctoral studies, the field work for which was conducted in the same general area where he had been working more than a hundred years ago.

My earliest years were spent in the small towns of Banjar and Tasikmalaya in the heart of West Java surrounded by the beauty of terraced rice fields, volcanos, rivers and the coastline. My father took me to many fascinating places for picnics when he was not on duty as a military man. Even in my childhood my dream was to climb all those mountains that surrounded me and to sail the ocean off the beach I played at not far from my house. Thus, my curiosity about nature was developed, which in turn led to my career in biology and ecology and my concern for protecting biodiversity.

My love of nature was reinforced after staying six months in the jungles of Borneo studying leaf-eating monkeys at the Orangutan Research Center at Tanjung Putting National Park in Central Kalimantan. My mentor, Prof. Dr. Birute Galdikas, inspired me to study more and to get involved in conservation after finishing my undergraduate degree. During my three years working at the National Biological Institute of the Indonesian Institute of Science (LIPI) I travelled to many forested areas in Sumatra, Kalimantan, Sulawesi and other smaller islands with and supervised by my mentor

Dr. Kuswata Kartawinata and later on Prof. Soekarja Somadikarta. I started writing about nature and wildlife. My first book, called "Poisonous snakes of Indonesia" was published by Brathara Karya Aksara in 1981. Once I became a lecturer at the University of Indonesia, I decided that my profession would be conservation biology and that I would tackle the problems of protecting endangered animals and understanding the ecology of their habitats and the role of the national park system in protecting them.

I wrote a book on tourism in Indonesia's national parks in 2016, published by Obor Foundation. When it became time to revise and update this book, and write it in English, Chris Margules agreed to help me. Chris has wide-ranging experience in Indonesia, with both its biodiversity and human community development issues. Chris and I worked together at Conservation International. We both travelled to many national parks in Indonesia, from Sumatra to Papua. Since 1995, Conservation International's Indonesia Program has supported the Ministry of Forestry, which at the time of writing was called the Ministry of Environment and Forestry, in many activities related to national parks in Indonesia.

The main reason for writing this book was to share our experiences travelling throughout Indonesia and studying its biodiversity, ecosystems and community development. A second reason was to explore the potential for the development of eco-tourism associated with national parks. I have written many books and scientific papers on the ecology and biodiversity of various national parks, but it is scattered throughout different publications and not readily accessible to lay readers. Thus, a third reason for writing is the many promises I have made to colleagues from the conservation community in Indonesia and internationally, to let the world know about the ecology and biodiversity of Indonesia and the strength of the commitment of many Indonesians to protect that biodiversity for the benefit of all. Finally, both of us as professors at the University of Indonesia, also want to answer the many questions raised by our graduate students and NGO friends who complain that too few books have been written on Indonesia's parks and their biodiversity.

Depok, Indonesia   Jatna Supriatna

INTRODUCTION

About this book

This book contains information on geographic location and park size, climate, topography, history of each park, and the biodiversity and ecosystems encompassed by them. It also contains information on local communities adjacent to the park and their culture, tourism opportunities, access, tourist facilities nearby and how to find the park office to get a visitor's permit.

All of this information was gathered during our own research, visits to the parks, from the literature, and from direct communication with the Director General KSDAE (Conservation of Natural Resources and Ecosystem) of the Ministry of Environment and Forestry, Mr. Wiratno and his Director of Conservation Areas, Mr Sujatno Sukandar. Both have endorsed this book and facilitated full access to many heads of national parks and their officers, who provided information on park history, maps and zoning systems, biodiversity status, local communities and problems within the parks. We are extremely grateful for this help.

Some general information was taken from the ecology of Indonesia series that has been published in English commencing in the mid-1970s, for example, *Ecology of Sumatra* (Whitten et al., 1987), *Ecology of Sulawesi* (Whitten et al., 1997), *Ecology Java and Bali* (Whitten et al., 2002), *Ecology of Kalimantan* (McKinnon et al., 2000), *Ecology of Nusa Tenggara and Maluku* (Monk et al., 2000) and the last one, published in 2004, the *Ecology of Papua* (Marshall et al., 2004). Many other sources were also used for general information on biodiversity in Indonesia such as *Megadiverse Countries* (Mittermeier et al., 1997) and *The Hotspots*, (Mittermeier et al., 1998), published by Conservation International and CEMEX. Indonesia's *Biodiversity Action Plan* published by Bappenas in 2003 has been cited in many places. Many field guides such as "Field Guide to the Birds of Java and Bali authored by MacKinnon and Ramsey (1988), Frogs (the Amphibians of Java and Bali was written by D.T. Iskandar (1998), Reptiles (Indraneil Das in 1998, Herpetological bibliography of Indonesia), Mammals (Payne et al., in 1985 for Borneo, Kitchener et al. in 1990 of Wild mammals of Lombok island, Nusa Tenggara, Indonesia, and Supriatna in 2019 on *Primates of Indonesia*. More information on biodiversity can be found in other books by Jatna Supriatna, *Primate Tourism in Indonesia* (Yayasan Pustaka Obor Indonesia, 2016), and *the Gunung Leuser National Park* (YABSHI, 1996)

Information on the communities living around each park was gathered from the park websites and newsletters or directly from the park officers or from anthropologists who have worked in the area. Some of the parks cross many community boundaries and even province boundaries. Information on tourism activities and facilities around or close to the parks was gathered during personal visits, from national park officers and from tourism offices at both the central and local government levels as well as from existing literature. Some field guides such as Lonely Planet, magazines and other tourist guides were also consulted but then verified in the field by the authors or with the help of our network of collaborators and friends.

The Creation and Management of National Parks

Management of National Parks in Indonesia is based on law that has been evolving since the Dutch era, culminating in Law no 5, 1990 and the many rules and regulations under that law established more recently. It is also based on the commitment of Indonesia to international treaties that have been signed such as the UN CBD (Convention of Biological Diversity), IUCN (International Union for Conservation of Nature), RAMSAR the convention on wetlands, CITES the convention on international trade in endangered species of wild fauna and flora and others. The most important of these for the management of protected areas is the CBD agreement.

In February 2004 in Kuala Lumpur, the CBD Parties agreed to the most comprehensive and explicit protected area commitments ever made by the international community by adopting the Program of Works for Protected Areas (PoWPA). It is a framework for cooperation between Governments, donors, NGOs and local communities. It is important because without such collaborations, programs cannot be successful and sustainable over the long-term. The PoWPA encourages development of participatory, ecologically representative and effectively managed national and regional systems of protected areas, where necessary stretching across national boundaries. From designation to management, the framework can be considered as a defining framework or "blueprint" for protected areas. To date, there are many good signs of progress and there is much to celebrate. Political will and commitments are clearly being catalyzed. For example, most countries have developed a system of protected area. The protected area network covers about 14.6% of earth's land surface. Globally, 11.5% of marine territorial waters were protected in 2017 (ourworldindata.org/biodiversity-and-wildlife, 2022).

While these are commendable achievements, there are still some areas that are lagging. Consideration of the social costs and benefits, the effective participation of indigenous and local communities and the diversification of various governance types need more commitment and resolute actions. The evaluation and improvement of management effectiveness, and the development and implementation of sustainable finance plans with diversified portfolios of traditional and innovative financial mechanisms both need more effort. Mitigation and adaptation responses to climate change need to be incorporated in management strategies. Strengthening the implementation of PoWPA will require concerted efforts and the combined strength of all sectors of society, as well as alliances at national, regional and international levels between policy makers, civil society, indigenous and local communities and the private sector.

Two ministries are responsible for national parks and other protected areas in Indonesia, the Ministry of Environment and Forestry for terrestrial areas and included some mixed marine and terrestrial parks and the Ministry of Marine and Fisheries for marine areas. Indonesia has surpassed all of its commitments to the CBD, almost doubling the area of terrestrial protected areas. Protected areas are the cornerstone of biodiversity conservation; they maintain key habitats, provide refugia, allow for species migration and movement, and ensure the maintenance of natural processes across the landscape. Not only do protected areas secure biodiversity conservation, but they also help secure the well-being of humanity itself. Protected areas provide livelihoods for nearly 1.1 billion people worldwide, are the primary source of clean drinking water for over a third of the world's largest cities, including Jakarta and are a major factor in ensuring global food security (www.cbd.int/protected/overview). Well managed protected areas with participatory and equitable governance mechanisms yield significant benefits far beyond their boundaries, which can be translated into cumulative advantages across a national economy and contribute to poverty reduction and sustainable development including achievement of the UN Sustainable Development Goals.

As the detrimental impact of climate change threatens the planet, protected areas provide a convenient partial solution to this inconvenient truth. Governments are increasingly likely to consider protected areas as a strategic investment in their national economies. A report estimates that investments in creating and managing protected areas will yield returns in societal benefits in the order of 25:1 to 100:1 (teeweb.org). Governments are also likely to view protected areas as a fundamental strategy to not only conserve biodiversity, but also to achieve some of the UN Sustainable Development Goals, secure vital ecosystem services, support local livelihoods, and support humans and nature alike to adapt to the impacts of climate change (www.cbd.int/protected/overview).

Similarly, national parks can play a role in income generation that is yet to be realized. There has been much discussion about the potential for direct revenue generated from tourism activities or indirect revenue from ecosystem services. The problem that has occurred with Indonesia surpassing its commitment and doubling the area of parks is that the budget to finance park management is only $4.5/ha/year. This is too low even compared to our neighbouring countries Malaysia ($20) and Thailand ($25). Parks will have to move towards generating revenue if they are to survive.

A Brief History of National Parks in Indonesia

Panji has written several books related to the history of Indonesia's national parks especially the six national parks established between 1980 and 1990 (Panji, 2016). Indonesia adopted the national park system in 1980 before hosting a national park conference in Bali in 1981. The first national parks established were in places that had been very popular destinations for tourists such as Gunung Gede Pangrango in West Java province, Baluran in East Java province, Komodo on Komodo island in East Nusa Tenggara Province and Gunung Leuser in North Sumatra and Aceh Provinces. In 1982, Indonesia added more parks including the island chain of Kepulauan Seribu in the Java Sea north of Jakarta, Meru Betiri in East Java, Tanjung Putting in central Kalimantan, Kutai in East Kalimantan, Manusela in Maluku and Bukit Barisan Selatan in South Sumatra and Bengkulu provinces.

Table 1. The History of the Establishment of Indonesia's National Parks and the areas they cover

Decade of decree	Name	Province	Marine (ha)	Terestrial (ha)	Total (ha)
1980-1990	Gunung Gede Pangrango (1980)	West Java	-	15,196	15,196
	Ujung Kulon (1980)	Banten	44,337		122,956
	Baluran (1980)	East Java		25,000	25,000
	Bali Barat	Bali	6,280		19,002
	Komodo (1980)	East Nusa Tenggara	112,500		173,300
	Kutai (1980)	East Kalimantan		198,629	198,629
	Merubetiri	East Jawa		58,000	58,000
	Karimun Jawa	Central Java	111,625		104,562
	Bromo Tengger Semeru	East Java		50,276	50,276
	Way Kambas	Lampung		130,000	130,000
	Bukit Barisan Selatan	Lampung & Bengkulu		365,000	365,000
	Manusela	Maluku (Seram island)		189,000	189,000
	Tanjung Putting	Central Kalimantan		425,040	425,040
	Leuser	North Sumatra & Aceh		1,094,692	1,094,692
	Kepulauan Seribu	Jakarta	106,963		107,489
	Bunaken	North Sulawesi	89,065		89,065
	Lore Lindu	Central Sulawesi		229,000	229,000
	Rawa Aopa	Southeast Sulawesi		105,194	105,194
	Wasur	Papua		413,810	413,810
	Gunung Rinjani (1990)	West Nusa Tenggara (Lombok Island)		40,000	40,000

Period	Name	Location	Col A	Col B	Col C
1991-2000	Bukit Tiga Puluh	Jambi and Riau		127,698	127,698
	Gunung Palung	West Kalimantan		90,000	90,000
	Gunung Salak Halimun	West Java & Banten		113,357	113,357
	Alas Purwo	East Java		43,420	43,420
	Betung Kerihun	West Kalimantan		800,000	800,000
	Bukit Baka Bukit Raya	West and Central Kalimantan		70,500	70,500
	Kayan Mentarang	North Kalimantan		1,360,050	1,360,500
	Kelimutu	East Nusa Tenggara (Flores Island)		5,000	5,000
	Siberut	West Sumatra (Siberut island)		190,500	190,500
	Berbak	Jambi		162,700	162,700
	Danau Sentarum	West Kalimantan		129,700	129,700
	Laiwangi Wanggameti	East Nusa Tenggara (Sumba island)		47,014	47,014
	Manupeu Tanah Daru	East Nusa Ttenggara (Sumba Island)		87,984	87,984
	Lorentz	Papua			2,505,600
	Bogani Nani Wartabone	Gorontalo		287,115	287,115
	Teluk Cendrawasih	West Papua	1,453,500		1,453,500
	Kerinci Seblat	Jambi, Bengkulu and West Sumatra		1,375,349	1,375,349
2000-2010	Sembilang	South Sumatra		205,750	205,750
	Takabonarate	South Sulawesi	530,765		530,765
	Gunung Merapi	Jogjakarta and Central Java		6,410	6,410
	Gunung Merbabu	Central Java		5,725	5,725
	Sebangau	Central Kalimantan		568,700	568,700
	Aketajawe Lolobata	North Maluku		167,300	167,300
	Gunung Ciremai	West Java		15,589	15,589
	Kepulauan Togean	Central Sulawesi	292,000		362,605
	Batang Gadis	North Sumatra			108,000
	Bantimurung Bulusaraung	South Sulawesi		43,750	43,750
	Teso Nilo	Riau		35,576	35,576
2010-2016	Wakatobi island	Southeast Sulawesi	1,390,000		1,390,000
	Gunung Tambora	West Nusa Tenggara (Sumbawa Island)		71,645	71,645
	Zamrud	Riau		39,480	39,480
	Gandang Dewata	West Sulawesi		214,201	214,201
	Gunung Maras	Bangka Belitung (Bangka Island)		16,806,91	16,806,91
2002	Moyo Satonda	West Nusa Tenggara (Moyo and Satonda island)	25,200.15	6,000	31,200.15

The National Parks of Indonesia

After these early parks were established, the number increased steadily reaching 41 in 2003. In a major expansion in 2004, nine new national parks were created, raising the total number to 50. By 2017 the number had increased to 54 national parks and by 2022 became 55. The latest addition was Mount Tambora National Park. Of the 55 national parks, 6 are World Heritage Sites, 9 are part of the World Network of Biosphere Reserves and 5 are wetlands of international importance under the Ramsar convention. A total of 9 parks are largely marine national parks. Table 1 lists all of these national parks (KLHK pers.com 2022).

Biodiversity in Indonesia

Encompassing major parts of the Sundaland and Wallacea Global Hotspots, roughly half of the New Guinea Tropical Wilderness Area and spanning much of the Coral Triangle, Indonesia undoubtedly harbors many of the most biologically rich and diverse ecosystems on Earth. This archipelagic country spans 17,508 islands along the Equator, including three of the worlds six largest, Sumatra, Borneo, shared with Malaysia and Brunei, and New Guinea, the western half of which belongs to Indonesia. Java and Sulawesi are also considered to be big islands while many others are categorized as medium to small. In 2019, Indonesia consisted of 34 provinces, 514 regencies and approximately 75 thousand villages. The archipelago covers 1,922,570 km^2 of land and 3,257,483 km^2 of ocean. The total population of Indonesia in 2016 was around 259 million people, increasing at less than 1% per year. There are 1,068 ethnic groups and 746 different languages (BPS, 2016).

Although occupying only 1.3% of the world's land surface, Indonesia is a megadiverse country, ranking very high in species diversity and endemism and holding almost 10% of the world's biodiversity (Mittermeier et al., 1998). At the heart of the Coral Triangle, Indonesia's 17,000 islands encompass 81,000 km of coastline and more than 80,000 km^2 of coral reefs. Its waters hold the richest marine biodiversity in the world, including 14% of the Earth's coral reefs (IBSAP, 2003), more than two-thirds of the world's hard coral species (more than 80 genera and 600 species; Veron, 2000).

Indonesia also possesses at least 2.5 million hectares of mangrove ecosystems, occurring throughout Sumatra, Java, Bali, Kalimantan, Sulawesi, Maluku, and Papua. Mangroves have diverse ecological functions and provide economic benefits for coastal fisheries, serving as nurseries, spawning and feeding grounds for fishes, shrimps and other marine organisms. Seagrass ecosystems also occur in many parts of Indonesia, particularly areas with shallow waters. At least 12 species of plants inhabit Indonesian seagrass beds, which are closely associated with mangrove and coral reef ecosystems. Seagrass beds produce nutrients, bind sediments and stabilize soft substrate, serving as nurseries and feeding grounds for inshore fishes. Seagrass beds are home to numerous threatened species such as dugong, *Dugong dugon* and green turtle, *Chelonia mydas*. Seagrasses also provide materials for local livelihoods, food and medical industries (IBSAP, 2003).

Historically, the economic benefits of Indonesia's biodiversity have not been shared directly by the country and its people. The ideal of an equitable share for the source country of the profits from a successful new product derived from its biodiversity should be established as a principle of every bioprospecting effort. However, the potential for commercial development and profit-making is only one part of the value of Indonesian biodiversity. A larger return is derived from the protection extended to watersheds and ecosystems and to the support this lends to the cultivation of natural products consumed by local people. The expertise arising from the traditional medicinal uses of plants and the vast regions not yet fully explored floristically are additional reasons to be optimistic about the prospects of ventures in drug discovery and development. The Indonesian domestic herbal medicines ("jamu") industry is strong and the demand in developed countries for natural products is increasing. It is reasonable for Indonesia to expect to derive increasingly more benefits from the usefulness of its natural resources. However, this can only be done with the support of the local communities in partnership with the private sector and international organizations involved in research and development programs. In order for the benefits to be distributed and shared equitably, they should contribute to the sustainability of all parties involved (Supriatna and MacManus, 1997).

Indonesia is first in world mammalian diversity with approximately more than 700 species, and first in the world in endemic mammals with about 280, fourth in reptile diversity with 411 species, 150 of them endemic and fifth in amphibian diversity with 270 species, 100 of them endemic (LIPI, 2021). In bird diversity, Indonesia ranks fifth after Columbia, Peru, Brazil, and Ecuador. In total diversity of non-fish vertebrates, Indonesia, with 2,906 species, ranks third after Columbia and Brazil, and second after Australia in total non-fish vertebrate endemism with 927 species (Mittermeier 1999). Indonesia has the highest marine fish diversity, on a par with Australia. While in freshwater fish diversity Indonesia is either second or third, behind Brazil and very similar to Columbia.

The number of plant species in Indonesia is between 30-35,000, making it fifth in the world for plant diversity. The Provinces of Papua and West Papua, in the western part of the island of New Guinea, contain 19,000-20,000 species of flowering plants. They are followed by Kalimantan, which has approximately 14,000-15,000 species, almost the same number as Sumatra at 14,000 species. Other islands have smaller numbers of species; Java with 10,000, Sulawesi with 9,000 species, Maluku with 6,500 and Nusa Tenggara with 6,500. Indonesia also tops the world list for palm diversity with 477 species, 225 of which are endemic. Indonesia has over half of the 350 species of the economically very important *dipterocarpaceae* family of trees, with 155 being endemic to Borneo alone.

Although Indonesia is the sixth largest country on Earth, it occupies only 1.3% of the world's land surface, but has roughly 12% of the world's mammals, 16% of the world's reptiles and amphibians, 17% of the world's birds, and 25% of the world's fishes. As is the case for most tropical countries, these figures are surely underestimates. (Mittermeier et al., 1998).

The Provinces of Papua and West Papua, the western half of the island of New Guinea, covering an area of over 416,000 km^2, contain a significant portion of the planet's remaining intact tropical forests, as well as some of the most pristine coral reefs on Earth. It is one of the largest remaining tropical wilderness areas in the world, with over 85% still covered in forest. Nearly one half of Indonesian biodiversity is contained within these two provinces. The flora and fauna of the island combines elements of two major bioregions, southeast Asia and Australia. This, combined with its enormous altitudinal range (from sea level to the highest mountains in the Asia-Pacific region) helps give rise to an extraordinary array of terrestrial ecosystems. These include equatorial glaciers and surrounding alpine valleys, a variety of montane forests in the many rugged ranges throughout the province, a diverse mix of lowland rainforest types, swamps, savanna forests, and mangroves (Marshall et al., 2004).

The biodiversity of Indonesia has changed considerably since the first humans arrived in the archipelago. Hunting, habitat loss and population pressure are some of the factors that have accelerated the rate of transformation. Sadly, these changes have already resulted in the extinction of several species. Among the most spectacular creatures that have already been lost are the Balinese and Javanese subspecies of the tiger, *Panthera tigris balica* and *P. tigris sondaica* and several species of birds, such as the Caerulean paradise flycatcher, *Eutrichomyias rowleyi*. Unfortunately, the list is likely to increase in the future if strong conservation action is not taken. Indonesia holds the grim record of having the greatest number of vertebrate species threatened with extinction (128 species of mammals and 104 species of birds) (Whitten et al., 2000).

Indonesia is a country with a well-established infrastructure to enact and enforce laws regarding the protection of natural ecosystems and species. This is done under the authority of the Directorate General of Conservation of Natural Resources and Ecosystem in the Ministry of Environment and Forestry who, recognizing the need to conserve its rich biological diversity, has made a commitment to protect more land Indonesia's land area. Also, it is planned that some 200,000 km² of coastal and marine habitats will be set aside as conservation areas.

Status of Indonesia's National Parks

The protected area (PA) system in Indonesia is considered to be the most comprehensive in all of Asia encompassing national parks, game reserves, grand forest parks, strict nature reserves, hunting reserves, protected forests and recreation parks or forests. The entire 23.2-million-hectare PA system consists of 18.4 million hectares of terrestrial and 4.8 million hectares of marine protected areas. The recreation parks or forests are generally of little biological significance. The national parks, however, are designed to protect biota of national and international significance and are assuming increasing importance as recreational resources for both domestic and international tourists; the latter group contributing increasingly to Indonesia's foreign exchange revenue. Figure 1 is a distribution map of the national parks of Indonesia.

Figure 1. The National Parks of Indonesia

Biogeographic Representation

Biogeography is the study of the distribution patterns of plants and animals through time and across space. A biogeographic region has characteristic plants and animals that are different to those of other regions, though some species may be shared. Indonesia's broad biogeographic regions are Sunda, Sahul and Wallacea. Sunda is affiliated with Asia, and covers Java, Kalimantan and Sumatra. Sahul is affiliated with Australia and New Guinea and covers the Aru islands and Papua. Between these two, where the biota includes a mixture of both as well as its own endemics, lies Wallacea, consisting of Sulawesi, Maluku and Nusa Tenggara.

MacKinnon & MacKinnon (1986) and MacKinnon & Artha (1982) further divided these three into seven; Sumatra, Jawa-Bali, Kalimantan, Sulawesi, Nusa Tenggara, Maluku and Papua. At least 47 distinct natural ecosystems can be recognized within these seven biogeographic zones. It would be desirable for the national park network to represent each of these ecosystems at least once. Currently that has not been achieved, but each of the seven biogeographic zones are well represented. Scientists would urge that any new national parks should be aimed at gaps in the coverage of ecosystems by the existing network.

Jatna Supriatna and Chris Margules

Sustaining National Parks and Other Conservation Areas in Indonesia

According to the theory of island biogeography, the number of species on an island is a function of island size and distance to a source, such as the mainland. Small far islands have fewer species than near large islands. This has been applied to the situation in which natural habitat has been cleared or degraded by uses such as agriculture, logging etc. Remnant habitat 'islands' are expected to sustain fewer species than the same area before isolation and the more distant they are from substantial tracts of natural habitat, the fewer species they will retain in the long term. This is still a matter of debate because the theory only deals with size and distance while many other factors determine the actual number of species, such as environmental heterogeneity, ability of some species to survive in disturbed habitat (it is not an actual sea), and so on. However, it is true to say that when habitat remnants, including protected areas, become isolated by land use changes surrounding them, they become susceptible to changes brought about by this isolation. For example, populations of species are separated from neighboring populations and these sub-populations are smaller than the previous contiguous larger population, which means they may suffer in-breeding depression. Other changes may also occur. The physical environment is altered, especially near the edges, which become exposed to desiccating winds and altered hydrological conditions. All of this points to the common sense conclusion that protected areas should be as large as possible in order to mitigate the potential impacts of isolation.

A set of criteria and principles have been developed to help guide the selection of protected areas, including national parks. Ideally, a national park should include the entire watershed from mountain tops to the lowlands. This would ensure that the entire range of ecological processes is maintained. The area should be large enough to support viable populations of the species it encompasses. The selection and management of conservation areas—in particular the preservation of ecosystems—should take the following into account (Groom et al., 2006):

1. The protected area should be as large as possible and should ideally contain thousands of individuals of each species, even if they are species with low population densities. The shape should consider relevant biogeographic boundaries. For example, watershed boundaries are better than tenure boundaries because they represent natural landscape changes. Buffer zones between protected areas and agricultural areas should be established,

where some activities that would not be allowed in the protected area, may be carried out. An example is selective logging. Something that would not be allowed in the protected area. Whenever possible, a protected area should meet that habitat needs of the species it contains the whole year round.
2. The protected area should include a broad array of ecological communities, which are mutually ecologically sustainable for example the communities of different forest height classes. This is because very few species are tied to a single community, and there are very few communities that do not depend on other communities nearby.
3. Total isolation of a protected area should be avoided. Whenever possible, it is better to group protected areas rather than have them scattered, and to link them with semi-natural habitat corridors, which may help form the buffers referred to above.

Why are these criteria needed? Because we know that the biodiversity is not evenly distributed. Different species occur in different habitats on different geological substrates in warmer or colder areas at low or high altitudes. Not all of this natural variation is usually covered by the protected area network. The general strategy to help mitigate this problem is to ensure that at least one protected area is established in each major ecosystem. We know that currently there are many ecosystems that do not have protected areas within them, such as on the islands of Bangka and Belitung, Enggano, Selayar Island and many more.

The Effectiveness of National Parks

Experience shows that the success of management depends greatly on the good will and support of local communities. In such cases, the managing authority will need to have good consultative and communications skills, and effective mechanisms which may include incentives, to secure compliance with management objectives. In addition, we are only beginning to understand the value and varied applications of indigenous knowledge. Because of this, legislative and policy frameworks should be formulated that allow local communities to participate in the regulation of access to parks, and to maintain a degree of sovereignty over their natural resources as well as their indigenous and traditional knowledge.

The real issue with local communities is very simple: local communities view conservation areas as a deterrent to their economic development. Without a share of tangible benefits from conservation programs, communities will not participate in conserving areas. Some 20 ICDP programs have been carried out in Indonesia, with only very limited success in increasing support from communities close to the parks. The economic benefits have either not been linked directly with conservation in and around the park and/or have reached too few people (Wells, et al., 1997).

Many believe that overall funding levels for the park system in Indonesia are adequate. In comparison with the other countries, park budgets are relatively high, and the average levels of staffing are better than the global average (MacAndrew and Saunders 1997). However, funding is not allocated evenly among regions and parks. Most of the funding in Indonesia is going to parks in Java, despite relatively low levels of species richness and endemism. While many parks in Kalimantan, Sumatra and Papua with much higher concentrations of biodiversity face many problems such as poaching, hunting and illegal logging with only limited budgets.

Financing for protected areas management has occurred primarily through the Ministry of Forestry, now the Ministry of Environment and Forestry, budget, which is provided by central government. Revenue for parks comes only from entrance fees that are unrealistically very low and have not changed for decades (Rp5000-7500 = USD 0.30-0.50/person/day for domestic and USD150/person/day for international visitors). None of the funds are earmarked for management in the park in which they are collected but go to the central government. Further, the distribution system of revenues generated from user fees does not provide adequate incentive for aggressive collection of those fees. Indonesia needs to reassess user fee collection and distribution in order to support the management of park, local governments and people nearby.

Other possible fees, such as from wildlife-tourism/ecotourism, mining, film making etc. have not been tapped due to the rigidity of the present fiscal system. In many countries, such as Costa Rica and Kenya, ecotourism contributes significant revenues to the country and to some protected areas. In Indonesia, potential tourism in parks has been poorly understood both by the park managers and by the tourism industry. Of the total USD26 billion gross output from the tourism industry in Indonesia in 1997, very little was generated from tourism related activities in parks. Indonesia should change its approach to future tourism development from traditional tourism destinations based on mass-tourism of culture, beach and landscape towards sustainable tourism that includes ecological and wildlife tourism.

Important financial resources are also provided by the international donor community, NGOs, and some multinational corporations. Most of the investment by donors, however, is too short-term while conservation projects need long-term programs and on-going processes. Conservation management also requires extensive institutional strengthening and human resources development. Therefore, conservation programs can only be achieved if they occur over long periods of time. Since conservation management is an adaptive and on-going process, it cannot rely solely on foreign funding agencies. For that reason, the Indonesian government should create trust funds generated from reforestation fees. Such financing could offset routine expenses covered in the GOI (Goverment of Indonesia) budget and provide flexibility for more adaptive and multi- stakeholder management.

Nevertheless, several long-term programs have been initiated by international agencies. Several NGOs and bilateral donors have established conservation trust funds and financing mechanisms using Debt for Nature Swap (MacManus, et al., 1999) and carbon emission offsets through Joint Implementation (JI) (Cannon et al., 1999) and also REDD (Reduce Emission from Deforestation and Degradation). Those initiatives have involved collaboration among institutions, i.e. NGOs (WWF, CI, TNC, Kehati) and the USAID-NRM program.

We hope that the publication of this book will help stimulate a more inclusive approach that involves local communities as key stakeholders in national parks. In many cases such people are among Indonesia's poorest. National parks provide significant opportunities for tourism, which is increasing throughout the world. Indonesia's national parks are among the most spectacular and contain many charismatic species that many people will pay a lot of money to see. National parks will become more secure and remain available for the benefit of future generations if local people value them.

ACKNOWLEDGMENT

We are very grateful to the Minister of Environment and Forestry, Prof. Dr. Siti Nurbaya, for writing the Foreword for this book. We also thank our many friends at the Ministry of Environment and Forestry, especially Vice Minister Dr. Alue Dohong, Secretary General Dr. Bambang Hendroyono, Senior Advisor to the Minister KLHK Ir. Wahyudi Wardojo MSc, former Director General of KSDAE Dr. Wiratno, and the many directors at the KSDAE and heads of national parks in Indonesia who have shared their knowledge and pictures. We thank Prof. Abdul Haris, Vice Rector UI, Dr. Dede Djuhana, Dean of the Faculty of Mathematics and Natural Sciences, the University of Indonesia (FMIPA-UI), Dr. Budiawan and Dr. Tito Indra Latif, Vice Deans FMIPA-UI, Head of the Dept. of Biology, Dr. Anom Bawolaksono and colleagues in the Department of Biology, FMIPA-UI, who supported us but are too numerous to be named individually, as well as Friends of the WILD research group, Department of Biology FMIPA UI. Special thanks to senior colleagues Prof. Soekarja Somadikarta and Prof. Indrawati Gandjar who have been Jatna's mentors for decades and have supported and encouraged him to become a better researcher.

We are also indebted to the willingness of our many friends who shared knowledge on biodiversity, local communities, and tourism activities around the national parks. Special thanks to Dr. Frans Teguh (Special Advisor to the Ministry of Tourism) and our students at the Dept of Biology and School of Environmental Science of the University of Indonesia. Special thanks to our friends who help design and collect data for the book, Dr. Nurul Winarni, Dr. Asri Dwiyahreni, Indartono Sosro Wijoyo, Maya Dewi, and to Liz Poon for advice and encouragement.

PROFILE OF JATNA SUPRIATNA AND CHRIS MARGULES

Jatna Supriatna completed his Master of Science in 1986, and his Doctorate in 1991 from the University of New Mexico, Albuquerque- USA. He also completed pre and post-doctoral studies at Columbia University in New York. He studied management at the Sloan School of Management, Massachusetts Institute of Technology in Boston and at Tsing Hua University in Beijing. He served as Lecturer and is now Professor of Conservation Biology in the Faculty of Mathematics and Sciences, University of Indonesia. He was Country Director and then Vice President of the major international non-government organization based in Washington DC, Conservation International (CI) from 1994-2010. Since January 2011 to date, he has served as Chairman of the Research Center for Climate Change and Institute for Sustainable Earth and Resources of the University of Indonesia. He was co-director of the Association of Pacific Rim Universities (APRU, based at the National University of Singapore) of CMAS (Climate Change Mitigation and Adaptation Strategy) based at the University of San Diego, USA. He also serves as Country chair of the United Nations Sustainable Development Solutions Network (UN SDSN) for Indonesia. He became a member of the Indonesia Academy of Science (AIPI) in 2011. He also served on the Board of Trustees of the Universitas Indonesia 2014-2019, Board of Trustee member of UID Foundation, Belantara Foundation, Indonesia Climate Change Trust Fund of BAPPENAS, Bornean Orangutan

Society Foundation, CSF Indonesia, TFCA Kehati (Indonesia Biodiversity Foundation), and Konservasi Indonesia Foundation. His research interests are in Primatology, Wildlife Tourism, Landscape conservation, Biodiversity Conservation, Climate change Mitigation and Adaptation and linking these to Environmental policy.

For his dedication to the environment and biodiversity works, he received the Most Excellence Order of the Golden Ark from his Royal Highness Prince Bernhard of the Netherland in 1999. In 2008, he received the prestigious award from President B.J. Habibie of Indonesia, or Habibie Award, for outstanding achievements in research on Natural Sciences. In 2010, he received the Terry MacManus Award of the United States of America for his dedication to conserving nature. In 2011, he received the Achmad Bakrie Award for Science for his commitment to developing Field Biology and Conservation in Indonesia. In 2017, he received Lifetime Achievement in the Field of Biodiversity Conservation from Conservation International. The SSC-Primate Specialist Group of IUCN named a new species of tarsier, a primate from Gorontalo, North Sulawesi, after him; *Tarsius supriatnai*. Other species named after him are the Togean flying lizard, *Draco supriatnai* from Togian Island in Central Sulawesi, and Jatna's gecko, *Cyrtodactylus jatnai* from Bali. He has served as an editorial board member of several international journals such as IUCN Parks Journal, Asian Primate Journal (IUCN-Primate Specialist Group), Tropical Conservation Science Journal, American Journal of Wildlife Policy and Law, Biosphere Conservation, and Climate Resilience and Sustainability. He has published 30 books, mainly on Indonesia's environment and biodiversity. He has published more than 190 papers in reputable International Journals such as Science, Nature, Conservation Biology, Evolution, Scientific Reports. Current Biology, Primates, International Journal of Primatology, and Primate Conservation, among others. According to Google scholar his H-index was 37, as September 2022.

Chris Margules is Adjunct Professor in the Faculty of Mathematics and Natural Sciences, University of Indonesia, Kampus Depok, and in the Centre for Tropical and Environmental Sustainability Science at James Cook University, Cairns Campus, Australia. His current research focuses on integrating conservation and development at the landscape or seascape scale. Previously, he designed and implemented a large-scale experiment on the ecological effects of habitat fragmentation and played a key role in discovering and then implementing the idea of complementarity in systematic conservation planning.

Chris completed a Bachelor of Applied Science at the University of Canberra, Australia, in 1973, a Graduate Diploma in Recreation Planning, also at the University of Canberra, in 1978 and a Doctorate in Biology from the University of York, UK, 1981. Chris was a research scientist and later research manager at the Commonwealth Scientific and Industrial Research Organisation (CSIRO) for 32 years, where he led programs on landscape management and sustainable development in the tropics. He joined the US based charity, Conservation International, in 2006 and later became Senior Vice-President and leader of the Asia Pacific Division until 2011. He led Country Programs in China, the Philippines, Cambodia, Indonesia, Papua New Guinea, Solomon Islands, Fiji, New Caledonia and Samoa

Chris was a fellow of the Wissenschaftskolleg zu Berlin, 1993-1994. He was a Web of Science Highly Cited Researcher for publications between 1982 and 2002. He received Order of Australia honours in the General Division (AM) for services to science in 2005. He has published one textbook and over 100 refereed papers and book chapters.

CHAPTER 1
SUMATRA

For the porposes of the book, Sumatra consists of the very large island of Sumatra as well as the many smaller adjacent islands including the Mentawai Islands as well as Nias, Simeuleu, Batu, and Enggano islands off the west coast and Bangka, Belitung and Riau Islands off the east coast. The large island of Sumatra itself is 1,800 kilometers long and 400 kilometers across at its widest point. It encompasses an extraordinary wealth of natural resources and diverse natural habitats. Stretching along the western side, the Bukit Barisan Mountains form the backbone of the island. The majority of this mountainous area is protected by nature reserves and national parks. The western slopes are very steep, while the eastern side slopes gradually to the plains and swamps of eastern Sumatra, where some of the most important wetlands in Indonesia can be found. The island was almost completely forested until the end of the 19th century. Today, approximately a quarter of the original jungle remains.

Sumatra's 476,000 square kilometers support an estimated 10,000 or more species of higher plants, mostly in lowland forests. The number of tree species per unit area in the lowland forests is likely to be greater than in similar areas of West Africa or South America. The large number of plant species is partly due to the nearperfect growing conditions: warmth, adequate moisture, and tall trees providing a framework for a wide range of structural niches. These are filled by smaller trees, shrubs, herbs, scramblers, climbers, epiphytes, parasites, and a complementary diversity of fauna (Whitten et al., 1997).

The fact that simple unstructured surveys still reveal locations of species outside of their previously known range — and even new species — reflects the richness and importance of the area for biodiversity conservation. During 1997, botanists carrying out a one-month survey of bamboo found five new species of three genera and nine new records for the Leuser Ecosystem in northern Sumatra. A survey of fish in one river revealed three new species.

In Sumatra, much of the remaining forest is on the hills and mountains along the western side of the island, some of which is of sedimentary origin and some of which is volcanic. The highest mountain, Mt. Kerinci, reaches 3,804 meters above sea level. There are also major areas of forest on the deep peat swamps of the eastern margins. The steep hills of the northern tip of the island have abundant forests, and further south is the huge volcanic caldera now occupied by Lake Toba. To the west are the biologically and anthropologically unique Mentawai Islands. In the south there are are some large rivers which, together with the island's abundant natural resources, have been the foundation for major industrial developments. In the far south there is very little forest left.

Sumatra and its nearby islands have at least 15 distinct ecosystems: sub-alpine forest, tropical montane forest on limestone, tropical montane forest, tropical pine forest, lowland forest on limestone, lowland semi-evergreen forest, lowland evergreen rainforest, ironwood forest, peat swamp forest, freshwater swamp forest, rivers, lakes, mangrove forest, sandy beach forest, and coral reefs and reef lagoons. A study by Leksono (2000) found that lowland and lower-level montane forests had been shrinking at an alarming rate for the previous five years. The lowland semi-evergreen forest covered only 81,000 hectares, 15% of its original area. There is relatively little disturbance to the ecosystems of the high mountain and sub-alpine zones. The remaining sub-alpine ecosystem covered 159,000 hectares, 94% of its original area because steep topography and limited access make it difficult to exploit.

Of these 15 ecosystems, eight are under-represented in protected areas, with less than 10% of their remaining area protected: heath forest, 0% protected; lowland semi-evergreen forest, 1%; tropical moist forest, 5%; ironwood forest, 2%; peat swamp forest, 2%; mangrove forest, 4%; tropical pine forest, 7%; and freshwater swamp forest, 6%. Unfortunately, these ecosystems are in lowland areas already under enormous land conversion pressure. The ecosystems that could be said to be adequately represented in conservation areas include subalpine forest, 21%; limestone montane forest, 29%; moist montane forest, 14%; lowland forest on limestone, 15%; sandy beach forest, 15%; and coral reef, 30% (Suroso, 2001).

Fauna

Sumatra has one of the most diverse faunas in Indonesia. With 210 species it has the most mammals and with 580 species of birds, is second only to New Guinea. This great natural wealth is due to Sumatra's large size, its diversity of habitats, and its past links with the Asian mainland. Sumatra is home to 31 species of endemic mammals, 16 of which are endemic to the isolated Mentawai Islands group. Sumatra is the last place on Earth where elephants, rhinoceroses, tigers, clouded leopards, and orangutans are all found together.

Table 1. Endemic Mammals of Sumatra

Pagai Islands horseshoe bat	*Hipposideros breviceps*
Herman's mouse-eared bat	*Myotis hermani*
Mentawai macaques	*Macaca pagensis*
	Macaca siberu
Thomas' langur	*Presbytis thomasi*
Mentawai langur	*Presbytis potenziani*
Siberut langur	*Presbytis siberu*
Black Sumatran langur	*Presbytis sumatrana*
Mitred Langur	*Presbytis mitrata*
Black-and white Black langur	*Presbytis bicolor*
Black-Crested langur	*Presbytis melalophos*
East Sumatra bonded langur	*Presbytis percura*
Snub-nosed monkey	*Simias concolor*
Mentawai gibbons	*Hylobates klossii*
	Hylobates agilis
Sumatran orangutan	*Pongo abelii*
Tapanuli orangutan	*Pongo tapanuliensis*
Sumatran rabbit	*Nesolagus netscheri*
Aceh squirrel	*Callosciurus albescurus*
Loga squirrel	*Callosciurus melanogaster*
Soksak squirrel	*Lariscus obscurus*
Mentawai black-cheeked flying squirrel	*Iomys sipora*
Mentawai orange-cheeked flying squirrel	*Hylopetes sipora*
Mentawai civet	*Paradoxurus lignicolor*
Giant Mentawai rat	*Leopoldamys siporanus*
Mentawai forest rat	*Maxomys pagensis*
Mentawai rat	*Rattus lugens*
Hoogerwerf's rat	*Rattus hoogerwerfi*
Kerinci rat	*Maxomys hylomyoides*
Kerinci rat	*Maxomys inflatus*
Mentawai pencil-tailed tree mouse	*Chiropodomys karlkoopmani*

| Sumatran shrew-mouse | *Mus crociduroides* |
| Sumatran gymnure | *Hylomys parvus* |

(Adapted from *The Ecology of Sumatra*, Whitten; Supriatna 2019, Nater et al., 2017)

Eight endemic mammals in Sumatra and Mentawai are listed in the IUCN Red Data Book of species in danger of extinction and in the appendix of the Convention on International Trade in Endangered Species (CITES). The Sumatran tiger, *Pathera tigris sumatrae*, numbers no more than 500 in highly fragmented populations and is the last of Indonesia's three tiger subspecies (out of eight worldwide). The Bali tiger became extinct in the 1940s, followed by the Javan tiger in the 1980s (Seidensticker et al., 1999). Some scientists hold that the Sumatran tiger comprises the only true tiger subspecies separate from the tigers of Mainland Asia. The main threats to the survival of the Sumatran tiger are poaching to supply the commercial demand for its parts (skin, bones, whiskers, teeth, and claws) and habitat destruction.

Sumatra is a home to six other wild cat species, clouded leopard, *Neofelis diardi*, golden cat, *Pardofelis temincki*, fishing cat, *Prionailurus viverrinus*, marbled cat, *Pardofelis marmorota*, leopard cat, *Prionailurus bengalensis sumatrans*, and flat-headed cat, *Prionailurus planiceps* (Nowell and Jackson 1996; Sunquist and Sunquist 2002; McDonald et al., 2010). Of these, only the leopard cat is considered to have a low risk of extinction (Sanderson et al., 2008). In addition to habitat loss and fragmentation, poaching and hunting are the main threats to wild cats in Sumatra (Nowell and Jackson 1996, Sunquist ans Sunquist, 2007, McDonald, 2010).

All wild cat species are protected by Indonesian law through Government Regulation No. 7/1999 on the preservation of animal and plant species. With the exception of the Sumatran tiger, conservation efforts and initiatives to protect wild cat species are limited (McCarthy et al., 2015). Most studies of wild cats have been conducted in Borneo (Azlan and Sharma, 2006; Wilting et al., 2006; Rajaratnam et al., 2007; Wilting *et al.* 2010; Cheyne and Mcdonald 2011; Hearn et al., 2013; Wearn et al., 2014) and Thailand (Grassman, 2000: Grassman and Tewes, 2002; Grassman et al., 2005a; Grasman et al., 2005b; Ngoprasert et al., 2012; Lynam et al., 2013). In Sumatra, ecological research on wild cats has only recently been conducted (Sunarto et al., 2015; Pusparini et al., 2014; McCarthy et al., 2015). Given the growing threats to those species, more studies on the biology and ecology of the wild cats of Sumatra are clearly needed as a basis for the development of comprehensive conservation and management plans (Ario, 2010).

Other notable Sumatran mammals include tapir, *Tapirus indicus*, dhole or forest dog, *Cuon alpinus*, bearded pig, *Sus barbatus*, sun bear, *Helarctos malayanus*, flying squirrels, *Iomys sipora* and *Hylopetes sipora*, mountain goat, *Capricornis sumatraensis*, as well as other primates, rodents, otters and porcupines (Whitten, et al., 2000).

Sumatran elephant, *Elephas maximus sumatranus*, as with other members of the elephant genus, live in groups. They need significantly large areas to roam. However, conflict of interests occurs in many elephant habitats, which are targeted for conversion into large scale plantations, settlements and agriculture for transmigration schemes, logging activities, etc. At the beginning of this century, elephants could be found in almost all forests across the island. These days, they are mainly confined to smaller, degraded forests. Many elephant – human conflicts occur throughout the island because of development and poor land-use planning. Consequently, many people argue that elephants should be captured and sent to six Elephant rehabilitation centers on the island, where they would be protected. However, it is costly to maintain them in the centers and it will reduce the size and therefore viability of wild populations.

The Sumatran rhinoceros, *Dicerorhinus sumatrensis*, just as other rhinoceroses is an endangered species. As a result of poaching for the lucrative trade in rhino horn, numbers have declined by 50% since 1980. Few Sumatran rhinos survive in very small and highly fragmented populations in South-East Asia with Indonesia having the only significant numbers. In Malaysia, only 3 Sumatran rhinos survive in Sabah and are too old to breed anymore. In the last few decades, Sumatran rhinos have mostly disappeared from Kalimantan, although there may be a few remain in North Kalimantan Province. They are now extinct in Kutai National Park in East Kalimantan which was originally created specifically to protect this species. In Sumatra, remaining viable population occur in only two places, Gunung Leuser National Park in northern Sumatra and Way Kambas National Park in southern Sumatra. Experts estimate that the total population in Sumatra stands at between 185 and 250 animals, but some pessimistically put the number at fewer than 100 (Payne, 2017). Anti-poaching efforts appear to have largely halted the hunting of these animals. However, illegal poaching to supply the demand for rhino horn used in traditional medicine has undoubtedly had a devastating effect.

There is a lack of information on populations of the sun bear, *Helarctos malayanus* in Sumatra. Experts assume that sun bears occur wherever there is sufficient forest cover in both Sumatra and Kalimantan. Sun bears are omnivores eating fruit, grubs, birds' eggs, fledglings and honey. They are expert climbers and will climb to great heights to reach the nests of bees and termites. Given both declines in forest habitat and direct captures and kills, the species is considered to be threatened. The trade in bear claws, canine teeth and gall bladders continues. An Indonesian government environmental impact assessment investigation in 1995 found that they were regularly hunted and fell prey to snares. Their canine teeth and claws were found for sale in many shops in Medan, the capital of North Sumatra. Other shops sold gall bladders for use in medicine. Although the international trade in sun bear parts is banned (the species is listed on Appendix I of CITES), it is thought that some threat remains from the demand for gall bladders from other Asian countries. The sun bear is thought to be particularly sensitive to logging operations. Researchers in Central Kalimantan learnt of a cub which had been captured by villagers after its mother had been killed by logging workers.

Of the 40 primate species found in Indonesia, Sumatra is home to 17 (Supriatna & Hendras, 2000), ranging from the smallest, a tarsier, *Tarsius bancanus* to two of the largest apes, the orangutans, *Pongo abelii* and the recently described *Pongo tapanuliensis*. Flagship species are those selected as icons or symbols representing biodiversity in a given habitat, or to promote a conservation campaign or environmental cause. Orangutans are examples of flagship species and the primates of the Mentawai Islands are all endemic flagship species. They include the Mentawai gibbon or bilou, *Hylobates klossii*, the two subspecies of Mentawai pig-tailed macaque or bokkoi, *Macaca pagensis pagensis* and *Macaca pagensis siberu*, the Mentawai leaf monkey or joja, divided into two subspecies, *Presbytis potenziani potenziani* and *Presbytis potenziani siberu*, and the distinctive pig-tailed langur or simakobu, again with two subspecies, *Simias concolor concolor* and *Simias concolor siberu*. In all, there are seven distinct primate taxa endemic to this small island group, including the genus *Simias*. This results in a level of endemic primate diversity per unit area unmatched anywhere on Earth.

Currently, 1,531 bird species are recorded in Indonesia (based on Peter's taxonomy, 1934-1986). With 580 of these, Sumatra is the second-richest region for birds after Papua, with 647 species. Only 14 are endemic, most of which are found in hill and mountain forest. Only a few, such as owls, are found in lowland areas. Many of those endemics occur in northern Sumatra, for example, Scheneider's Pitta, *Pitta schneideri*, bulbuls, *Pycnonotus leocogrammicus and P. tympanistrigus*, babblers, *Tricastoma vanderbilti* and *Napothera rufipectus*, Sumatran niltava, *Niltavaa sumatrana* and bronze tailed peacock, *Lophura inornata*. Some of them are restricted to certain islands such as the Mentawai Scops owl, *Otus mentawi* and the Island Scops owl, *Otus umbra*, which is found on Enggano and Simeuleu islands. Hornbills, Family *Bucerotidae*, reach their highest diversity in Sumatra, where 10 species are found. According to BirdLife International, there are 34 Important Bird Areas (IBAs) in Sumatra., of which, 18 are found outside the existing protected areas system, and 6 are in the highly threatened lowland forests.

Sumatra has at least 300 reptile and amphibian species, 69 (23%) of which are endemic. However, based on undescribed specimens from the Indonesian Institute of Science, there may be as many as 88 endemic species. Most of these occur in West Sumatra and in the Kerinci Seblat lowlands, which are under very severe threat from logging. Sumatra's regionally endemic reptiles and amphibian genera include *Aphaniotis, Gonocephalus, Harpesaurus, Phoxophrys*, and other arboreal agamid lizards, *Larutia* skinks, *Xenophidion, Hydrablabes* and *Macrocalamus* snakes, *Staurois* frogs, and *Leptobranchella, Ansonia,* and *Leptophryne* toads. Diversity of Sumatra's herpetofauna may be substantially underestimated, and the number of known species is likely to increase dramatically through further surveys and taxonomic work.

Reptiles include several flagship species, notably two large river terrapins, the mangrove terrapin, *Batagur baska* and the painted terrapin, *Callagur borneoensis*. The false gharial, *Tomistoma schlegelii*, is another flagship reptile. At least 12 endemic amphibian species are found in Sumatra, including the frog *Bufo sumatranus*. Another species, *Rana blythii*, is the largest species of ranidae (true frogs) found in West Sumatra and is protected by Indonesia law.

Sumatra has 270 species of freshwater fishes, 42 (15%) of which are endemic (Kottelat and Whitten, 1996). One new species, *Gymnochanda limi*, was recently found in Bukit Tigapuluh National Park. New species of fish are still being discovered in rivers, lakes and swamps.

Flora

Sumatra has 17 endemic plant genera, fewer than some other parts of Indonesia. But at the species level, it is rich, with more than 150 out of the 1,100 Indonesian endemics. Most endemic species are found below 500 meters in lowland forests. Sumatra has 105 species of the family *dipterocarpaceae*, of which 11 are endemic. Sumatra's most notable endemic plant is *Rafflesia arnoldi*, a flagship for Sumatra's lowland forests. It has the largest flower in the world and is the most specialized of all parasitic plants. Only about 15% of the plant species believed to occur in Sumatra have ever been recorded. Of the 49 Centers of Plant Diversity and Endemism in the Sundaland global hotspot (Davis et al., 1995), six are in Sumatra.

The National Parks

The table below lists the national parks of Sumatra and the provinces they occur in. Detailed decriptions follow.

Table 2. The National Parks of Sumatra

Name	Province
Gunung Leuser	Aceh & North Sumatra
Batang Gadis	North Sumatra
Tesso Nilo	Riau
Bukit Dua Belas	Jambi
Bukit Tigapuluh	Jambi & Riau
Kerinci Seblat	West Sumatra, South Sumatra, Jambi & Bengkulu
Siberut	West Sumatra
Berbak	Jambi
Sembilang	South Sumatra
Bukit Barisan Selatan	Lampung & Bengkulu
Way Kambas	Lampung
Zamrud	Riau
Gunung Maras	Bangka Belitung

The National Parks of Indonesia

River at Sembilang National Park & River at Tesso Nilo National Park (Photo by Jatna Supriatna) & Lake at Zamrud National Park (Photo by Sri Mariati)

Forest at Zamrud National Park (Photo by Sandy Leo); View of forest path at Berbak Sembilang National Park (Photo by KLHK) & Landscape of Gunung Leuser National Park (Photo by Ahtu Trihangga)

Anak Dalam Tribe at Bukit Dua Belas National Park (Photo by Sri Mariati) & *Raflesia sp.* at Gunung Leuser National Park (Photo by Darwin)

Dicerorhinus sumatrensis at Gunung Leuser National Park ; (Photo by Adhi Nurul Hadi) & *Pongo abelii* at Gunung Leuser National Park (Photo by Misdi)

Elephas maximus sumatranus at Way Kamnbas National Park (Photo by Nurul Winarni) & *Panthera tigris sumatrae* at Sembilang National Park (Photo by Subagyo)

The National Parks of Indonesia

Sus scrofa at Bukit Barisan Selatan National Park (Photo by Nurul Winarni) & *Tapirrus indicus* at Kerinci Seblat National Park (Photo by Wilson)

Presbytis melalophos & Symphalangus syndactylus at Bukit Barisan Selatan National Park (Photo by Nurul Winarni)

Neofelis sp. ,& Pardofelis badia, at Kerinci Seblat National Park (Photo by Wilson)

1. GUNUNG LEUSER NATIONAL PARK

Geographic Location and Size

Gunung Leuser National Park is located between $2^0 55'$-$4^0 05'$ North Latitude and $96^0 30'$-$98^0 35'$ East Longitude. It straddles the border of two provinces, Aceh and North Sumatra and includes parts of five regencies, Southeast Aceh, South Aceh, North Aceh, Langkat and Tanah Taro.

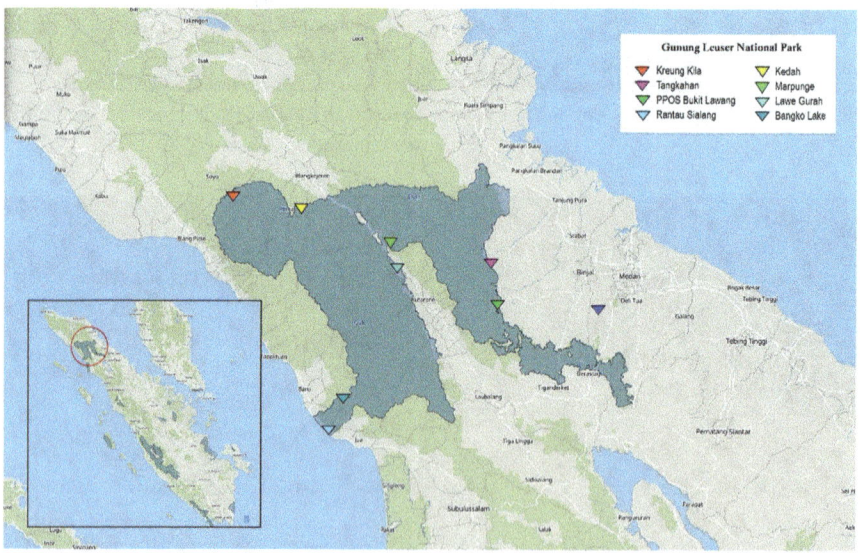

Gunung Leuser National Park covers 1,094,692 ha and has a border 850 km long. It spreads over 100 km along the Bukit Barisan Mountain Range, from the west coast of Sumatra in the southwestern tip to less than 25 km from the north east coast at the northeastern tip.

Climate and Topography

The mean minimum temperature is approximately 21°C and the mean maximum is approximately 28°C. The highest annual rainfall recorded is 4,600 mm in the mountains towards the center of the park. Rainfall drops off to between 3,000 and 3,500 mm towards the coast in the west. The lowest, at 1,300 mm is in the Alas Valley. March and April and September and October are the four wettest months, while February, June and July are the driest months, but it is wet year-round. The rainfall is never less than 100 mm per month. Humidity ranges between 62% and 87% in the drier months and reaches 100% during the wettest months. In areas higher than 1,700m asl, humidity can be very high with the air becoming saturated for long periods of time.

The north end of Bukit Barisan, which is part of the park, is characterized by separate mountain ranges in the northwest and the northeast, consisting of some of the highest mountain peaks in Sumatra. Not all are volcanoes. Local relief is very high. The southern part of the park has lower peaks and local relief becomes very low on the Bengkong plateau.

History

In 1924, Aceh's local leaders asked the Dutch Colonial Government to protect forests in the Alas area to halt plans to log them.

In May 1928, F.C. Van Heurn proposed to the Netherlands government the creation of a National Park in west Aceh of 928,000 ha, covering the land between the Alas, Kluet and Tripa rivers and all ecosystems from the coast to the the mountains

In 1934, after A. Ph. Van Aken was elected governor of Aceh, the first part of what would eventually become Gunung Leuser National Park was declared as a *Wildceservaat Goenoeng Leoser*, of 142,800 ha.

Between 1934 and 1938 further conservation areas such as the Mount Leuser Wildlife Reserve of 583,310 ha (No. 317/35), Kluet Wildlife Reserve of 20,000 ha (SK. ZB. No. 122/AGR), the Langkat and Sikunur Wildlife Reserves were added

In December 1976 three more reserves were added, Kappi Wildlife Reserve of the Decision Letter of Ministry of Agriculture Number. 697/Kpts/Um/12/1976, Sikundur National Park and Lawe Gurah National Park

In March 1980 the letter of Agriculture Ministry Number 736/Mentan/X/1980 declared that all the conservation areas of and around Mount Leuser would be gathered together, along with an area of restricted and production forests totaling 292,707 ha to become Gunung Leuser National Park, with a total area of approximately 792,675 ha. Four other national parks in Indonesia were announced at the same time to demonstrate the willingness of the Indonesian government to further the IUCN World Conservation Strategy that was announced at that date

In 1981 this area was announced as a Biosphere Reserve by UNESCO

In 1989 TAHURA (Grand Forest Park) Bukit Barisan was established including Sinabung and Sibayak Mountains (volcanoes) as well as Langkat Selatan, East of the Wampu River, all within the Leuser area.

The TAHURA area was also declared as a World Heritage Site as well as a Sister Park of one in Malaysia, part of an agreement between Indonesia and Malaysia

On the 23rd of May 1997 the Decision Letter of the Ministry of Forestry was released, Number. 276/Kpts-VI/1997, to announce the total area of Leuser National Park had become 1,094,692 ha.

Biodiversity and Ecosystems

Gunung Leuser National Park includes the complete range of ecosystems from coastal mangroves through lowland tropical rain forests and tropical mountain forests to sub-alpine mountain ecosystems of low trees and shrubs. Most of the lowland forest areas are dominated by species of the family *Dipterocarpaceae*, such as meranti, *Shorea spp.*, keruing, *Dipterocarpus spp* and kapur, *Dryobalanops spp*. The kapur *Dryobalanops aromatica* is a favored furniture species. There are three famous rare plants that characterize the Mount Leuser area, the massive umbrella tree, *Johanesteisjmania altifrons*, *Rafflesia arnoldi* var *atjehensis* and a parasitic lily with a flower up to 1.5m in diameter, *Rhizanthes zippelii*. There are also many orchids including orchid shoes, *Paphiopedilum liemianum* and many different tropical carnivorous pitcher plants, *Nepenthes* spp.

The lowland *Dipterocarpaceae* forests below 600 m asl, cover approximately 12% of the park. Sub-montain rainforest covers around 485,000 ha areas and is located between 600 and 1,500 m asl. This type of forest has a lower crown, usually less than 30m, with plants such as *Pirola sumatrana, Swerina bimaculatus, Valeriana sp., Ranunculus sp., Aenemona sp.* and *Gentiana sp.* Above 1,500 m asl there is real Montain forest, which is characterized by trees with many branches, rarely more than 20m high. Over 1,200 m asl, there is moss forest, in which the ground and trees are covered by moss. Over 2,500 m asl and in several open valleys, there is sub alpine forest and low shrubs. The trees become very small and the understory is characterized by short grasses and sedges such as *Carex* spp., accompanied by small trees and very small bushes such as *Rhododendron* spp., *Vaccinium* spp., *Parnassia* spp. and *Gentiana* spp. At the western edge of the park near Kluet there is beach sand, swamp vegetation and freshwater swamp forest.

There are more than 4,000 plant species, some of which are consumed by local people and could be valuable sources of germ plasm for further commercial development. These include forest durian, *Durio exyleyanus* and *Durio zibethinus*, forest orange, *Citrus macroptera*, rambai, similar to a menteng fruit, *Baccaurea montleyana*, menteng, *Baccaurea racemosa*, duku, *Lansium domesticum*, rukem, *Flacourtia rukem*, rambutan hutan, *Nephelium lappaceum* and mangoes, *Mangifera feotida* and *Mangifera guardrifolia*.

More than 380 bird species have been recorded, 350 of which are residents. This constitutes 85% of Sumatra's 438 resident breeding bird species. This park also protects 36 Sumatran endemic birds out of 50 Sundaland endemic bird species (van Marle and Voous, 988, Holmes, 1996, Wind 1996, Buij et al., 2016). The park includes more than 175 species of mammals, including 15 species of rat, 13 species of bat and 17 species of squirrel. At least 89 rare species have been recorded at Gunung Leuser National Park. The 129 species of mammals out of 205 species found in Sumatra, represents 65% of all Sumatran mammals (Van Strein, 1996).

Rare species that can be found in the park include Sumatran elephant, *Elephas maximus sumatranus*, Sumatran rhinoceros, *Dicerorhinus sumatrensis sumatrensis*, Sumatran tiger, *Panthera tigris sumatrae*, Sumatran mountain goat, *Capricornis sumatraensis*, great hornbill, *Buceros bicornis*, leopard cat, *Prionailurus bengalensis sumatrans*, clouded leopard, *Neofelis diardi*, golden cat, *Pardofelis temincki*, fishing cat, *Prionailurus viverrinus*, marbled cat, *Pardofelis marmorota* and flat-headed cat, *Prionailurus planiceps*. Other endemic Sumatran mammals include Kloss's squirrel, *Callosciurus albescens*, Sumatran striped rabbit, *Nesolagus netscheri* and Hoogerwerf's rat, *Rattus hoogerwerfi*. Primates found in the park are mawas or Sumatran orangutan, *Pongo abelii*, Sumatran siamang, *Symphalangus syndactylus syndactylus*, Lar gibbon, *Hylobates lar*, Thomas' leaf monkey or kedih, *Presbytis thomasi*, silvery leaf monkey, *Trachypithecus cristatus*, pig-tailed macaque, *Macaca nemestrina*, long-tailed macaque, *Macaca fascicularis* and slow loris, *Nycticebus coucang*.

Approximately 50% of the remaining Sumatran orangutan habitat falls inside the park directly managed by the Ministry of Environment and Forestry, and 78% lies within the boundaries of the wider vast Leuser Ecosystem Area that includes the park (Wich et al., 2008). Thus, Gunung Leuser National Park is very important habitat for the critically endangered Sumatran orangutan. Rijksen and Meijard (1999) estimated that there were 85,000 Sumatran orangutans in 1900. By 2017 only 6,600 were thought to exist, all in North Sumatra and Aceh provinces (Wich et al., 2008). However, recent surveys have been conducted beyond the previously known range, especially at higher altitudes and it is now estimated that there may be more than 14,000 (CBSG and MoEF, 2017). Similarly, 70 Sumatran tigers out of the approximately 500 individuals in Sumatra, are found in this park. But the number goes up to 250 individuals, or almost half of the Sumatran tiger populations, for the wider Leauser ecosystem. The park is also a key location for the critically endangered Sumatran rhino, of which only 100-150 individuals are left in three locations, Gunung Leuser, Way kambas and Bukit Barisan National parks. But it is Gunung Leuser that has the largest block of suitable habitat for Sumatran rhinos. Similarly, this park and the wider Leuser Ecosystem have the largest blocks of suitable habitat for the Sumatran Elephant.

The two provinces, Aceh and North Sumatra, have witnessed total forest losses of 22.4% and 43.4%, respectively from 1985 to 2009. The highest rate of forest loss in both provinces was on peatlands (Tripa swamp and Singkil swamp forests), mainly due to draining and burning for oil palm plantations, a practice that also releases Greenhouse gases into the atmosphere (Perbatakusuma et al., 2009).

The park harbors more than 95 species of herpetofauna. There are at least seven species of turtles and tortoises, seven species of agamids, three species of geckonids, three species of scincids, two species of varanids, and many species of snakes and frogs (Supriatna and Sidik, 1996).

Local Communities and Culture

Originally there were two ethnic groups, Alas and Gayo. In the north of the Alas valley and the mountains at the northern end of the park, the people are mostly Gayo. Alas people traditionally occupied southern areas, especially the main Alas valley. In recent decades there have been immigrations, especially from Batak Karo, Mandailing and Singkil, areas close to the park, as well as from Java. The communities on the west coast are mainly Minangkabau who have migrated north along the coast of the Bukittinggi area in West Sumatra. Ethnic Kluet people are found along the Kluet River, near the border in the west. Other ethnic peoples in the park include the Jamu and Aceh in south Aceh, and the Melayu in the Langkat area to the north.

Tourism

Tourism is already well established in the Bohorok area and at Berastagi with the Sibayak volcano. The Berastagi area is one of the most visited areas on weekends in North Sumatra. It is only 30 km from Medan, the third largest city in Indonesia, with a population of more than 4 million people. From Berastagi, you can climb for an hour to the Sebayak mountain with an amazing volcanic crater. Similarly, thousands of people, mostly Indonesian but also many foreign tourists visit Bohorok, approximately 96 km to the south of Medan to see orangutans and other wildlife and to bathe in the beautiful clear water of the river there. A few hundred kilometers to the south of the park, there is Lake Toba, also an important tourist destination. This lake is in the largest caldera in the world. Formed by a super-volcano, the lake is more than 100 km wide and approximately 700 m deep.

Existing tourism locations and areas with tourism potential in Gunung Leuser National Park are:
- Gurah Tourism Forest. This is an area of about 9,200 ha at Balailitu, Kutacane-Aceh Tenggara, Aceh Province. The attractions are the beautiful landscape, the Alas River, a waterfall, hot springs, many species of wildlife and plants including *Rafflesia*, wild orchids, orangutans, monkeys, and snakes as well as many species of birds and butterflies. It is possible to see

orangutans each day, in the morning or in the afternoon. Gurah Tourism Forest is 250 km from Medan. It can be reached by public bus from Pinang Baris bus Terminal of Medan city everyday and takes approximately eight hours. Alternatively, minibus taxis are available from Padang Bulan and Simpang Kuala Medan, which take around 6 hours to Kutacane. Then, to the forest gate of Gurah Tourism Forest by public transportation from Kutacane Bus Terminal takes around 30 minutes.

- Sekundur Tourism Forest. This is an area of 18,500 ha, at the Sekundur part of the park, Langkat, North Sumatra Province. The forest harbors many elephants, deer, monkeys, birds and natural caves as well as a beautiful landscape. There are camping grounds between Bohorok and Langkat as well as in the Sekundur area itself.
- Kluet Preserve Sanctuary. This is an area of 20,000 ha in South Aceh, Aceh Province. The attractions are kayaking on the river or lake, beach forest views and natural caves. This location includes Sumatran tiger habitat.
- The Bohorok Orangutan Rehabilitation Center. This is an area of 200 ha at Bohorok Bukit Lawang, Langkat, North Sumatra Province. The attractions include the rehabilitation center and the orangutans there, monkeys and released orangutans in the nearby forest, views of nature including the river and a variety of birds. Accommodation is available at Bukit Lawang. The gate is at Bohorok/Bukit Lawang, around 96 km or 2.5 hours by public bus from Pinang Baris Terminal in Medan, or about the same time by hire car.
- White-water rafting on the Alas River can be accessed from Muara Seulan village.
- Mountain climbers could try Mount Kemiri, Mount Simpali, Mount Mamas, and Mount Bendahara, as well as Mount Leuser itself.
- The beauty of nature in the Tiga Lingga Dairi, Lawe Gurah and Tapak Tuan areas can be enjoyed while walking.
- At Mount Tuan, a stone tomb can be found. 2 meters
- The Lau Pengurukan area has many natural caves such as Pintu Air, Pintu Angin, Patu, Rizal, Palong-long, Pamuite, Pasar and Pasugi. At 600 m, Pintu Angin Cave is the longest. Lau Pengurukan can be reached from Medan by bus or hire car to Bukit Lawang, then continued using a four-wheel drive hire car such as a Jeep or Land rover through Tanjung Naman Village for 1 hour. Then walk for 2 hours to Lau Pengurukan.

Access to the National Park

The best time to visit the park is from June to October. Gunung Leuser National Park has entrances at Lawe Gurah-Ketambe, Bohorok-Bukit Lawang and Sikundur-Besitang. From Jakarta Medan can be reached by land, sea or air. The routes from Medan to the three entrances are reachable by car:

- The Medan-Kutacane-Lawe Gurah-Ketambe route (± 275 km) can be traveled by public transport (bus, taxi) and will take 6-7 hours. Buses depart from the Medan Pinang Baris bus terminal to Kutacane around 15 times/day. Lawe Gurah Tourism Park is 43 km from Kutacane and buses depart twice each day. In Kutacane, there is an orangutan field research station.
- The Medan-Bohorok-Bukit Lawang route (± 91 km) can be traveled by bus, which will take ± 2.5 hours.
- To go to Tangkahan village of Langkat, in the edge of park in North Sumatra, take a bus to Langkat from the Bus Terminal in Medan for 2 hours. Then rent a car to the village.

Facilities

There are cities close to the park with good hotels, such as Medan and Kutacane, Berastagi and others. But many resorts can also be found close to national park, such as Bukit Lawang and Tangkahan in North Sumatra and Ketambe in Aceh.

Park Office

Balai Besar Taman Nasional Gunung Leuser
Jl. Selamat no.137, Kel Siti Rejo 3
Medan Amplas, Medan 20219
Tel: +62-61-7872919
Fax: +62-61-7864540
Call Center: 0812 6354 6663
Instagram: @bbtn_gn_leuser
Email: balai-tngl@dephut.go.id

2. KERINCI SEBLAT NATIONAL PARK

Geographic Location and Size

Kerinci Seblat National Park is located between $1°07'$ and $3°26'14''$ s and between $100°31'18''$ and $102°44'01''$ E. It is the largest National Park on the island of Sumatra covering 1,389.509 ha. It is spread along 345 km of the Bukit Barisan mountain range in the middle part of western Sumatra. The park extends into four provinces and nine regencies and includes 139 villages.

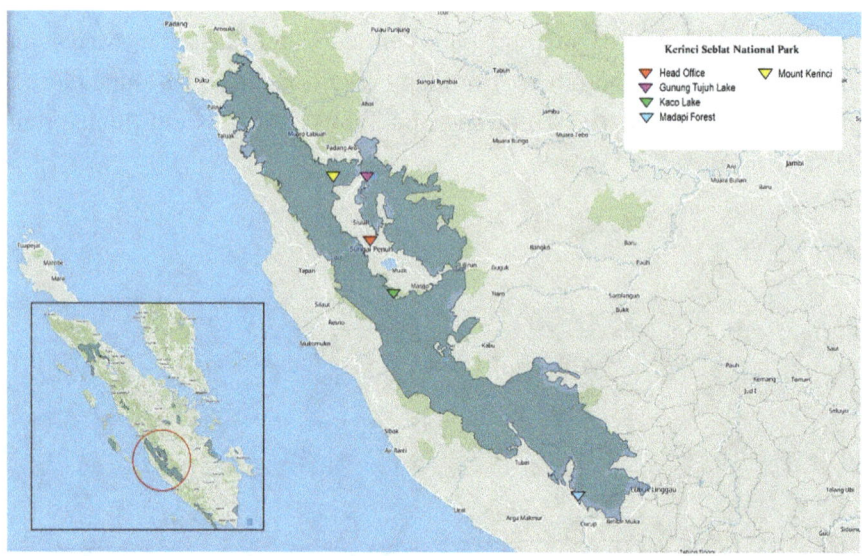

Jambi Province contributes 588,462 ha, which is 40% of the park area, covering parts of Batanghari, Sorolangun, Bangko and Kerinci Regencies and includes 48 villages. West Sumatra Province contributes 375,934 ha, 25% of the park area, covering parts of Pesisir Selatan and Solog Regencies and includes 39 villages. Bengkulu Province contributes 310,579 ha, 21% of the park area, covering parts of Bengkulu Utara and Rejanglebong Regencies. South Sumatra Province contributes 209,675 ha, 14% of the park area, covering part of Musi Rawas Regency and includes 18 villages.

Climate and Topography

Rainfall is high averaging 3,086 mm/year. Rainfall is highest on the west coast and in the north with April and November being the wettest months. Valleys have lower rainfall, due to the orographic effect of the mountains that surround them. At lower elevations the average maximum temperature is 28^0C with little monthly fluctuation. In valleys at higher elevations, the average maximum temperature is 23^0C. At highest elevations temperatures can be as low as 7^0C. Average humidity is around 80%-100%.

The topography is generally steep with local relief of around 200-300m. There are many high mountains such as Kerinci (3,805 m, the highest mountain on Sumatra), Tujuh (2,604 m), Seblat (2,383 m), Raya (2,543 m), Nilo (2,400 m), Masurai (2,600 m) and Sumbing (2,500 m). Bukit Barisan mountain range in Sumatra is part of the volcanic arc that spreads along the south of Indonesia from Sumatra in the west to Java and on to the islands of Nusa Tenggara in the east. The middle of the Bukit Barisan range, with active volcanoes, is part of this park. It contains a hidden valley of about 140,000 ha, all sides of which are surrounded by mountain peaks. The view to the north of the valley is dominated by the cone of Kerinci volcano which is still active. Just a little south and west of Mount Kerinci is the Gunung Tujuh crater lake.

Topographically, the area has very high local relief with normally steep and rugged slopes, with marked spurs to the north and west from the north-south spine of the Bukit Barisan range. Three main streams flow out of the National Park, but the most important hydrological feature is the small streams that flow to the north and southeast, which form the headwaters of the Batanghari and Musi Rivers, the latter being the largest and longest river in Sumatra.

History

Kerinci Seblat National Park, along with Gunung Leuser and Bukit Barisan Selatan national parks together have been named by UNESCO as its Southeast Asia Heritage Site.

Originally, the area of the park was covered by numerous small wildlife reserves, protected forests and limited production forests until late in 1999, when these were amalgamated and together with some other new areas, declared to be Kerinci Seblat National Park,

The history is as follows

In 1921 the Netherlands government declared the forests at Bayang area, Batanghari I, Kambang, Sangir I and Jujuhan to be protected forests

In 1926 the forests at Sangir Ulu, Batang Tebo and Batang Tabir were also declared protected forests

In 1929 an area including Mount Indrapura was declared a Wildlife Reserve

In 1978 Bukit Tapan Wildlife Reserve was declared

In 1979 Rawas Hulu Lakitan Wildlife Reserve was declared

In 1980 Kambang Wildlife Reserve was declared

On October 14th, 1982, according to the Statement Letter of the Agriculture Minister Number 736/Mentan/X/1982, all these areas were gathered together and declared to be Kerinci Seblat National Park. On January 5th 1996, according to the Decision Letter of the Forestry Minister Number 192/Kpts-II/96, areas including Mount Nilo (2,400 m), Mount Masurai (2,600 m) and Mount Sumbing (2,500 m) were included in Kerinci Seblat National Park, bringing it to a size of 1,368,000 ha.

On Februari 27th 1998, the Decision Letter of the Forestry and Plantations Minister Number 280/Kpts-II/1998 was realeased allocating a 348,125.10ha area located in Pesisir Selatan Regency, Solok, Sawahlunto/Sijunjung, in West Sumatra Province to Kerinci Seblat National Park.

In 1999 the Decision Letter of the Forestry and Plantations Minister Number. 46/Kpts/VII-3/1999 was released, part of several blocs of forest in South Sumatra Province was included into Kerinci Seblat National Park.

On April 14th 1999, the Decision Letter of the Minister of Forestry and Plantations Number 200/Kpts-II/1999 was released included some of forests in Jambi province into Kerinci Seblat National Park

On October 14th 1999, the Decision Letter of the Forestry and Plantations Minister Number. 90/Kpts-II/1999 was released, confirming the status of Kerinci Seblat National Park, located in West Sumatra, Jambi, South Sumatra and Bengkulu provinces, covering a total area of 1,375,349.867 ha.

Biodiversity and Ecosystems

There are at least 4,000 species of plants in the park. Most of these are found in lowland forests, which are dominated by the families Dipterocarpaceae, Leguminosae, Lauraceae, Myrtaceae and Bombacaceae. More than 300 species of Orchids as well as bamboos, cinnamons, rattan, *Calamus pilosellus*, *Calamus exilis*, and the rare edelweis daisy, *Anaphalis* sp. are found in the park. In the understaory of the lowland forests there are the very large flowers of *Rafflesia arnoldi*, and *Rafflesia hasseltii* and the tallest flower in the world *Amorphophallus titanium*, as well as the carnivorous pitcher plants, *Nepenthes* spp. The most commercially important tree species are members of the Dipetrocarpaceae family, located in lowlands and hills below about 1,000m altitude. Examples are *Shorea parvifolia, Dipterocarpus spp., Parashorea spp., Koompassia malaccensis* and *Dialium* spp. Between 1,000-1,500m altitude low mountain forest is dominated by other species of Dipterocarpaceae such as *Hopea sp.* and *Shorea platyclados.* Other tree species include *Litsea* sp., *Rhodamnia cinerea*, as well as species from the *Euphorbiacea* and *Leguminosae* families. The understory includes the palms, *Livistona altissima* and *Areca catechu*, epiphytic plants such as *Asplenium* spp., *Bulbophyllum* spp., *Dendrobium* spp. and *Eria* spp., and pitcher plants, *Nepenthes* spp. Above 1,500m altitude, the mountain forests are dominated by members of the *Lauraceae* and *Ericaceae*, families, such as *Podocarpus amarus, Castanopsis* sp., *Ficus variegata* and *Cinnamomum parthenoxylon*. The flora at high altitudes on Mount Kerinci goes from *Schima-Symigtonia* forest between 1,800 and 2,250 m, through to *Quercus-Engelhardtia* and *Symplocos-Myrsine* forests between 2,250 and 3,000 m, and then to *Vaccinium-Rhododendron* shrublands above 3,000 m.

In Kerinci Regency there are two swamps, Rawa Ladeh and Rawa Bento, located at 1,950 m altitude covering 150 ha. These are the highest swamps in Sumatra. Bento swamp in the headwaters of the Sangir River is characterized by the grass *Leersia hexandra*, and the woody shrubs, *Glochidion* sp. and *Eugenia spicata*.

Plants with limited ranges can be found in this area, such as Kerinci pine, *Pinus merkusii kerinci*, leech wood, *Harpullia arborea*, jelutung, *Dyera costulata*, Rafflesia, *Rafflesia arnoldi*, Mersawa, *Anisoptera* sp., and flesh flower, *Amorphophallus titanium*.

The park covers a continuum of habitats from lowland forests to mountain forests, including natural tropical pine forest to alpine vegetation communities. Interspersed there are swamps and swamp forest and open water lakes. Thus, it includes the habitats of most Sumatra birds. There are more than 139 species of birds (nine of them hornbills), but no field guides have been produced and there is no definitive bird list for the park, but it does include the small Sumatra Kuau, also known as bronze-tailed peacock-pheasant, *Polyplectron chalcurum*, and Kerinci owl, *Otus stressmanni*.

Sumatran endemic wildlife that can be found in Kerinci Seblat National Park include Sumatran rhinoceros, *Dicerorhinus sumatrensis* (unfortunately this species was not found during the last survey in 2010), tapir, *Tapirus indicus*, Sumatran tiger, *Panthera tigris sumatrensis*, Sumatran elephant, *Elephas maximus sumatranus*, wild goat, *Capricornis sumatrensis*, Sumatran rabbit, *Nesolagus netscheri*. Many different cats can be found including leopard cat, *Prionailurus bengalensis sumatrans*, clouded leopard, *Neofelis diardi*, golden cat, *Pardofelis temincki*, fishing cat, *Prionailurus viverrinus*, marbled cat, *Pardofelis marmorota*, and flat-headed cat, *Prionailurus planiceps*.

The primates found in the park are siamang, *Sympalangus syndactylus*, agile gibbon, *Hylobates agilis*, grizzled langur, *Presbytis melalophos*, silvered langur, *Trachypithecus cristatus*, pig-tailed macaque, *Macaca nemestrina*, long-tailed macaque, *Macaca fascicularis*, coucang, *Nycticebus coucang* and Bangka tarsier, *Tarsius bancanus*.

Local Communities and Culture

The communities in the area come from the Minangkabau, Kerinci, Komering, Rejang, Kubu, Siak and Sakai ethnic groups. There are also now ethnic groups that have transmigrated from Java. Each ethnic group has its own unique and characteristic culture, which includes customs, traditional dance, art, food, language, traditional architecture, traditional clothes, etc.

Forests which are managed by local communities include Temedak and Muara Air Dua.

The valley and slopes between Mount Kerinci in the north and Lake Kerinci in the south is the Kerinci Regency, which is entirely within the park. Here there is a famous traditional festival, *Kenduri Seko*, which is held to celebrate the establishment of the communities. The people who live here believe that short ground dwelling upright bipedal creatures covered in short grey fur inhabit the forest. Sightings have been reported over the past 100 years and more by villagers The local name is "orang pendek" (short person) the mysterious ruler of the jungle.

Tourism

There are many tourist attractions and considerable tourism potential in Kerinci Seblat National Park. There is a natural pathway, which adventurous tourists can walk from Aman Estuary in Bengkulu to Mandras-Bangko in Jambi over 3 days. A summary of other tourist attractions and potential attractions follows.

Jambi Province:
- Around Sungai Penuh city, Lempur, Mount Kerinci and Masurai, there are beautiful views and hot springs. Hot springs can also be found at Semurup.
- Mount Kerinci (3,805 m) is an active volcano. Many people climb the mountain. The peak can be reached in around 10-12 hours from Kersik Tuo Village.
- Lake Gunung Tujuh (1,996 m) consist of 1,000 ha of fresh water surrounded by 7 mountains. It is the highest freshwater lake in Asia and can be reached by foot in about 3 hours from Polempek. Other lakes in this National Park are Lake Kerinci, Lake Duo and Lake Nyalo. Each of them is a volcanic lake surrounded by hills and mountains.
- Near Mount Tujuh visitors can ride elephants belonging to an elephant safari business.
- The Tea Garden at the edge of Mount Kerinci. Here there is volcanic ash and at night the red color of fire or lava on the top of the mountain can be seen.

- Kayu Aro is famed as a tourist destination where the river is perfect for white-water rafting. The famous Talun Barasok waterfall can also be found here. Other waterfalls in the area include Pancaro Rayo, Tujuh Tingkat, Siulang Bersisik Emas, Sungai Medan, Pincuran Gading, Pauh Tinggi and Nyai Meh Kupak.
- Padang Satwa Ladeh Panjang and Bukit Tapan Wildllife Reserve in West Sumatra province are habitats of the Sumatran tiger, deer, tapir and gibbon. They can be reached by walking for around 10 hours from the Kayu Aro tea plantation.
- Between Mount Masurai, Lake Depati Ampat and Padang Penggembalaan Inumraya (Renah Kemumu) there are enclaves of traditional communities. There are also rare vegetation types such as the homogeneous bamboo forest, a variety of spiky plants and carnivorous pitcher plants, *Nephentes* spp. It can be reached in around 10 hours walking from Sungai Penuh city.

West Sumatra Province:
- In Bukit Tapan Wildlife Reserve and at Mount Indrapura, there are beautiful natural landscapes and habitat for wildlife especially primates. It can be reached by public transportation or bus from Padang city or Sungai Penuh in Jambi province.

Bengkulu Province:
- Mount Seblat (2,383 m) is the habitat of *Rafflesia arnoldi*, many species of primate and reptiles such as python. It can be reached in around 12 hours by walking from Aman Estuary.
- Sambar, Mount Payung, Mount Gedang Seblat and Kayu Embun are areas that have rhinoceros, elephant and Sumatran tiger habitat. They can be reached in around 10-12 hours by walking from Muko-muko

South Sumatra Province:
- At Rawas Ulu Lakitan there are the Sungai Keruh waterfall, Sungai Kerali, Sungai Koten, and the river and waterfall at Sungai Ampar with an extremely rapid flow as well as steep gorges. It is perfect for rafting.
- The Ulu Nasal area has many caves and many unique stones. It can be reached in around 10 hours from Lubuk Linggau by land

Access and Transportation

Kerinci Seblat National Park can be reached through entrances in four provinces, as follows:

Jambi Province

From Jambi to Sungai Penuh by public transportation (bus) takes 10 hours. The trip passes by Bukit Dua Belas National Park, home of of the Anak Dalam Ethnic group, and the Batanghari River. Forests within the national park can be seen from the road. There are handicrafts available from local communities and many species of fruit will be encountered. From Sungai Penuh the park can be reached by car

West Sumatra Province

There are two alternative routes from Padang to Sungai Penuh, through Tapan or Muara Labuh. Both routes are serviced by public transportand and the road is sealed. It will take 7 to 8 hours. The route through Tapan runs beside the National Park and the route through Muara Labuh passes by a tea plantation at Kayo Aro beside Mount Kerinci.

Bengkulu Province

Curup can be reached by public transport or hire car from Bengkulu in 2.5 hours. To Danau Tes and Muara Aman/Ketaun will then take another 3-4 hours by rental car. The road, part of the trans Sumatra road, is sealed and passes West Sumatran beaches, coconut palms and rubber plantations, as well as the Elephant Training Centre and the and Sumatra rhinoceros breeding centre

South Sumatra Province

From Palembang to Lubuk Linggau takes 5 hours on public transport. Muara Rupit, Surulangun and Napal Licin can then be reached by rental car in another 4 hours. Muara Rupit-Napal Licin can also be reached by water transport in 2 hours.

Facilities

- At Sungai Penuh, Kerinci Regency, there are sufficient local hotels
- At Kersik Tuo there are home stays and tour guides for visitors who want to climb Mount Kerinci.
- At Curup, Lubuk Linggau and Bangko there are large hotels and restaurants. Cars can be rented in these places.
- There is a guest house near Mount Tujuh which can accommodate 10 people, for visitors who want to climb the mountain or climb to Lake Gunung Tujuh. The lake can be reached by walking for 3-4 hours from the guest house
- At Lake Gunung Tujuh there is a waterfall and pathway. Visitors can enjoy views of the lake and waterfall as well as wildlife including gibbons and many species of bird. Kayaking is also available.
- There are airports at Padang and Jambi cities.
- At Curup, Painan and Lubuk Linggau there are National Park Offices, Information Centres and ranger post (Jagawana pos).
- There are about 20 security posts in the area of Kerinci Seblat National Park:
  - Jambi Province 8 posts
  - West Sumatra Province 5 posts
  - Bengkulu Province 4 posts
  - South Sumatra 3 posts
- At Bangko there is a Ranger Post

Support facilities in Kerinci Seblat National Park include guest house, National Park museum, tourism cottages, information center, work cottages, stores, parking areas, camping grounds, shelters, observation towers and pathways.

Park Office

Balai Besar Taman Nasional Kerinci Seblat
Jl. Basuki Rachmat No. 11 Kotak Pos. 40, Sungai Penuh, Jambi 37101
Telp: (+62-748) 22250
Fax: (0748) 22300
Call Center: 082287971791; 08217811 5812; 0812 2351 8491; 08112009 778; 082288516729; 085274710371; 081398952211
Instagram: @ bbtn_kerinciseblat
Email: bbtnks@gmail.comwebsite: www.kerinciseblat.dephut.go.id

3. SIBERUT NATIONAL PARK

Geographic Location and Size

Siberut National Park is located on Siberut Island in the Mentawai Islands Regency, West Sumatra Province. Siberut Island is located between $0°55'$ and $1°49'$ S and between $98°35'$ and $99°18'$ E. The national park runs from north to south and almost divides Siberut Island into two parts. The park is bordered by the Indian Ocean in the west and Mentawai strait in the north.

The total size of Siberut National Park is 190,500 ha

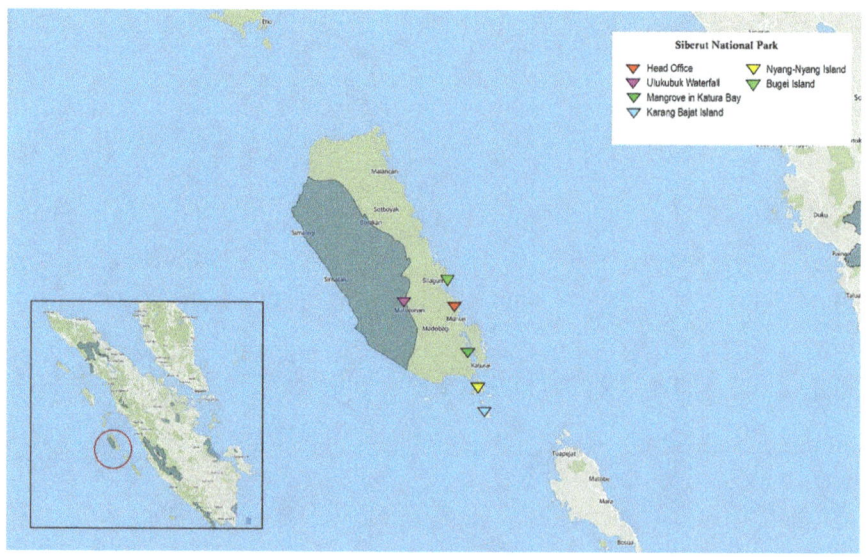

Climate and Topography

Being equatorial and adjacent to the sea the park is always hot and humid, with temperatures between 22^0-31^0C. Rainfall is around 2,900-3,750 mm/year with monthly rainfalls of more than 200 mm. The number of rain days per year is approximately 300. The highest rainfall is in November, between 400-500mm. Other periods of high rainfall are April (290 mm) and October (390 mm). February (220 mm) and June (220 mm) have the lowest monthly rainfalls.

At 430,300ha, Siberut Island is one of the largest in the Mentawai archipelago. It is located between 100-155 km west of Padang, the capital city of West Sumatra Province. The elevation of the park is from 0-384 m. The Mentawai Islands are separated from Sumatra Island by the Mentawai basin that reaches a depth of 1,500 m. The islands were isolated about 500,000 years ago during the Pleistocene era. This isolation has given rise to a unique flora and fauna derived from a different evolutionary pathway than that followed on the Sunda shelf to the east.

The island is characterized by hilly topography. Subsidence from tectonic activity has made the east coast irregular with many bays, small islands, capes and coral reefs. The west coast is more stable. It is straight and bordered by a seemingly endless wide sandy coast.

History

On October the 25th 1976, Decision Letter of Agricultural Minister Number 670/Kpts/Um/10/1976, declared the forest of Teiteibati on Siberut Island a Wildlife Conservation reserve of 6,500 ha.

On December the 5th 1978, according to the Decision Letter of Agricultural Minister Number 758/Kpts/Um/12/1978, the status was changed to Teiteibati Wildlife Conservation reserve of 56,500 ha.

In 1981 this reserve was designated a Biosphere Reserve.

On August the 23rd 1982, according to the Decision Letter of Agricultural Minister Number 623/Kpts/Um/8/1982, the area was expanded to consist of Teiteibati Wildlife Conservation reserve of 82,900 ha on the northern end and Siberut II Wildlife Conservation reserve of 50,000 ha on the southern end.

On October the 8th 1993, Decision Letter of Forestry Minister Number 407/ Kpts-II/93, declared Siberut National Park, with an area of 109,500 ha, consisting of Wildlife Conservation reserves, Restricted Forest, Limited Production Forest and Production Forest.

Biodiversity and Ecosystems

Mixed forests, swamp forest, beach and mangrove forests, spread from the hills to the beach. The dominant trees in this park are from the *Dipterocarpaceae* family, the species of which become tall trees up to 60 meters high. At the northern end of the park dominant species include *Dipterocarpus gracilis* and *Dipterocarpus sublamellatus*, while at the southern end, the dominant species include *Shorea* sp. and *Hopea* sp. Common species from other families are *Palaquium* sp. and *Hydnocarpus* sp. Mixed lowland forest with high species diversity occurs from valleys to hill slopes. Characteristic species are *Aporusa quadrilocularis, Diospyros brevicalyx, Drypetes subsymmetrica, Horsfieldia macrothyrsa, Mesua catharinae, Dipterocarpus elongates, D. gracilis, D. retusus, D. sublamellatus, Shorea pauciflora, Hopea dryobalanoides, Pentace* sp., *Durio* sp., *Dialium* sp., and gaharu, *Aquilaria malaccensis*. This forest has many lianas that reach high into the crown, as well as epiphytes and palms such as rattan manau, *Calamus manau*, other rattans, *Calamus caesius, Calamus* spp.and *Daemonorops* sp.

Freshwater swamp forest occurs in valleys on the east side of the park. Trees in this forest have pneumatophores, aerial roots that emarge from the water to facilitate root respiration. Sago trees, *Metroxylon sagu* are common, as well as palms such as *Licuala spinosa*. Other species that can be found here are *Alstonia spathulata, Stemonurus secundiflorus, Oncosperma* sp., *Terminalia phellocarpa* and *Artocarpus kemundo*

Beach forest can be found on the south west coast that are not exposed by the ebb and flow of the sea. The characteristic species are *Barringtonia* sp., cottonwood, *Hibiscus tiliaceus,* Pandan, *Pandanus tectorius*, nipa palm, *Nypa fruticans* and *Calophyllum inophyllum,* cemara laut, *Casuarina equisetifolia* and *Syzygium pseudoformosum* form pure stands. A *Eugenia grandis* and *Calophyllum inophylloides* community occurs on the coral reef substrate at the south west coast. Other species at this site are *Dipterocarpus* sp., *Shorea* sp., *Durio* sp., *Musa* sp., *Solanum* sp. as well as *Ipomoea* sp.

Mangrove forest spreads along the south west coasts of Siberut island and adjacent islands. Most mangrove forests are ound in the west of Siberut island, outside of the park. Mangrove forest grows well on sand, mud, and coral as well as in river estuaries which are affected by the ebb and flow of the tide. This forest cannot be found on the west coast due to strong winds and waves as well as the big offshore waves favoured by surfers. The dominant mangrove species are *Rhizophora apiculata, Rhizophora mucronata, Rhizophora stylosa, Bruguiera gymnorrhiza, Bruguiera sexangula, Ceriops candoleana, Sonneratia alba, Sonneratia caseolaris*, with *Nypa fruticans* on the edge of mangrove formations. Secondary forest, disturbed forest, is found close to communities living around the park. It results from community disturbances like burning or the remnant forest after logging activities.

Besides a high diversity of plants, this park has a diverse array of wildlife. There are about 31 species of mammals, 17 of which are endemic. There are four endemic primates, four species of endemic squirrels, four species of mice, one of which is endemic and 105 species of bird with one endemic species and 13 endemic sub-species. Endemic primate species found in Siberut National Park are Siberut macaque or bokkoi, *Macaca siberu,* Siberut langur or joja, *Presbytis siberu*, dwarf gibbon, or Kloss' gibbon, *Hylobates klossii* and simakobu or pig-tailed snub-nose langur, *Simias concolor*. Altogether, there are six endemic Mentawai Island primates. The two found outside of the park are on the Pagai island, Pagai langur, *Presybtis potenziani*, and Pagai macaque, *Macaca pagensis.*

Simias is a monotypic genus with two subspecies: *Simias concolor concolor* (Miller, 1903) inhabits Sipora, North Pagai, and South Pagai islands and several small islets off South Pagai; *Simias c. siberu* (Chasen and Kloss, 1927) is restricted to Siberut Island (Zinner et al., 2013). *Simias concolor* is classified as Critically Endangered on the IUCN Red List (Whittaker and Mittermeier, 2008), threatened mainly by heavy hunting and commercial logging (Whittaker, 2006). The Pagai island populations are threatened by forest conversion to oil palm plantations, and forest clearing and product extraction by local people (Whittaker, 2006). Although hunting appears to be declining and opportunistic in many areas of the Pagais, where it still occurs it has devastating effects on *S. concolor*, which is the preferred game species (Mitchell and Tilson, 1986; Fuentes, 2002; Paciulli, 2004). Tenaza (1987) estimated that twice as many individuals were killed by hunters each year as were born in the Pagai Islands. In a multi-population study, Erb et al. (2012) found that hunting pressure reduced group size, resulting in the formation of male-female pairs, which is atypical for Asian colobines, which normally form small one-male groups with around five females.

The Mentawai owl, *Otus mentawi*, is also endemic to these islands, but little is known of its ecology or behavior.

Local Communities and Culture

The indigenous population of the Mentawai islands may represent the first inhabitants of Indonesia, Proto Melayu people, who were the ancestors of the people now occupying Malaysia, parts of the east coast of Sumatra and parts of central west Sumatra. There are 15 ethnic groups with different languages in 50 villages, including Simatalu, Sagulubbe, Sirileleu, Sarareiket, Sempungan, Saibi and Sarabua.

The Siberut people hunt by using a bow and arrow and respect the forest and wildlife. Originally, they practiced animism, and some still do. The Social political and religious life of a village is centered on the *Uma*, the long house that is inhabited by a number of families who are related to each other. The *Sikerei* is the guardian of the clan's traditions and the spiritual leader.

Tourism

- From the east coast of Siberut some of the best views of Sumatra can be had
- In Simabugai area, western part of the park, endemic primates and other species of wildlife can be seen easily. In this area, the German Primate Center and Bogor Agricultural University established a research station.
- The ocean off the southern part of the park is often full of surfers from all over the world. The Mentawai Islands are some of the most favorite surfing sites in Indonesia.

Access and Transportation

There is a ferry that runs three times each week and takes about 10 hours from Padang to Muara Siberut and Muara Sikabaluan estuaries.

From Muara Siberut or Muara Sikabaluan estuaries there are canoes or motor boats up the rivers to the park, or walking to the park is also possible

Facilities

In Siberut Estuary there is a small hotel and guesthouses.

Park Office

Balai Taman Nasional Siberut
Jl. Khatib Sulaiman No. 46 Padang 25433
Telp: (+62-754) – 7059986
Fax: (+62- 754-7050585
Call Center: 0853 7747 2240
Instagram: @ btn_siberut
Email: Taman-nasional_siberut@yahoo.com

4. BATANG GADIS NATIONAL PARK

Geographic Location and Size

Batang Gadis National Park (BGNP), which covers 72,150 ha, is located in Mandailing Natal (Madina) regency, North Sumatra Province. It is located between 99° 12' and 99° 47' E and between 0° 27' and 1° 02' S. The name of the National Park comes from the name of the main river, Batang Gadis, that originates in and flows through the regency.

The area is hilly between 300 m and 2,145 m asl with its highest point on the peak of the volcano, Mount Sorik Merapi. BGNP was formed from areas of Restricted Forest, Limited Production Forest and Permanent Production forest.

History

The story of establishing the park began with the Conservation International Indonesia (CII) approaching the most influential person in the community, the Islamic school's top spiritual leader, or imam. CII staff described the potential pollution problem facing the Batang Gadis river running through his school's property. The river would become polluted with tailings if a proposed gold mine upstream in the Batang Gadis forest went ahead as planned. The river would also become clogged with silt and could possibly dry up if logging around the river was permitted. It was explained to the imam that the Batang Gadis water that he and thousands of his students depended upon to wash themselves five times before prayer, would become very dirty and contaminated with poisonous chemicals.

The imam was skeptical of these predictions, so he was taken to see what the mining and logging activities in the upper part of the river were doing to the water quality. The imam was surprised, and he agreed that the logging and mining activities were indeed contaminating the river. He knew that polluted water would not be good to use for the 'ablution" or cleansing ritual. The imam began to discuss these issues with his students, and his students talked to their parents, and so the potential hazards to the quality of the water spread to the entire community. The village council was also approached to remind them of their traditional practices of "naborgo-naborgo" and "lubuk larangan." Lubuk larangan is a pool of fishes in the deep river protected by community while naborgo-naborgo is where community harvests those fishes together as part of a holy day celebration.

As awareness of the potential pollution problems spread, concern grew from the bottom up. The communities took the issue to the bupati (the regency head, at that time an appointed position) and asked for the forest to be protected. The bupati agreed to consider this request but was worried about how he could compensate the people who profited from the logging and mining activities. It was explained to him that the Batang Gadis River was the source of irrigation water for 42,000 hectares of local rice fields, all of which would be imperiled if the river was polluted by new industries established upstream.

With a lot of support from different elements of society, the Governor of North Sumatra, through his letter Number 050/1116 on the 2nd of March 2004, formally give support for the establishment of BGNP. This was followed by the release of the decision letter from Forestry Minister on the 29th of April 2004 No.l26/Menhut-11/2004 declaring 108,000 hectares in Mandailing Natal Regency of North Sumatra as Batang Gadis National Park. The ultimate support was given by the President of the Republic of Indonesia, Megawati Sukarnoputri, by inaugurating BGNP through the signing of an inscription in Panyabungan in May 2004. Conflict between BGNP and the gold mine continued until 2012 when the Ministry of Forestry revised the total area of the park from 108,000 ha down to 72,150 ha, so that 35,850 ha was given to the gold mine Sorik Merapi (Tanjung and Dewi, 2014).

Biodiversity and Ecosystems

Satellite imagery shows that 88% of the forest in the park is still in a good condition. Only 7% is degraded forest and 5% is agricultural land. From a research plot of 200m^2 242 species of vascular plants were recorded. This is about 1% of the approximately 25,000 vascular plants found in Indonesia. There is also a rare undescribed new species of *Rafflesia*. The biodiversity in this park represents a blending of north, south and east Sumatra.

Using camera trapping and from observations during surveys, some rare species of wildlife have been found, such as Sumatra tiger, *Panthera tigris sumatrae*, wild goat, *Naemorhedus sumatraensis*, tapir, *Tapirus indicus*, golden cat, *Catopuma temminckii*, mouse deer, *Tragulus javanicus*, binturong, *Arctitis binturong*, honey bear, *Helarctos malayanus*, deer, *Cervus unicolor*, Indian muntjac, *Muntiacus muntjac* and porcupine, *Hystrix brachyura*. The survey team has also found a legless amphibian, *Ichtyopis glutinosa*, and the three horns toad, *Megophrys nasuta*, which are endemic to Sumatra (CI, 2004).

The primates in this park are siamang, *Sympalangus syndactylus*, agile gibbon, *Hylobates agilis*, Grizzled langur, *Presbytis melalophos*, possibly black Sumatran langur, *Presbytis sumatrana*, silvery leaf monkey, *Trachypithecus cristatus*, Pig-tailed macaque, *Macaca nemestrina*, Long-tailed macaque, *Macaca fascicularis*, and Sumatran coucang, *Nycticebus coucang*.

At least 247 bird species are found in BGNP. Of this number, 45 are protected by law, 8 are globally endangered and 11 are threatened. Examples include Sumatran ground cuckoo, *Carpococcyx radiceus*, Salvadori's pheasant, *Lophura inornata*, and Crested fireback, *Lophura ignita*. Two species of bird listed as data deficient by IUCN were also found. There are 13 species categorized as range restricted birds that contribute the identification of Endemic Bird Areas and Important Bird Areas as defined by Burung Indonesia. The park is also as a transit site for migrant birds that come from northern countries. Salvadori's pheasant and Schneider's pitta, *Pitta schneiderii*, are rare and endemic Sumatran species. They had been deleted from the Sumatra Bird Species List for almost a decade. A specimen of Sunda ground cuckoo, *Carpococcyx radiceus*, was found for only the second time in almost a decade.

BGNP has six hornbills from the family Bucerotidae, which is 60% of the hornbills found on Sumatra Island. This indicates that the tropical forest of BGNP is still in good condition for fruit eating wildlife.

Local Communities and Culture

The park is mostly surrounded by Batak Tribes of Mandailing, Natal, and Angkola and several tribes from Toba. Mandailing tribes are mostly muslim although some are christian. These people still use the traditional adat system of law on space and land, which here is known as Na Mora Na Toras (nobles and elders). The local leader is called Pamusuk King. This system governs the ways in which each tribe utilize the resources of their territories.:

The Batang Gadis River flows like a major artery through the forests of six different watersheds, spreading out into smaller streams and creeks for a total of over 137.5 km. This watershed system supplies almost all of the domestic water needs of 400,000 people in the regency, including water for irrigating 42,100 hectares of paddy fields and 108,320 hectares of commercial crops such as coffee, cinnamon, cacao, palm oil, cloves, ginger, and others. Economists estimate an additional amount up to $24.8 million per year in indirect values from this extensive water system (Midora and Anggraeny 2006).

For the people of Mandailing Natal, a local Islamic community, the river provides water for drinking and bathing, sanitation needs, irrigation, and additional socio-cultural, religious, and economic functions. These people traditionally protect places called "naborgo-naborgo" through customary law.

In the 1970s, local committees formalised these laws into river protection (lubuk larangan) schemes (Lubis, 2001). An Islamic practice known as a harim (meaning zone) prohibits the harvesting of fish from rivers close to human settlements for 6 to 12 months each year. The community leaders have the power to decide when the best harvesting time is and a small fee is charged to residents or visitors who desire to fish. The income generated is used to pay for the development of local social facilities such as schools, roads, and mosques, and some of it goes to provide educational scholarships and administrative salaries and grants to orphans, poor families and invalids (Lubis, 2001).

The river also has cultural value to the people here. Students attending Islamic boarding schools (pesantren) use the river to bathe and to perform the ritual "wudhu," the washing of their bodies before prayer, which they do five times a day. One of the largest boarding schools in the region, Al Mustafawiyah, has more than 7,000 students from surrounding villages, regencies and other provinces, and even from abroad. The school itself has been established for many years, and is the best known school in Sumatra for studying Islamic teachings.

The exploitation of Batang Gadis for logging displaced almost all traditional customs and regulations by shifting control, including the ownership of the forest, to central government, diminishing the role of traditional cultural values. For centuries local communities have wisely managed and depended on the natural forest and rivers, but then access to these resources became more and more restricted. This situation triggered several social problems causing conflicts between the central government, local governments and traditional communities.

For many years local communities were suppressed and unable to appeal to any government system regarding land use issues. Only after the old regime was replaced, and authority over the forests was decentralized to the regency level in early 2000, were those conflicts brought up and begun to be addressed. As a result, a new local government was established that enabled the traditional institution of Na Mora-Na Toras to be revived and to play a critical role in the management of local affairs including the protection of forests and rivers. This helped the Mandailing people to recover their tradition of consultative governance and encouraged them to challenge the Bupati of the regency to promote a participatory planning and decision-making process (Lubis, 2001.)

Tourism

The tourism sites are centered in the villages of Sibanggor Jae, Sibanggor Tonga and Sibanggor Julu. Mount Sorik Marapi at 2,145 m asl is a stratovolcano with a very active and interesting crater, which can be climbed from Sibanggor Julu village.

Potential tourist attractions in this park are Bagas Godang (mansion of the king in traditional Mandailing villages), a former Japanese soldier tunnel, Portuguese cannon and others. Other potential tourism sites are:
- Sibanggor Hot Springs, at Tambangan
- Sampuraga Hot Springs, at Panyabungan Regency

- Siabu Hot Springs, at Siabu Regency
- Lake Siombun, Panyabungan Regency
- Lake Marambe, Panyabungan Barat Regency
- Batang Gadis Dam, Panyabungan Regency
- A group of habituated macaque monkeys, at Siabu Regency
- Putusan Hot Springs, at Panyabungan Selatan Regency
- Sitaut Waterfall, at Kotanopan Regency
- Sopotinjak View, at Batang Natal Regency
- Natal Beach, at Natal Regency
- Sikara-Kara Beach, at Natal Regency

Access

Fly from Medan, the biggest city in North Sumatra, to Aik Godang, the small airport in South Tapanuli. Then continue by rental car to the national park, which takes about 3-4 hours.

Facilities

At Panyabungan City, the nearest city to the national park, there are several fine hotels plus other smaller hotels. This city is on the main road from North Sumatra to West Sumatra, so the bus is available 24 hrs either to Padang or to Medan.

Park Office

Balai Taman Nasional Batang Gadis
Jl. Wilem Iskandar no.1 Kel. Pidoli Dolok
Panyabungan - Sumatera Utara
Telp. (062-636) 321670, 321675
Fax. (062-636) 321670
Call Center: 08116250555
Instagram: @ btn_batang gadis
Email: btnbtggadus42@gmail.com

5. BUKIT TIGA PULUH NATIONAL PARK

Geographic Location and Size

Bukit Tiga Puluh National Park is located in two provinces, Jambi and Riau. In Jambi this park is bordered by seven regencies, Jabung Regency, Tungkal Ulu Regency, Merlung Regency, Bungo Tebo and Tebo Ilir Regencies, Sumai Regency and Tujuh Koto Regency. In Riau it is bordered by Indragiri Hulu and Indragiri Hilir regencies. The villages near the park are Pemayungan, Semambu, Muaro Sekalo, Suo-suo, Lubuk Mandarsah, Lubuk Kambing, Suban and Lubuk Bernai.

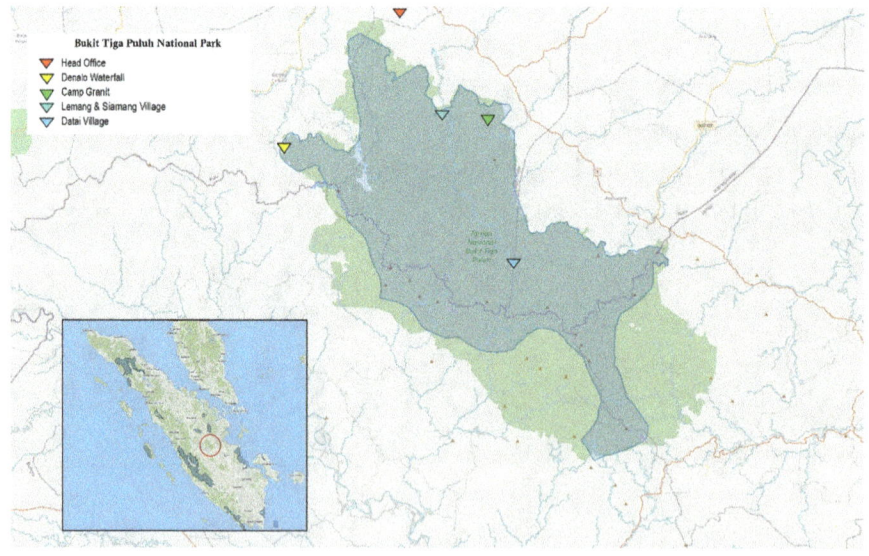

Bukit Tiga Puluh National Park covers 33,000 ha in Jambi and 94,698 ha in Riau and additional forest was added in 2002, therefore the total national park area is 144,223 ha.

Climate and Topography

The temperature range is from 28^0-37^0C. The park covers lowland to highland tropical rain forest from 60-843 m asl level. There are 30 hills, hence the name Tigapuluh. The highest peak is Bukit Hulu Supin (843 m). From this park, two big rivers flow into Jambi, the Indragiri and Batanghari Rivers.

History of the park

Before they were amalgamated into a national park, there was Seberida Nature Preserve in Jambi province covering 33,000 ha, an area of 57,448 ha of Limited Production Forest and 37,250 ha of Restricted Forest, both in Riau Province.

On the 5th of October 1995, they were collectively declared to be Bukit Tiga Puluh National Park, of 127,698 ha, according to Decision Letter of Forestry Minister Number 539/Kpts-II/95.

In 2002, additional forest was added to the park based on Ministry Forestry decree no. 6407/Kpts-II/2002, when the size of Bukit Tiga Puluh National Park was increased to 144, 223 ha.

Biodiversity and Ecosystems

The ecosystems in this park are lowland rain forest, swamp forest and upland rain forest. It protects the catchment of the Kuantan Indragiri River. Characteristic plants include jelutung, *Dyera costulata*, red sap, *Palaquium* spp., pulai, *Alstonia scholaris*, kompas, *Koompassia excelsa*, dipterocarps, *Shorea* spp., rafflesia, *Rafflesia hasseltii,* jernang or blood dragon palm, *Daemonorops draco*, and various species of rattan.

There are 59 species of mammals, six species of primates, 151 species of birds, 18 species of bats and many different species of butterflies. Iconic species include the Sumatra tiger, *Panthera tigris sumatrae*, tapir, *Tapirus indicus*, honey bear, *Helarctos malayanus malayanu*s, sempidan biru or crested fireback, *Lophura ignita* and kuau, *Argusianus argus*. Many wild cats are found in the park such as leopard cat, *Prionailurus bengalensis sumatrans*, clouded leopard, *Neofelis diardi*, golden cat, *Pardofelis temincki*, marbled cat, *Pardofelis marmorota* and flat-headed cat, *Prionailurus planiceps*.

The primates that occur here are siamang, *Sympalangus syndactylus*, agile gibbon, *Hylobates agilis*, grizzled Langur or simpai, *Presbytis melalophos*, black and white langur, *Presbytis bicolor,* silvered langur, *Trachypithecus cristatus*, pig-tailed macaque, *Macaca nemestrina*, long-tailed macaque, *Macaca fascicularis*, Sumatran coucang or slow loris, *Nycticebus coucang* and western tarsier, *Cephalopachus bancanus*.

Local Communities and Culture

There are 345 families from eight villages inside Bukit Tigapuluh National Park. They are all indigenous Sumatran people. In the north of the park in Riau Province there are people of Talang Mamak ethnicity Proto-Malays (Melayu Tua), which are indigenous to Indragiri Hulu, a regency in Riau Province. They are also known as "Tribe Tuha", which means the first tribe to arrive. This means they have more rights to the natural resources in Indragiri Hulu than later arrivals.

Another tribe is the Kubu ethnic group (known as Jungle Men) who live nomadic lives. The Kubu people spread from the Alim and Cindaku headwaters downstream in Indragiri Regency along the Sumai River and its streams, such as Langgas, Cempogan, Gelumpang, Rotan, Sentano, Salak, Serut, Sako, Peladangan, Tembulun, Kubu Jahat, and many others in Bungo Tebo regency of Jambi Province.

On the borders of the park there are also scattered tribal communities of Anak Dalam ethnicity, as well as Talang Mamak and traditional Malay communities. Here there may be many villages suitable for the development of cultural tourism, with many different types of handicrafts, traditional people who still wear traditional clothes, the agriculture slash and burn or swidden agriculture (burning, cropping and then moving on). There are also sacred and historical places such as the traditional houses of the proto-Malay tribes.

Local communities have a high level of dependence on local forest products. Some examples are the use of jernang, honey, rattan, jelutung sap, resin, bamboo, wood, medicinal plants, wildlife and many others.

Tourism

Current and potential tourist attractions are described below.

- In Pemayungan village there are waterfalls, and *sialang* trees, which are favoured for the honey that bees produce from their flowers. It is interesting to watch the local people climb high into the trees to collect the honey.
- In Semambu village there are waterfalls, indigenous residences and sialang honey
- In Suo-suo village there are waterfalls, indigenous residences and sialang honey
- In Lubuk Mandarsah village there are the ancestor tombs, which are a hundred years old and sacred, as well as natural caves.
- There are also waterfalls in Lubuk Kambing, Lubuk Bernai and Suban villages
- The best season to visit this park is between March-July.

Access and Transportation

Bukit Tiga Puluh National Park is located near Rengat city, which is on the highway, on the east side of Sumatra. The entrance is in Rantau Langsat City, which can be reached from Siberida village on the highway south of Rengat. From Pekanbaru to Siberida by motorbike is around 4 hours (± 285 km), then continue to the park on the road of the former of logging concession.

Facilities

There is a path in the National park 12 km long and 20 m wide running from Sanggi to Bengkunat. Small hotels are available in Rengat city. Some of them are very good and reliable.

Park Office

Balai Taman Nasional Bukit Tigapuluh
Jl. Lintas Timur Km. 3 Rengat Barat, Indragiri hulu, Riau
Telp. : +62-769 2341008
Fax: +62-769-341727
Call Center: 0811 7675 733
Instagram: @ btn_bukittigapuluh
Email: btnbt2003@yahoo.com

6. TESSO NILO NATIONAL PARK

Geographic Location and Size

Tesso Nilo National Park is located in Riau Province between $1°05'$-$1°35'$ South Latitude and $104°05'$-$104°30'$ East Longitude. Originally this park was various logging concessions in the regencies of Pelalawan and Indragiri. The size of the national park is 81,793 ha.

Climate and Topography

The climate is humid with annual rainfall between 2,000 and 3,000 mm. The park protects some of the last remaining forest of a low flat plain that is extremely high in biodiversity. Most has now been lost to agriculture and timber extaction followed by plantations. The park is between 100-200 m asl and local relief is very low.

History

In August 2003, a logging concession of 38,576 ha of PT. Inhutani IV was cancelled by the Minister of Forestry and was reserved for developoing a conservation area

In 2004, the Forestry Minister declared the area to be a ntional park located in Pelalawan and part of Indragiri Regencies Riau Province

Biodiversity and Ecosystems

Tesso Nilo National Park contains forest representing the transition from low to high plain ecosystems with very high biodiversity. Studies by Gillison et al (2001) found 360 plant species in one hectare of forest. Protected and endangered plants include bata wood, *Irvingia malayana*, kempas, *Koompasia malaccensia*, jelutung, *Dyera costulata*, kulim wood, *Scorodocorpus borneensis*, tembesu, *Fagraea fragrans*, gaharu, *Aquilaria malaccensis*, ramin, *Gonystylus bancanus*, keranji, *Dialium* spp, meranti, *Shorea* spp, keruing, *Dipterocarpus* spp and different species of durian, *Durio* spp.

There are also at least 82 species of medicinal plants in this national park. Pasak Bumi, *Eurycoma longifolia*, is famous for bestowing stamina. Other species of medicinal plant include kunyik bolai, *Zingiber purpureum*, jarangau, *Acorus calamus*, white ginger, *Alpinia galanga*, bulu root, *Argyreia capitata*, sundik langit, *Amorphopalus* sp. and yellow wood root, *Lepionurus sylvestris* that is used to treat jaundice

In each hectare of the park, on average, there are 360 plant species from 165 genera and 57 families. The park also harbours at least 107 species of bird, 23 species of mammal, seven of which are primates, 50 species of fish, 15 species of reptile and 18 species of amphibian. The primates are siamang,

Sympalangus syndactylus, agile gibbon, *Hylobates agilis*, banded langur, *Presbytis siamensis*, silvery leaf monkey, *Trachypithecus cristatus*, pig-tailed macaque, *Macaca nemestrina*, long-tailed macaque, *Macaca fascicularis*, and Sumatran coucang, *Nycticebus coucang* (Supriatna and Mariati, 2014). Many wild cats are found in the park, including Sumatran tiger, *Panthere tigris sumatrae*, clouded leopard, *Neofelis nebulosa*, marbled cat, *Pardofelis marmorata*, leopard cat, *Prionailurus bengalensis*, and flat-headed cat, *Prionailurus planiceps*.

The lowland forest in Tesso Nillo is an important habitat for endangered species such as the Sumatran tiger, *Panthera tigris sumatrae* and the Sumatran elephant, *Elephas maximus sumatranus*. Unfortunately, most of the local forest has been converted into oil palm plantations, pulp wood plantations, and human settlements. This has led to conflict between land use and biodiversity protection, which is increasing exponentially (Margono et al., 2012). WWF (2012) found that serious conflicts between oil palm plantation and protecting wildlife began after the construction of two roads, Baserah and Ukui, commenced in 2000 and 2012 respectively. Illegal logging in the Tesso Nilo Forest increased after the roads were built, primarily in the park, since there is better timber potential than production areas, most of which had already been logged.

A survey of the elephant population in Riau carried out by WWF (2013) found that numbers had declined very sharply compared to a survey in the 1980's. The cause is habitat loss due to the expansion of agricultural plantations as well as farmers killing elephants directly when they stray onto their farms.

Local Communities and Culture

There are traditional communities inside the park from at least 19 tribes including Pana kampung, Batin Palabi, Datuk Songgan, Monti Gola, Datuk Lelo Bunsu, Datuk Mangkuto besar, Datuk Tua Banjo, Ninik Mamak, Datuk Raja Melayu, Datuk Momat, Panglimo Garang, Datuk Majo, Datuk Rajo Bilang Bonsu, Monti Rajo, Batin Mudo Gondai and others. Those tribes are called Batin and have claimed the forest inside the park as theirs (Mariati 2014).. Then, in spite of the declaration of the national park, they sold the forest to people who migrated into the area. These migrants easily got acceses through a road built by companies with a government permit. By the end of 2015, much forest in the park had been converted illegaly into oil palm plantations, and only in certain areas was it still in good condition.

Tourism

- The main purpose of this national park is to protect Sumatra elephant habitat. Tourism is still not common because suitable facilities have not yet been established, even for such a very high-profile attraction as wild elephants.
- Park rangers ride elephants as a part of a patrol called Flying Squad. During the weekend, it is possible for tourists to ride these elephants. There is an observation tower located near Camp Flying Squad in the north central part of the park. It provides a good overview of the park and a magnificent vantage point for seeing sunsets and sunrises.
- Fresh tracks from wild tigers are frequently seen near Camp Flying Squad. In other parts of the park, rangers will probably be able to locate tracks made by Malayan tapir, wild pigs and deer. There are also trees that have been clawed by hungry Malayan sun bears seeking food. Visitors can also walk through that section of the park. Driving through the park with the rangers, there is also a good chance of spotting primates on the road and in the trees that were planted for tree farms before the creation of the park.
- Charter boats costing 300,000 rupiah go up the Nilo River to just past where it joins with the Tesso River. The rangers and guide will indicate that part of the jungle is the most intact in the park.
- It may be possible to arrange to see the ceremony that local people conduct before sustainably harvesting wild honey from sialang trees. The ceremony includes casting of shadows of the hands and head of prospective climbers. If the shadow of a person's hand shows only four digits or if a person's head is not connected to the rest of their body, that person is not allowed to climb the tree and harvest the honey.

Access and Transportation

Access to Pekanbaru, the capital city of Riau Province, is easy and there are many alternatives. It can be reached by driving or taking public transport, by plane and by ship. From Pekanbaru, Tesso Nilo National Park is a ± 5 hour drive on logging roads. From Pelalawan, the town with the park headquarters, it is a 2 hour drive to the park.

Facilities

The closest best hotel facilities are in Pekanbaru. There are many good accommodation options in that city, from bed and breakfast to 5 star hotels. Pelalawan also has a number of good hotels.

Park Office

Balai Taman Nasional Tesso Nilo
Jl. Langgam km 4, Kotak Pos 1, Pangkalan Kerinci, Pelalawan, Riau
Telp./Fax. (062-761) 494728
Call Center: 0811 7513 086
Instagram: @ btn_tessonilo

7. BERBAK NATIONAL PARK

Geographic Location and Size

Berbak National Park is located in Jambi Province, approximately 50 km east of Jambi City, between $1°05'$-$1°35'$ South Latitude and $104°05'$-$104°30'$ East Longitude. It covers Kumpeh District of Batanghari Regency and Nipah Panjang and Sadu Districts of Tanjung Jabung Regency. The southern boundary follows the Benu River west to the Simpang Kecil River (a tributary of the Benu River), before following the border with South Sumatra Province. On the west side the boundary passes through forest and beside agricultural land to the north, before following the Air Hitam (black water) River east to the beach. The eastern boundary goes from the Benu River in the south to the Remau River in the north.

Berbak National Park covers 162,700 ha and is the largest protected area of freshwater and peat swamp forest in southeast Asia.

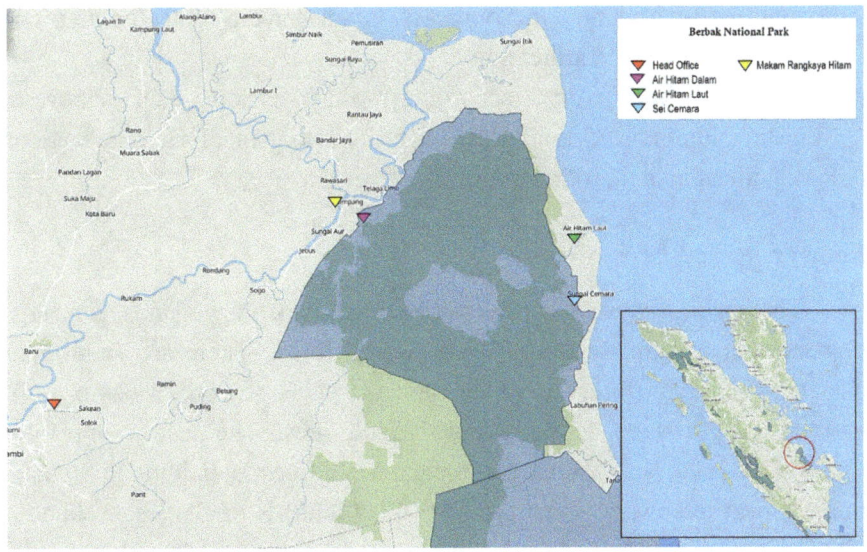

Climate and Topography

The rainfall is variable ranging from 1,500 to 2,500 mm/year. There is a long dry season from April to October. A dry cycle is expected to occur every 3-6 years, and a very dry cycle every 80 years, an effect of the *El Nino* southern oscillation. The average annual temperature is about 22^0C. The humidity is around 85%, which increases from November to February, when northwest winds bring humid air to the east coast of Sumatra.

The park protects part of the large alluvial plain that occupies about one quarter of the island of Sumatra. Local relief is between 0 and 12.5 m. The plain is dissected by meandering rivers that drain east and northeast towards the coast. Along the low coastal plain there are mud flats and mangroves.

History

On the 29th of October 1935, the government of the Netherlands, with Provision Letter Number 18 GB. Stbl. 521, declared the site to be a Wildlife Conservation reserve of 190,000 ha

In 1985 the area was reduced by 15,000 ha to175,000 ha

On the 19th of October 1991 according to the President Decision Number 48/1991, the site was declared a *World Natural Heritage Site*, protected by the multi-lateral Ramsar Conservation. With this decision, Indonesia became the first country in Southeast Asia to have a protected area with Ramsar Convention Ratification

On the 26th of February 1992, according to the Decision Letter of Forestry Minister Number 285/Kpts-II/92, the site was named Berbak National Park with an area of 162,700 ha.

Biodiversity and Ecosystems

The major ecosystems in Berbak National Park are wetlands, consisting of freshwater swamp forest and peat swamp forest. There are small areas of mangrove, beach forest, savanna, savanna palm, seasonal swamp forest, seasonal wet grassland, seashore and riverine forest. The peat swamp forest (110,000 ha) is the largest intact example of such forest remaining in Sumatra. The non-peat freshwater swamp forest (60,000 ha) is the largest wetland of this kind in southeast Asia. This 27, this forest has the most palm species found anywhere in Indonesia. At least 260 plant species found in the park include 10 species of pandan, *Pandanus* spp. Typical plant species are durian, *Durio carinatus*, jelutung, *Dyera costulata* and ramin, *Gonystylus bancanus*. There are also rare plants such as the umbrella leaved palm, *Johannesteijsmannia altifrons*.

The average high tide on the coast is 2 to 2.5 m, which creates estuarine conditions up to 10 km inland. At the Benu River on the southern border, the coastal beach forest is dominated by nipah, *Nypa fruticans* and pandan, *Pandanus tectorius*. Inland, the forest is dominated by *Mammea* spp.

At least 250 species of bird can be found in the park, including 22 species of migrant birds and water birds, which are protected in Indonesia. Three other notable birds found in the park are crested goshawk, *Accipiter trivirgatus*, wreathed hornbill, *Aceros undulatus* and eastern curlew, *Numenius madagascariensis*. White-winged wood duck (*Asarcornis scutulata*) can be seen at the park. This duck is one of the largest living species of duck next only to the steamer duck which are heavier. This secretive duck is only known to feed at night. Its diet consists of seeds, aquatic plants, grain, rice, snails, small fish and insects.[6] It inhabits stagnant or slow-flowing natural and artificial wetlands, within or adjacent to evergreen, deciduous or swamp forests, on which it depends for roosting and nesting, usually in tree holes

The charismatic wildlife in Berbak National Park include the binturong, or bear cat, *Arctictis binturong*, honeybear, *Helarctos malayanus*, Sunda pig-tailed macaque, *Macaca nemestrina*, banded langur, *Presbytis melalophos*, agile gibbon, *Hylobates agilis,* siamang, *Symphalangus syndactylus*, silvered leaf monkey, *Trachypithecus cristatus* and many different cats including the Sunda clouded leopard, *Neofelis nebulosa diardi*. There are also Sumatran Tigers, *Panthera tigris sumatrensis*, and Tapirs, *Tapirus indicus*.

There are more than 100 species of freshwater and brackish water fish, such as arowana, *Scleropages formosus*.

Local Communities and their Culture|

The original occupants of Berbak National Park were ethnic Kubu people, but they have been displaced in many areas by migrants. Air Hitam Laut Village, close to the park, is mainly occupied by Bugis people originally from Sulawesi. Villages on the riverbanks are normally occupied by Melayu people and migrants from Java and West Sumatra.

Tourism

Berbak National Park is known for its adventure tourism opportunities, such as boating along the rivers or walking through swamp forest. This peat swamp forest is very acidic with a high humus content.

To see Suak Kandis Dock on the Sriwijaya River, it will take 1.5 hours from Jambi city by rental car. The Air Hitam Dalam River can be reached from Suak Kandis Dock by motor boat (1 hour) where it is possible to see Rang Kayo Hitam Tomb which is 400 years old.

Air Hitam Dalam: Tourists can spend the night in existing huts. In the morning or evening, several activities can be carried out including walking along the trail in the forest, canoeing along the Air Hitam Dalam River (while looking at the animals on the left and right of the river), fishing in the provided place (shelter/pier), and look around the arboretum, which is a collection of various types of plants in Berbak National Park that are deliberately planted.

Air Hitam Laut: Tourists can spend the night in the provided cottages. In the morning or evening, you can walk along the trail in the forest, canoe along the black sea water river, fish in the places provided, and see the arboretum.

Sei Cemara: Is a beach where the birds look for food. At this location, a bird-watching hut equipped with binoculars and reference books about shorebirds will be provided

Access and Transportation

Berbak National Park can be reached through two main gates, one at Air Hitam Dalam and the other at Nipah Panjang. The gates can be reached by motor boat or public transportation such as a *pompong*, a medium sized boat, from the harbor near Angso Duo market in Jambi City, in the following ways

From Jambi city follow the Batanghari and Berbak Rivers to Telagalima/Telagalung village, then on to the Air Hitam Dalam River by motorboat, which will take 2.5-3 hours. Near Telagalima there is a park security post where visitors should report to the park ranger.

From Jambi city take the motorboat on the Batanghari River to Nipah Panjang, which will take 4-5 hours, then continue to Air Hitam Laut Village across the open sea by motorboat or security patrol ship for 5-8 hours. From Air Hitam Laut Village continue up the Penuh River, which is on the border between Jambi and South Sumatra Provinces.

Facilities

In Nipah Panjang there is a conservation office in the Natural Resources Building. At the office there are posters and photos about Berbak National Park

In Simpang Datuk or Sungai Lokan there are some huts

In Jambi city there are several good hotels.

Park Office

Balai Taman Nasional Berbak
Jl. Yos Sudarso km 4, Box 122 Sejinjang, Jambi Timur
Telp./Fax. (+62-741) 31257
Call Center: 0822 8853 9111
Instagram: @ btn_berbaksembilang
Email: berbak@ja.mweb.co.id and berbak@palsa,com

8. BUKIT DUA BELAS NATIONAL PARK

Geographic Location and Size

Bukit Duabelas National Park is a small park of 54,780 ha located south-west of Jambi City in the Province of Jambi, at 1 44` - 1 58 S and 102 29` - 102 49` E, with an elevation of 50 to 400 m asl. It covers parts of three regencies, Sarolangun Bangko, Bungo Tebo and Batanghari.

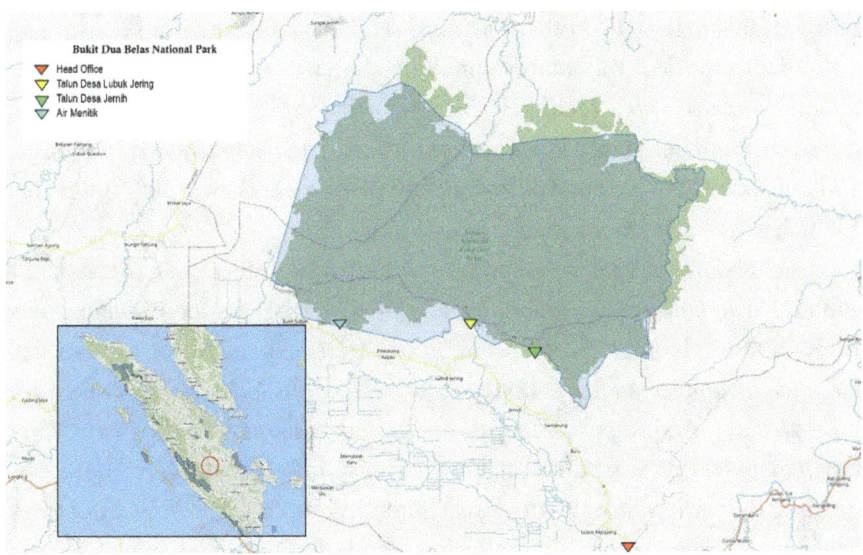

History

Bukit Duabelas National Park was established in 2000 by a Decree of the Minister of Forestry and Plantation Number 258/Kpts-II/2000.

Climate and Topography

The temperature ranges between 20-30°C.

There are a lot of rivers and creeks which originate from this area (seen on a map like fibrous roots). As a result, it is the most important catchment for the watershed of the Batanghari River.

The topography is flat to moderately rugged with hills and low mountains, such as Suban Hill (164 m asl), Panggang Mountain (328 m asl), and Bukit Kuran (438 m asl).

Biodiversity and Ecosystems

Bukit Duabelas National Park contains lowland and some hilly tropical rain forests. Originally this area consisted of a production forest, limited production forest and smaller areas with other designated uses, which were combined into a National Park. Primary natural forest is located in the north section of this park, while other parts are secondary forest.

Plant species of interest include bulian, *Eusideroxylon zwageri*, meranti, *Shorea spp.*, menggeris/kempas, *Koompassia excelsa*, jelutung, *Dyera costulata*, jernang, *Daemonorops draco*, damar, *Agathis sp.*, and rattan, *Calamus spp*. There are approximately 120 plant species including fungi some of which are used as medicines.

This National Park is a habitat for rare and protected animals such as siamang, *Symphalangus syndactylus*, pig-tailed macaque, *Macaca nemestrina*, Sunda clouded leopard, *Neofelis nebulosa diardi*, kanchil, or mouse deer, *Tragulus javanicus kanchil*, honey bear, *Helarctos malayanus malayanus*, barking deer, *Muntiacus muntjak montanus*, Leopard cat, *Prionailurus bengalensis sumatrana*, Sumatran lutra, or hairy-nosed otter, *Lutra sumatrana*, wild dog, *Cuon alpinus sumatrensis*, Sumatran rabbit, *Nesolagus netscheri*, cheela eagle, *Spilornis cheela malayensis*, and many others.

Local Communities and Culture

The indigenous community is the Anak Dalam tribe (Orang Rimba, or son of the forest), which has inhabited the Bukit Duabelas forests for many years. The Anak Dalam call the forest of Bukit Duabelas National Park, the wandering area; where they interact with nature, mutually giving, nurturing, and providing. To meet the necessities of life, Anak Dalam people hunt pigs, fish, seek honey, and tap rubber, which they sell.

The Kubu tribe, also known as Orang Rimba, is one of the minority ethnic groups living on the island of Sumatra, especially in Jambi and South Sumatra. The majority live in the province of Jambi, with a population estimated at around 200,000.

Access and Transportation

Travel from Jambi to Pauh, by bus or hire car passing through Muara Bulian. This will take approximately 3 hours. From Pauh continue to Lubuk Jering and Pematang Kabau by hire car.

This national park has only recently been established and does not yet have facilities for visitors within it. The best season to visit is from June to October.

Facilities

The best hotels are in Jambi city, the capital city of Jambi province. In Muara Bulian there are also small hotels.

Park Office

Balai Taman Nasional Bukit Duabelas
Jl. Lintas Sumatera km 4, Sorolangun, Bangko, Jambi 36124
Telp. (+62-745) 7002069
Fax: +62-745-91368
Call Center: 0812 9523 9036
Instagram: btn_bukitduabelas
Email: tnbukit12@yahoo.co.id

9. SEMBILANG NATIONAL PARK

Geographic Location and Size

Sembilang National Park is located on the coast of Musi Banyuasin Regency in South Sumatra Province, at 1°38` - 2°25` S and 104°12` - 104°55 E. It has an elevation of 0 – 500 m asl., although most of the area is a flat plain with an elevation of 0 to 225 m asl. In the northwest, the park shares a border with Berbak National Park, in Jambi Province. The total area is 205,750 ha, and 60% of it consists of swamps, scrub and former fields.

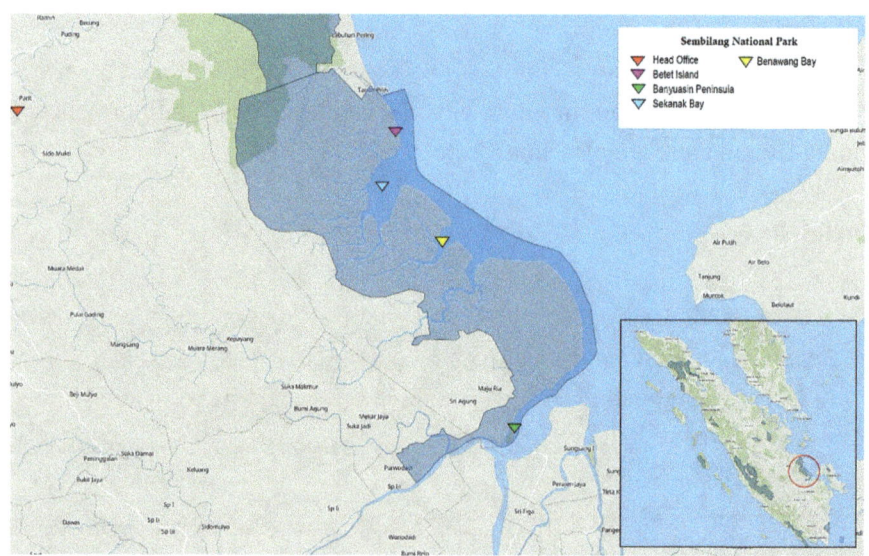

Sembilang National Park is a major international bird migration site recognized by the Ramsar Convention. It is a resting and feeding place for birds migrating from Siberia to eastern Indonesia and Australia, and back again

History

This park was established by Decree of the Minister of Forestry, 76/Kpts-II/2001. It unites the areas designated historically as Sungai Sembilang 1 protected area, Sungai Sembilang II protected area, Pulau Alang Gantang protected area, Terusan dalam protected area, Sungai Sembilang Protected forest and around 17,820 ha of coast and ocean.

Climate and Topography

Sembilang National Park has a long wet season from November until March. It is somewhat drier from July to September. The average annual rainfall is approximately 2,500 mm/year, with air temperatures ranging between 20°-31°C.

The park is flat. Only a few thousand hectares are non-watery land.

Biodiversity and Ecosystems

Sembilang National Park is essentially a large wetland consisting of peat forests, freshwater swamp forests, mangrove forests, riparian forests and mudflats. This park contains the largest mangrove forest in Sumatra and has been named Ramsar Site no. 1,945.

Plants typical of inundated areas include nipah, *Nypa fruticans*, *Melaleuca spp* and mangroves with *Rhizophora apiculate* and *R. mucronata* close to the beach. Further back there are many other mangrove species including *Sonneratia alba* and *Bruguiera gymnorrhiza*. On somewhat higher ground there is cemara laut, *Casuarina equisetifolia*, pandan, *Pandanus tectorius*, waru laut, *Hibiscus tiliaceus*, nibung, *Oncosperma tigillaria*, jelutung, *Dyera costulata*, menggeris, *Koompassia excelsa* and gelam tikus, *Syzygium inophylla*. More than 43% of Indonesia's mangrove species are found in this park.

Birds migrating from Siberia can be seen in Sembilang starting in October, reaching peak numbers in November and lasting through until

February. They return on their way back to Siberia from mid March to mid May. Thousands or even ten of thousands of birds flying in formation can be heard calling above the rumbling waves of Bangka Strait. Birds migrating to eastern Indonesia and north-western Australia find food in this park. The bird list compiled by the park staff is given in the table below

The forests of Sembilang and the Banyuasin peninsula generally are habitats for the Sumatran tiger, *Panthera tigris sumatrae*, Sumatran elephant, *Elephas maximus sumatranus*, tapir, *Tapirus indicus*, golden cat, *Catopuma temminckii temminckii* and sambar deer, *Cervus unicolor equinus*. Primates found in this park are siamang, *Symphalangus syndactylus*, agile gibbon, *Hylobates agilis*, white mitred langur, *Presbytis mitrata*, silvered leaf monkey, *Trachypithecus cristatus*, pig-tailed macaque, *Macaca namestrina*, long-tailed macaque, *Macaca fascicularis*, Coucang, *Nycticebus coucang* and Western tarsier, *Cephalopacus bancanus*.

In the lakes and rivers, there are crocodiles, *Crocodylus porosus, Crocodylus siamensis, Tomistome schlegeli*, monitor lizard, *Varanus salvator*, sembilang fish, *Plotusus canius*, giant freshwater narrow-headed softshell turtle, *Chitra indica* and Irrawaddy dolphin *Orcaella brevirostris*, among many others.

Bird species list for Sembilang National Park (Source: TN Sembilang 2017)

No	Engish Name	Scientific name
1	Common Redshank	*Tringa tetanus*
2	Spotted Redshank	*Tringa erytropus*
3	Black-tailed Godwit	*Limosa limosa*
4	Bar-tailed Godwit	*Limosa lapponica*
5	Far Eastern Curlew	*Numenius madagascariencis*
6	Eurasian Curlew	*Numenius arquata*
7	Wimbrel	*Numenius phaeopus*
8	Asian Dowitcher	*Limnodromus semiphalmatus*
9	Spotted Green Shank	*Tringa guttifer*
10	Greater Sand Plover	*Charadrius leschenaulti*
11	Lesser Sand Plover	*Charadrius mongolus*
12	Kemtish Plover	*Charadrius alexandrinus*
13	Green Plover	*Pluvialis squatarola*

14	Black Winged Stilt	*Himantopus himantopus*
15	Marsh Sandpiper	*Tringa stagnatilis*
16	Common Greenshank	*Tringa nebularia*
17	Teret Sandpiper	*Xenus cinereus*
18	Common Sandpiper	*Actitis hypoleucos*
19	Ruddy Turstone	*Arenaria interpres*
20	Red Knot	*Calidris canutus*
21	Curlew Sandpiper	*Calidris ferruginea*
22	Sanderling	*Calidris alba*
23	Great Knot	*Calidris tenuirostris*
24	Whiskered Tern	*Chlidonias hybrida*
25	White-winged Tern	*Chlidonias leucopterus*
26	Great Crested Tern	*Sterna bergii*
27	Common Tern	*Sterna hirundo*
28	Little Tern	*Sterna albifrons*
29	Lesser Crested Tern	*Sterna bengalensis*
30	Caspian Tern	*Sterna caspia*
31	Gull-billed Tern	*Sterna nilotica*
32	Sunda Teal	*Anas gibberifrons*
33	Lesser Adjuntant	*Leptoptilos javanicus*
34	Black-headed Ibis	*Thresciornis melanocephalus*
35	Cattle Egret	*Bubulcus ibis*
36	Little Egret	*Egretta garzetta*
37	Great Egret	*Egretta alba*
38	Striated Heron	*Butorides striatus*
39	Yellow Bittern	*Ixobrychus sinensis*
40	Javan Pond Heron	*Ardeola speciosa*
41	Grey Heron	*Ardea cinerea*
42	Great-billed Heron	*Ardea sumatrana*
43	Purple Heron	*Ardea purpurea*
44	Oriental Darter	*Anhinga melanogaster*
45	Lesser Whistling Duck	*Dendrocygna javanica*
46	Milky Stork	*Mycteria cinerea*

Local Communities and Culture

There are only a few villages inside the park, Sembilang Village being the largest. Most people in the park are local people from Banyuasin, although there are some migrant Javanese. Fishing is the primary occupation.

Tourism

The park is known mainly for its migrating birds, which can be seen in their thousands or tens of thousands from October to February, but peaking in November. They can be seen from the seashore at Bangka Strait on the Banyuasin Peninsula, and Teluk Benawan, Teluk Sekanak and Pulau Betet. There is a regular trip to see the migrant birds that costs around $200 for 3 days per person. The park also contains representative peat swamp forests, freshwater swamp forests and riparian forests, as well as the dolphins, crocodiles and primates that may be seen from boats.

Access and Transportation

To reach Sembilang National Park, it is possible to hire speedboats that can take up to 40 passengers. There are two routes that lead to this area, from Sungsang, in Banyuasin II District, which will take around 2 hours, or from Palembang which will take around 4 hours. A hire car from Palembang to Sungsang will take around 2 hours.

A permit from the National Park office is needed to travel in the National Park. Interesting sites and locations are Banyuasin Peninsula, Sembilang village and river, Benawan Bay, Sekanak Bay, and Betet Island. River and mangrove forests can be explored by boat, and there is fishing and animal watching, migrating birds and dolphins. The best time to visit is October-November during peak migrant bird arrivals.

Facilities

The best hotels you are in Palembang, the capital city of South Sumatra, the second biggest city on the island of Sumatra.

Park Office

Balai Taman Nasional Sembilang
Jl. Harun Sohar, Lorong Aster no.14 , Palembang 30152
Telp: (+62-711) 7839200, FAX: ^2-711-419737
Call Center: 0822 8853 9111
Instagram: @ btn_berbaksembilang
Email: Sptn2.tnbs@gmail.com

10. BUKIT BARISAN SELATAN NATIONAL PARK

Geographic Location and Size

Bukit Barisan Selatan National Park is in the southern park of the Bukit Barisan mountain range and covers parts of two provinces, Lampung (81.5% of the park) and Benkulu (18.5%). It is located between $4°31'-5°57'$ South Latitude and $103°24'-104°43'$ East Longitude. The total area is 374,081 ha.

Climate and Topography

The rainfall varies from 1,000 mm/yr in to 4,000 mm/year. The rainy season is from November to May and the dry season from June to August. The average annual temperature in the low land forest is about 28°C and, in the mountains, about 20°C.

The topography is varied rising from the coast to low sloping plains, then on to steep rolling hills and mountains, with altitudes from 0 to 1,964 m asl. The east side is generally steep, but on the west, where it borders the Indian Ocean the slopes are gentler. There are a lot of mountains. The highest is Mount Palung (1,964 m) on the western side of Lake Ranau. Others in the north are Mount Tompok Tunggal (1,181 m), Mount Lumtuk (1,473 m) and Mount Pandan (560 m) and in the south, Mount Kelerai (1,358 m), Mount Sekincau (1,183 m), Mount Meling (1,040 m) and Mount Penetok (1,156 m).

History

On the 24th of December 1935, according to Besluit Van der Gouverneur Van Nederlandsch Indie Number. 48 STB 621, the Government of the Netherlands established this site as Wildlife Conservation of South Sumatra I, with an area of 356,800 ha

On the 14th of October 1982, according to Decision Letter of Agriculture Minister Number 736/Mentan/X/1982, it was announced at the World National Park Congress in Denpasar, Bali, that the status and the name of the Wildlife Conservation of South Sumatra I was changed to Bukit Barisan Selatan National Park

On the 12th of May 1984, Decision Letter of the Forestry Minister Number 096/Kpts-II/1984, confirmed the status of Bukit Barisan Selatan National Park.

Biodiversity and Ecosystems

Bukit Barisan Selatan National Park covers many different ecosystems. There are mangrove forests, beach forests, lowland tropical forests, hill tropical forests, and low and high mountain forests. On the flater areas in the south and southeast areas there are open grasslands.

The typical plant species in the park are damar, *Agathis* sp., keruing, *Dipterocarpus* spp., meranti, *Shorea* spp. with nipah, *Nypa fruticans*, and mangroves, e.g. *Rhizophora* sp n riverine and coastal areas. Several Sumatran endemics such as flesh flower, *Amorphophallus decussilvae* and Rafflesia, *Rafflesia arnoldi* occur.

At least 276 species of bird are found in this park. They include tongtong heron, *Leptotilos javanicus*, bluwok heron or upeh bird, *Mycteria cinerea*, mountain eagle, *Spizaetus alboniger*, kuau raja or great argus, *Argusianus argus*, rhino hornbill, *Buceros rhinoceros*, tusk hornbill, *Buceros vigil*, red-chested kepudang or black and crimson oriole, *Oriolus cruentus* and black and yellow broadbill, *Eurylaimus ochromalus*.

The mammals include Sumatran rhino, *Dicerorhinus sumatrensis* Sumatran elephant, *Elephas maximus sumatranus*, honey bear, *Helarctos malayanus*, hairy-nosed otter, *Lutra sumatrana*, Sunda pangolin, *Manis javanica*, tapir, *Tapirus indicus*, sambar deer, *Cervus uniclor*, wild boar, *Sus scrofa*, barking deer, *Muntiacus muntjak*, napu or mouse deer, *Tragulus napu* and Lesser mouse deer, *Tragulus javanicus*. Wild cats include Sumatran tiger, *Panthera tigris sumatrae*, clouded leopard, *Neofelis diardi and Neofelis nebulosa*, golden cat, *Catopuma temincki*, fishing cat, *Prionailurus viverrinus*, Leopard cat, *Pardofelis marmorota*, marbled cat, *Prionailurus bengalensis*, and flat-headed cat, *Prionailurus planiceps* (McDonald et al., 2010)

Primates inhabiting this park are agile gibbon, *Hylobates agilis*, siamang, *Symphalangus syndactylus*, mitred langur, *Presbytis mitrata*, silvered leaf monkey, *Trachypithecus cristatus*, pig-tailed macaque, *Macaca nemestrina*, long-tailed macaque, *Macaca fascicularis*, Sumatra coucang, *Nycticebus coucang* and western tarsier, *Cephalopacus bancanus*.

Other notable wildlife include green sea turtle, *Chelonia mydas*, leatherback turtle, *Dermochelys coriacea*, and hawksbill turtle, *Eretmochelys imbricata*.

Local Communities and Culture

The nearest village to Bukit Barisan Selatan National Park is Pamekahan Village in the Belimbing area. This village is administratively part of Way Haru village. This area is unusual in that most of the communities consist of migrant people, especially from Sukabumi and Pandeglang, West Java and Banyuwangi, East Java, as well as Lampung, Palembang and Padang.

Tourism

The Tampang/Tanjung Mas, Danau Menjukut, Belimbing areas that are less than 200 m asl, consist of beach ecosystem, swamp and low plain forest. The potential tourism attractions are:

- The light house (57 m) built by the Government of the Netherlands in 1879
- An airstrip built during World War II
- Panerusan Beach where there are white sands from the Indian Ocean. Its big waves attract surfers and there are coral reefs with pumice stone, lobster and seaweed as well as the beautiful landscape
- Lake Menjukut and Sulaiman. There is a *gazebo* on the shore of the lake. There are various species of water bird, as well as other animals like wild buffalo and deer. Lake Menjukut is a perfect place for swimming and fishing
- Blambangan beach near Lake Menjukut, is a turtle-nesting beach
- Way Sleman, Sawung Bajo, Teluk Belimbing have the beautiful natural panoramas

Sukaraja Atas. This is the nearest location to the National Park Office, which is in Agung City. It retains the original rainforest with its massive flesh flower, *Amorphophallus decussilvae*, which can be found on Mount Tanggamus. Wildlife includes the honey bear, many different species of bird and other animals. There are about seven waterfalls and beautiful views over Teluk Semangka. There is a 2 ha camp ground, which has a shelter and toilet. There is a pathway for hiking, and some caves where swifts roost, such as Way Paya, Way Monok, Way Babuta and Way Kerinci

Lake Ranau. This is the largest lake in south Sumatra and is surrounded by the mountains of Bukit Barisan. Visitors can fish, sail and kayak on the lake that has clear water and cool air.

Suwoh. Potential tourist sites in this area are hot springs from the former volcano, hot mud geysers that change place from time to time and four lakes close to each other.

Kubu Perahu. Located adjacent to the Liwa and Krui roads, there is original rainforest with at least 60 species of orchids, several waterfalls, a camp ground, several huts and a pathway for hiking.

Access and Transportation

Facilties in Bukit Barisan Selatan National Park are office, home stay, tour Guide, work station/security post, pathways for hiking, an observation tower, campgrounds, shelters, toilets and a guest house.

In the Liwa area, Belimbing and Kota Agung there are several simple home stays. There are several places for resting at ranger offices, work station/security posts or even in the villagers' houses.

Bukit Barisan Selatan National Park can be reached such as:

Departure	Arrival	Distance	Time	Transportation
Bandar Lampung	Kota Agung	93 km	2 hours	Bus
Bandar Lampung	Liwa	211 km	6 hours	Bus
Kota Agung	Tanjung Mas		6 hours	Motor boat
Kota Agung	Sukaraja Atas	30 km	1 hour	Bus
Kota Agung	Banding	15 km	0.5 hours	Bus
Kota Agung	Kuncoro	15 km	0.5 hours	Bus
Tanjung Mas	Belubuk beach	7,2 km	2 hours	Walking
Belubuk	Danau Menjukut	7 km	2 hours	Walking
Danau Menjukut	Way Sulaiman beach	8.3 km	2 hours	Walking
Way Sulaiman	Belimbing	7.4 km	2 hours	Walking
Belimbing	Pamekahan	3,5 km	1 hour	Walking
Sukaraja Atas	Bumi perkemahan	2 km	0.5 hour	Walking
Banding	Suwoh	40 km	2 hours	Bus/motor bike
Kuncoro	Suwoh	25 km	2 hours	Motor bike
Liwa	Kubu Perahu	7 km	0.5 hours	Bus
Kubu Perahu	Air terjun	3 km	2 hours	Walking

Facilities

The best facilities close to the park are in Bandar Lampung, the capital city of Lampung Province, where there are many good hotels. Flights from Jakarta to Lampung take only 30 minutes. A bus or rental a car from Jakarta across the Sunda Strait to Bandar Lampung takes approximately 4 to 5 hrs. Kota Agung has many small hotels bed and breakfasts.

Park Office

Balai Taman Nasional Bukit Barisan-Selatan
Jl. Ir. Juanda 19 Kota Agung, Tanggamus
Lampung Selatan 35751
Telp./ Fax. (+62-722) 21064
Call Center: 0852 6600 9917
Instagram: @ bbtn_bukitbarisanselatan
Email: btnbbs@gmail.com

11. WAY KAMBAS NATIONAL PARK

Geographic Location and Size

Way Kambas National Park (WKNP) is in Lampung Province between $4°37'$-$5°16'$ south Latitude and $105°33'$-$105°55'$ east Longitude. It is 125,621 ha in size. It covers six regencies, Labuan Maninggai, Way Jepara, Sukadana, Purbolinggo, Rumbia and Seputih Surabaya. The park is bordered by the Sumatran coastline in the east, the Way Penet River in the southeast, the Way Sukadana River in the west, the Way Pegadungan River in the southwest and the Way Seputih River in the north.

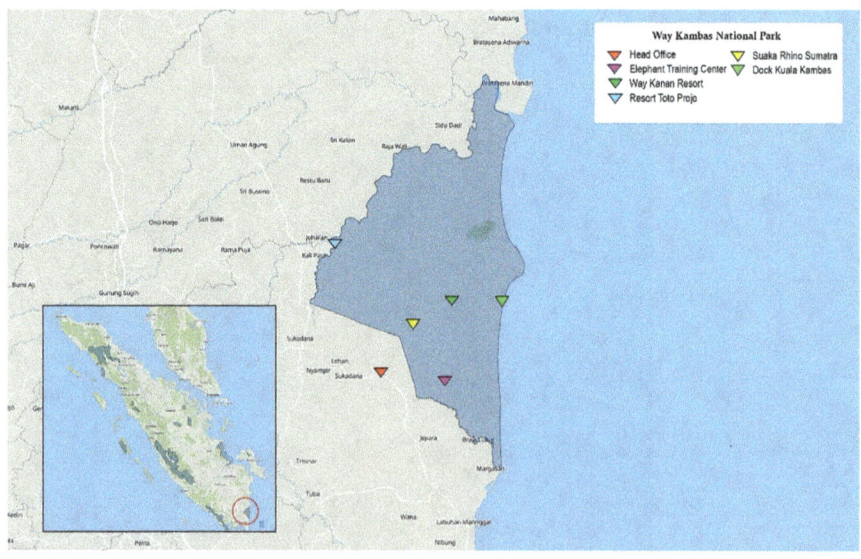

Climate and Topography

The rainy season occurs from November to March, and the drier season from July to September. Average rainfall is around 2,500 mm/year, with average temperatures from around 20^0 to 31^0C.

Most of the area is flat, ranging from 0 to 50 m asl. Annual rainfall is from 2,500–3,000 mm.

History

In 1924 Way Kambas and several areas around it were designated as Protected Forest areas

In 1936, Mr. Rookmaker, a resident of Lampung, proposed that the area should be a Wildlife Conservation reserve

In 1937 the Government of the Netherlands declared the Way Kambas Wildlife Conservation Reserve covering an area of 130,000 ha

In 1982 the Agriculture Minister of the Republic of Indonesian, through a Decision Letter, changed the status of this area to Way Kambas National Park

On the 1st of April 1989, Decision Letter of Agriculture Minister Number 444/Kpts-II/89, was released to officially announce Way Kambas National Park status, of 125,621 ha.

Biodiversity and Ecosystems

Way Kambas National Park contains a large area of non-peat freshwater swamp forest and covers a range of other ecosystems such as Dipterocarp lowland rainforest, beach forest, grasslands with low shrubs, as well as low mountain forest. There are also several types of wetlands, such as tidal mudflats, sand flats, mangrove swamps, rivers with slow contionus flows, riparian forest, seasonal wet grasslands, freshwater swamp forest and seasonal swamp forest.

Tree species of the canopy are dominated by the diptercarpaceae family, and include *Shorea ovalis, Shorea leprolusa, Dipterocarpus gracilis, Canarium littorale, Canarium denticulatum, Horsfieldia glabra* and *Albizia lebbeckiodes*. In the middle stratum, common trees are *Mallotus peltattus, Eurycoma longifolia, Baccaurea racemosa* and *Antidesma* spp (Soerianegara and Indrawan, 1988). Common swamp species include paperbark, *Melaleuca leucadendron*, Pandan, *Pandanus tectorius, Oncosperma tigilaria* and *Gluta renghas*.

The beach ecosytem here is a temporary refuge for birds migrating from Australia in the south to the plains of north China and Siberia in the north, and back again each year. Altogether, there are more than 280 species of bird, but among those seen easily are besra sparrowhawk, *Accipiter virgatus*, wreathed hornbill, *Aceros undulatus*, white bellied or pied hornbill, *Anthracoceros albirostris*, Storm's stork, *Ciconia stormi*, eastern curlew or gajahan, *Numenius madagascariensis*, iron woodpecker, *Dinopium javanense*, gold tiong or hill myna, *Gracula religiosa*, blue-tailed bee-eater, *Merops philippinus* and reddish scops owl, *Otus rufescens*.

WKNP is one of the key protected areas for in-situ wild cat conservation in Sumatra. It harbors six of the seven Sumatran wild cat species (Bastoni and Apriawan, 1997; Franklin et al., 1999). They are are Sumatran tiger, *Panthera tigris sumatrae*, clouded leopard, *Neofelis nebulosa* and Sunda clouded leopard, *Neofelis diardi*, golden cat, *Catopuma temincki*, fishing cat, *Prionailurus viverrinus*, leopard cat, *Pardofelis marmorota*, marbled cat, *Prionailurus bengalensis* and flat-headed cat, *Prionailurus planiceps*.

The mammals found in the park include Sumatran rhino, *Dicerorhinus sumatrensis*, Sumatran elephant, *Elephas maximus sumatranus*, honey bear, *Helarctos malayanus*, hairy-nosed otter, *Lutra sumatrana*, Sunda pangolin, *Manis javanica*, tapir, *Tapirus indicus*, sambar deer, *Cervus uniclor*, wild pig, *Sus scrofa*, barking deer, *Muntiacus muntjak*, napu, *Tragulus napu* and mouse deer, *Tragulus javanicus*. Primates occuring in this park are agile gibbon, *Hylobates agilis*, Siamang, *Symphalangus syndactylus*, mitred langur, *Presbytis mitrata*, silvered leaf monkey, *Trachypithecus cristatus*, pig-tailed macaque, *Macaca nemestrina*, long-tailed macaque, *Macaca fascicularis*, Sumatran coucang, *Nycticebus coucang* and western tarsier, *Cephalopacus bancanus*.

There are also two species of crocodile, senyulong crocodile or Sunda gharial, *Tomistoma schlegelii* and estuarine or saltwater crocodile, *Crocodylus porosus*, monitor lizards and many other reptilian and amphibian species including Sumatran endemics.

Local Communities and Culture

People around the park consist of local Lampung people and some migrants from Java and other parts of Sumatra. Local people can be found in Sukadana Village with Lampung traditional houses. Other villages are Wana Village, at Labuan Maninggai and Punggung Raharjo, where there are ancient sites

Tourism

- From the water, along the beach and the rivers Way Kanan, Way Kambas, Way Pegadungan, Kuala Penet, Rawa Wako, Rawa Kali Biru and Rawa Pasir, many species of bird and primate can be observed.
- At Karangsari - Kuala Penet - Kali Bathin - Way Kanan, it is possible to take an elephant safari ride and see wildlife like deer, wild elephant, honey bear and many species of bird
- The Centre of Elephant Training (PLG, Pusat Pelatihan Gajah) is an area of 400ha in Karangsari, Way Kanan, which is about 90 km to the east of Bandar Lampung city and 10 km from the park gate. Attractions include elephant training from $08.^{00}$-$16.^{00}$. Passengers of *Awani Dreams* cruiseship are regular visitors. PLG is crowded with visitors on Saturdays, Sundays and public holidays
- Around Way Kanan there is an area of 10,000 ha for the *in situ* conservation of the Sumatra Rhino.
- There is a pathway 13 km long between Plang Hijau and Way Kanan. It is about a 3 hour trip.
- Krakatau Island can be seen from the white sand beach at Merak Belantung

Access and Transportation

Way Kambas National Park can be reached by several routes.

By air

- From Jakarta to Lampung takes around 30 minutes, with many flights available every day. Flights from Padang, Batam and Medan to Palembang are also available every day. It is then possible to continue by car or train to Bandar Lampung.

- From Bandar Lampung to the park gate by public transport (minibus/taxi/rental transportation) takes 2-3.5 hours, through Sumatra Teluk Betung, Sriwabono, and Way Jepara
- Visitors who use their own transportation or taxi or hire car can go directly through the gate of Way Kambas National Park in Plang Ijo. Those who use alternative public transportation from Labuan Ratulama (Way Jepara) to Plang Ijo can rent a motorbike, which takes around 15 minutes

By land and sea crossing:

- From the intercity bus terminal of Kalideres, Jakarta to Merak Harbour by bus takes around 2-3.5 hours
- From Merak Harbour to Bakauheni (Lampung) Harbour by ferry takes around 2 hours, and then continue to Bandar Lampung and stop at Teluk Betung Bus Terminal or Rajabasa Tanjung Karang Bus Terminal. A private boat can go directly to the harbor of Labuan Maninggai
- From the Bandar Lampung bus terminal, Labuan Ratulama (Way Jepara) can be reached by public transport.

Facilities

In Way Kambas National Park there is an office, information center, Elephant Training and Attraction Center, observation tower, Flora and Fauna Research Center, pathways, security post and ommunication radio.

In Bandar Lampung there are many options for accommodation from bed and breakfast to 5-star hotels. There are many small bed and breakfasts in the villages close to the headquarters of the park. At the headquarters, there are also guest houses and small resorts.

Park Office

Balai Taman Nasional Way Kambas
Jl. Raya Labuhan Ratu Lama, Labuhan Ratu
Sukadana - Lampung Timur - 34196
Telp. (062-725) 7645024 Fax. (0725) 7645090
Call Center: 0823 7639 4648
Instagram: @ btn_waykambas
e-mail   : program@waykambas.or.id , kabalai@waykambas.or.id

12. ZAMRUD NATIONAL PARK

Geographic Location and Size

Zamrud National Park is located in the Siak Regency, Riau Province. It is a new park, only established in 2016. There are four protected areas on the Kampar Peninsula, three game reserves, Tasik Belat, Tasik Metas and Tasik Serkap and Zamrud National Park.

The park is peat swamp forest of 31,480 ha and consists of 2 lakes, Danau Pulau Besar (2,416 ha) and Danau Bawah (360 ha). In Danau Pulau Besar, there are 4 small islands, Besar, Tengah, Bungsu and Beruk islands. This park is surrounded by Industrial Plantation Forest.

Access from the west is the PT Bumi Siak Pusako/BOB (Joint Operations Agency) road

Active oil wells in the National Park surround the lower lake and to the east of the upper lake (the names Tasik Bawah and Tasik Atas are seen from the traditional access of the swamp river, the first small lake encountered is called Tasik Bawah. Tasik upstream is Tasik Atas). Tasik is the local Malay word for lakes

Climate and Topography

The topograhy of the park is almost flat from 0 to 200 m above sea level. Most is covered with peat forest, lakes with black water and mangrove forest behind the shoreline. Two rivers are found close to the park, Kampar and Siak. The climate is humid with annual rainfall between 2,000 and 3,000 mm. The park protects some of the last remaining forest of a deep peat land that is extremely high in biodiversity. Most has now been lost to agriculture and timber extaction followed by plantations.

History

In 1971 according to Presidential Decree No. 39 of 1971, PT. CPI (Caltex Pacific Indonesia) has obtained the determination of Mining Authorization Areas including those in the Lake Pulau Besar/Lake Bawah SM area, while field activities began in 1975. Julius Tahija, former Board of Commissioners of PT Caltex Pacific Indonesia who discovered two lakes in the CPI operating area in Siak Regency. Julius Tahija invited Emil Salim, who at that time served as State Minister for Development and Environment Supervision, to support the idea of conserving the area. Emil Salim supported the idea and issued a letter No. 812/MenPPLH/8/79. This letter later became the basis for the Decree of the Governor of Riau in November 1979, which designated the area as a protected forest. The Pulau Besar/Lake Bawah Lake area is known as the Zamrud field. The operational area in this area is 2,682 ha and an area of 2,288 ha including protected areas.

The Pulau Besar Danau Bawah Lake area was appointed by the Minister of Agriculture No. 846/Kpts/Um/II/1980 dated November 25, 1980 as one of the forest areas with a conservation function of the Pulau Besar/Lake Bawah Wildlife Reserve, covering an area of ± 25,000 ha. In 1982, oil drilling continued to be carried out by conducting oblique drilling where the initial drilling location was carried out as far as possible from the lake, with the aim of avoiding pollution that could occur from oil mining exploration activities.

In 1983, definitive boundary demarcation and bracelet gathering were carried out, resulting in an area of 28,237.95 ha. The area is determined through the Decree of the Minister of Forestry and Plantation No. 668/Kpts-II/1999 dated August 26, 1999 concerning the Designation of the Pulau Besar/Lake Bawah Lake Forest Group covering an area of 28,237.95 ha located in the District Level II Bengkalis, Riau Province Level I Region as a Forest Area with a Wildlife Reserve Function.

In 2002 the oil mining area changed its management from PT. CPI to the Joint Operations Agency (BOB) PT. Bumi Siak Pusako-Pertamina Hulu. Petroleum management continues and the function of the area as a wildlife sanctuary continues. In 2005 the Regent of Siak sent a letter to the Minister of Forestry No. 364/Dishut/205/2005 dated June 9, 2005 regarding the request for support from the Minister of Forestry for the expansion of Lake Pulau Besar/Lake Pulau Bawah from ± 28,237.95 ha to Emerald National Park covering an area of ± 38,500 ha. The letter is processed at the Ministry of Forestry in accordance with applicable regulations.

The idea of a national park was proposed in 2001 by the local government (Siak Regency) but it was not established until 2016 when it was declared by the Minister of Environment and Forestry with decree No.350/2016. It was launched by the Vice President of Indonesia, Mr. Yusuf Kala, during World Environment day in 2016.

On December 22, 2006 the Regent of Siak through letter No. 551/EK/133 addressed to the Minister of Home Affairs to request support for the inauguration of the Siak Bridge and other activities, one of which is the signing of the Zamrud National Park inscription. On July 16, 2007 the Regent of Siak submitted letter No. 100/TP/VII/2007/209 submitted an application for Lake Pulau Besar/Lake Bawah to become Emerald National Park as well as the inauguration/signing of the inscription by the sixth President of the Republic of Indonesia which coincided with the agenda of the President's visit to Siak Regency on 10 – 11 August 2007. Minister of Forestry on August 7, 2007 has given approval in principle to change the function of the Lake Pulau Besar/Lake Bawah SM into Zamrud National Park. Furthermore, during a working visit to Siak Regency, the President of the Republic of Indonesia signed the inscription declaring Zamrud National Park on August 11st, 2007.

On 12 May 2008, the Director General of PHKA wrote to the Head of Baplanhut No.S.216/IV-KK/2008, stating that in principle he supports the proposed expansion and change of function of the Lake Pulau Besar/Lake Pulau Bawah to become TN Zamrud covering an area of +38,500 ha. In March 2009, the Head of the Riau KSDAE Center wrote to the Director General of PHKA No.S.230/IV-17/TU.2/2009, requesting to immediately determine the area of the Pulau Besar/Lake Bawah Lake as a National Park covering an area of ±28,237.95 ha (or the same area as BC). The Integrated Team which involves various parties, including LIPI, Forestry Research and Development, Directorate General of Planology, Directorate General of PHKA, Siak Regency Government based on the Decree of the Minister of Forestry No. 463/Menhut-VII/2009 dated August 5, 2009 conducted a field review and study. The results of the study from the integrated team have been reported to the Minister of Forestry and are awaiting approval in principle from the Minister of Forestry.

On May 4, 2016, the Minister of Environment and Forestry of the Republic of Indonesia, through the Decree of the Minister of Environment and Forestry No. 350/Menlhk/Setjen/PLA.2/5/2016, has determined the change of function of the Pulau Besar/Lake Bawah Wildlife Reserve and the Permanent Production Forest Area of Tasik Besar Serkap to become the Emerald National Park in Siak Regency, Riau Province covering an area of ±31,480 ha with an area of 28,237 ha from the Lake Pulau Besar/Lake Bawah area and an area of 3,242 ha which is a Production Forest (HP) of Tasik Besar Serkap.

Biodiversity and Ecosystems

Biodiversity and Ecosystems have been studied on the Kampar peninsula, including Zamrud National Park, by the NGO Flora and Fauna International (FFI) in collaboration with RER (Restorasi Ekosistem Riau) of the APRIL-Pulp and Paper Company. These studies identified 220 bird species, 152 plants, 72 mammals, and 107 amphibians and reptiles. Sixteen of those are globally threatened, with two listed as critically endangered, the Sumatran tiger and the Sunda Pangolin. Surveys conducted by FFI and RER in the park found 549 species, 5 of which are critically endangered, 14 are endangered and 25 are vulnerable.

Primates are distributed widely in this peat forest. Two gibbons, siamang, *Sympalangus syndatylus* and agile gibbon, *Hylobates agilis* are found in the park. There are three leaf monkeys, lutung, *Trachypithecus cristatus*, banded leaf monkey, *Presbytis femoralis* and Sumatran surili, *Presbytis melalophos* and two macaques, pig-tailed macaque, *Macaca nemestrina* and long-tailed macaque, *Macaca fascicularis*. The Sunda slow loris, *Nycticebus coucang*, can also be found here. Cats include Sumatran tiger, *Panthere tigris sumatrae*, flat-headed cat, *Prionailurus planiceps*, leopard cat, *Prionailurus bengalensis*, Sunda clouded leopard, *Neofelis diardi* and marbled cat, *Pardofelis marmorata*. Other species include Malaysan sunbear, *Helarctos malayanus*, bintorong, *Arctictis binturong*, Sambar deer, *Rusa unicolor*, bearded pig, *Sus barbatus* and several rat species such as Whiteheads spiny rat, *Maxomys whiteheadi*, Rajah spiny rat, *Maxomys rajah*, and dark-tailed tree rat, *Niveventer cremoriventer*. Reptiles that occur in the park include freswater and saltewater crocodiles, monitor lizards and other lizards, and various snakes.

Dominant birds found in the park are Blekok cina (*Ardeola bacchus*), Dara-laut kumis (*Chlidonias hybrida*) and Kirik-kirik laut (*Merops philipinus*). Some fishes are very important and iconic such as arowana, *Scleropages fomosus*, a popular aquarium fish, which can be very expensive to buy. Many species of fishes found here have been adaptaed to peat swamp forest. Most plant species are also characteristic of peat swamps such as red meranti, *Shorea platycarpa*. Other species is *Vatica teysmanniana and Shorea teysmanniana*, both are Sumatran endemics. Both are endangered plants. On the beach, many mangrove species are found.

Local Communities and Culture

The local people in the park are from Malay tribes and are called Melayu Cuo in Bahasa. This traditional community in the kampar peninsula has a similar pattern of ownership and land rights as the people from Tesso Nilo National park.

Tourism

People who visit this park can experience the black water with beautiful trees in the two lakes; Danau Pulau Besar and Danau Bawah. Since there is no regular transporation to the park, visitors need to rent a boat from nearby communities. Since this is such a new national park, no facilities have been developed as yet.

Access and Transportation

This park is 180 km from the capital city of Riau Province, Pekanbaru. There are flights from Jakarta to Pekanbaru. A rental car can be used to get from Pekanbaru to Kampung Rawa Nekar Jaya, in the Sungai Apit District. From this village, it is possible to rent a boat called pompong to get to the park. One boat can take up to 15 people and it will cost around 400,000 rupiah or around USD25.

Facilities

Pekanbaru, the capital of Riau Province, has many hotels from the basic to 5-star.

Park Office

Balai Besar KSDA Riau
Jl. H.R. Soebrantas km 8.5., PO Box 1048, Tampan,
Pekan Baru,
Tel. +62-761-63135
Call Center: 081374742981
Instagram: @ bbksda_riau
Email: bbksdariau.admin@bbksdariau.id

13. GUNUNG MARAS NATIONAL PARK

Geographic Locataion and Size

Gunung Maras National Park, of 16,807 ha, is the newest national park in Sumatra, located in the Riau Silip and Kelapa Districts of Bangka Regency and Bangka Barat Regency, Bangka Belitung Province.

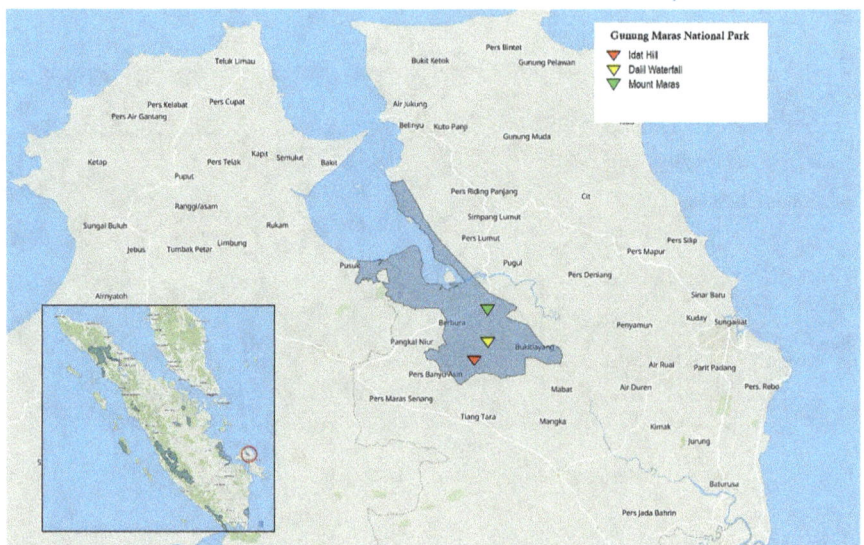

Climate and Topography

The rainfall averages about 3,000 mm per year, with a peak in December and January. July to September is drier. The natural vegetation under such conditions would be tropical rain forest but the extent of human activities has been such that little original vegetation remains and even those parts of the island that are not directly cultivated are covered with secondary vegetation.

The park is located in the center of Bangka island, a landscape of undulating hills, in several places rising to low granite mountains. Close to the park, there are several hills, such as Menoembing (448 m) near Muntok the Mangol or Mangkol Mountains (398 m), Permisan (457 m), and Pading/ Bebuluh (654 m).

History

The park was originally designated as a protection forest (hutan lindung) for the purpose of protecting the watershed. Based on Minister decree Nomor Sk. 576/Menlhk/Setjen/PLA.2/7/2016 on 27 Juli 2016, this block of forest was converted into national park.

Biodiversity and Ecosystems

The total number of bird species recorded from Bangka Island is 172. Two species, white breatsed waterhen, *Amaurornis phoenicurus* and hawk cuckoo, *Cuculus fugax* are represented by two subspecies, and one, hooded pitta, *Pitta sordida,* by three, making a total of 176 forms. Animals that can be found here are silvery leaf monkey, *Trachypithecus cristatus*, long-tailed macaque, *Macaca fascicularis*, and Bangka's tarsier, *Tarsius bancanus saltator*. Bangka's tarsier is endemic to Bangka Belitung. Other animals found in this park are wild pig, *Sus scrofa*, Sunda pangolin, *Manis javanica*, mouse deer, *Tragulus javanicus*, Asian palm civet, *Paradoxurus hermaphroditus*, and jungle fowl, *Gallus varius*.

There are at least 53 plant species. The forest is dominated by *Pithecelobium* sp, *Palaquium sp*, *Ficus* sp., *Trema orientalis*, *Eugenia polyantha*, *Arthocarpus integra*, *Arthocarpus chempeden*, *Havea braziliensis*, *Garcinia mangostana*, *Terminalia cattapa*, *Lancium domesticum*, *Lancium* sp, *Nephelium lapaceum*, *Castanopsis acuminate*, *Langestromia laudoni*, *Cassia siamea*, *Leucaena glauca* and many others. Three endemic plants of the genus *Tristaniopsis* are, water pelawan, *Tristania whiteana*, pelawan sungon, *T. obovate* and pelawan merah, *T. maingayi*.

Local Communities and Culture

The people of Bangka Island are similar to people from South Sumatra. However, Chinese immigrants settled in Bangka hundreds of years ago with their food, art and dances so there is now a blend between local and chinese tribes.

Tourism

Because the park is a very important watershed for the people of Bangka Island, the main attraction is the water itself. There are several waterfalls in the park. The most famous is located in Dalil village. Many people jump 6m from the waterfall into a natural pool. It is an approximately 2 km walk from the village but good trekking or biking across several landscapes, with traditional houses, gardens, and plantations to the forested land.

Access and Transportation

The closest city is Sungai Liat, approximately 70 km away. There is a smaller city, Belinyu, approximately 30 km away. Pangkal Pinang, where the airport is located, is the largest city on the island. Access to the park is very easy. From the airport drive to the north for approximately 90 km.

Facilities

Because this is a new national park there are no facilities for visitors near the park. However, you can easily drive to it in no more than 2 hours from Pangkal Pinang. In the city of Pangkal Pinang you can find many hotels from basic back packers to luxury hotels.

Park Office

Balai KSDA Sumatra Selatan
Jl. Kol II Berlian/Punti Kayu, km 6, no.79 Kode Pos 1288
Palembang, 30153, Sumatera Selatan
Email: bksdasumsel@yahoo.co.id
Tel.+62-711-410948 Fax: +62-711-411578
Call Center: 0812 7141 2141
Instagram: @ bksda_sumsel
Email: bksdasumsel@yahoo.co.id

CHAPTER 2
JAVA AND BALI

Java and Bali are treated together because they are close geographically and have similar ecological characteristics. Further east, Bali is separated from Lombok and the rest of Nusa Tenggara by Wallace's line, a biogeographic boundary proposed by Alfred Russel Wallace, the famous 19th century English biologist, who recognized that the composition of the flora and fauna was different either side of this line, although with some overlap. Wallace was particularly impressed with Java. He wrote *"Java is ... arguably the finest of all tropical islands"*. Java is known as the most fertile island in Indonesia, largely due to the volcanic ash and mud produced by its 30 volcanoes. This fertility varies across the island, following the pattern of rainfall. From the western end to the middle are high forested mountains with high rainfall, dominated by tree species of the *Dipterocarpacaea* family. To the east the landscape is drier with distinctive wet and dry seasons and monsoon forests. These geological and climatic factors have produced the very beautiful dramatic landscapes typical of Java and Bali.

They have also produced a great variety of local habitats and a diverse biota. Java is a relatively small island, yet originally supported many rare and endemic species, such as the Javan tiger, Javan rhino, Javan gibbon, Javan eagle and many more. Bali, an even smaller island, also supported a population of tigers, now thought to be the same sub-species as the Javan tiger, but that population has become extinct.

It is very unfortunate that 150 years after Alfred Wallace visited Java, the forest and its many species that amazed him are now found only on mountain peaks or in small remnants within agricultural landscapes. Java, with an area just over 10 million ha, is occupied by more than 120 million people. Central and local governments and government agencies, research centers and industry and business centres are found throughout Java and Bali, and there is a long human history represented by many historical sites.

For visitors who want to know details of the flora and fauna of Java and Bali, there are several useful books, such as the Ecology of Java and Bali (Whitten et al 2001), field guides to the primates (Supriatna and Hendras 1991; Supriatna and Ramadhan 2016; Supriatna, 2019) and birds (MacKinnon 1999).

Java and Bali together cover about 138,580 km^2, consisting of six provinces in Java and the single province of Bali. They are:
- Jakarta Special Capital: the capital city of Indonesia
- Banten Province: the capital city is Serang
- West Java Province: the capital city is Bandung
- Central Java Province: the capital city is Semarang
- Yogyakarta Province: the capital city is Yogyakarta
- East Java Province: the capital city is Surabaya
- Bali Province: the capital city is Denpasar

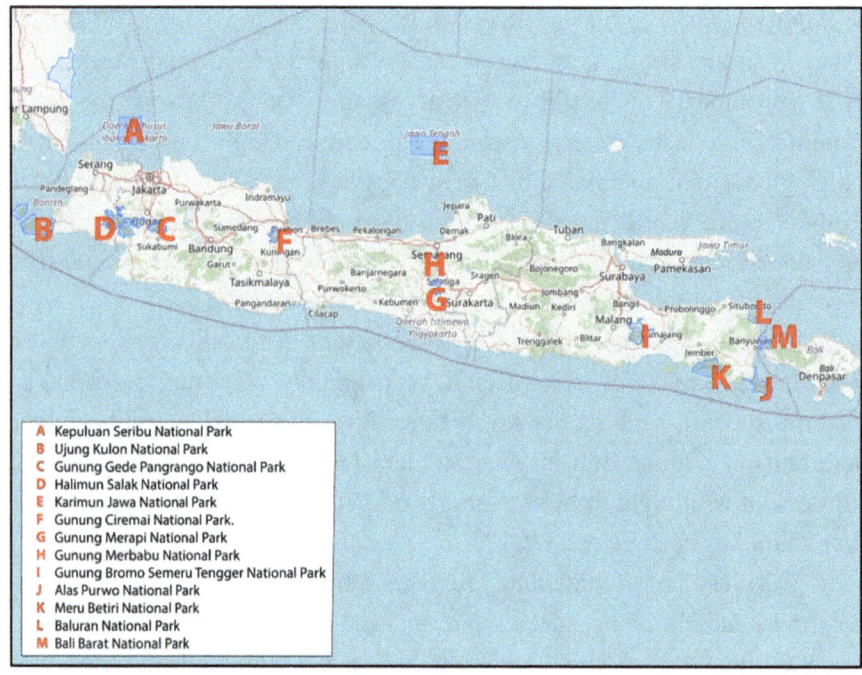

There are 12 national parks in Java and Bali, two of which are marine national parks. They are:

The National Parks of Indonesia

- Kepulauan Seribu Marine National Park, of 108,000 ha in Special Capital City, Jakarta
- Ujung Kulon National Park, of 122,956 ha in Banten Province
- Gunung Gede-Pangrango National Park, of 15,196 ha in West Java Province
- Gunung Halimun Salak National Park, of 40,000 ha in West Java Province
- Kepulauan Karimun Jawa Marine National Park, of 111,625 ha in Central JavaProvince
- Gunung Merapi National Park, of 6410 ha in Jogjakarta Province and Central Java Province
- Gunung Merbabu National Park, of 5,700 ha in Central Java Province
- Bromo Tengger Semeru National Park, in 50,276.20 ha in East Java Province
- Alas Purwo National Park, of 43,420 ha in East Java Province
- Meru Betiri National Park, of 58,000 ha in East Java Province
- Baluran National Park, of 25,000 ha in East Java Province
- Bali Barat National Park of 19,003 ha in Bali Province

Mount Baluran & Savannah (Photo by Nurdin Razak & Arif); Teluk Hijau at Meru Betiri National Park (Photo by Nurul Winarni)

Island, Mangrove, View of underwater at Kepulauan Seribu Marine National Park (Photo by Adhi Nurul Hadi)

View of underwater at Kepulauan Seribu Marine National Park (Photo by Adhi Nurul Hadi); Crater at Gunung Gede Pangrango National Park (Photo by Susmustika)

Landscape Mount Semeru & Bromo (Photo by Lili Adidjaja); Beach at Karimun Jawa National Park (Photo by Adhi Nurul Hadi)

Coral Reef & Mangrove in Karimun Jawa National Park (Photo by Adhi Nurul Hadi)

Mangrove in Karimun Jawa National Park (Photo by Adhi Nurul Hadi) &View of Gunung Merapi National Park (Photo by Jatna Supriatna & KLHK)

Leucopsar rothschildi at Bali Barat National Park (Photo by Michelle Angelina Sharon) and Badak Jawa (Rhinoceros sondaicus) at Ujung Kulon National Park (https://www.menlhk.go.id/)

Hylobates moloch in Gungung Halimun Salak & Gunung Gede Pangrango National Park (Photo by Nurul L. Winarni & Anton) & *Nisaetus bartels* at Java (Photo by Yopi Suhendar)

Bos javanicus in Meru Betiri National Park (https://merubetiri.id/) & Nycticebus javanicus at West Java (Photo by Anton Ario, Denny Setiawan – IAR, & Risanti - IAR)

Rusa timorensis Baluran National Park (Photo by Nurul Winarni) & *Panthera pardus melas* at Java (Photo by KLHK & CI)

14. KEPULAUAN SERIBU MARINE NATIONAL PARK

Geographic Location and Size

Kepulauan Seribu Marine National Park was the first Marine National Park to be declared in Indonesia. It is located between $5°24'$-$5°45'$ South Latitude and $106°25'$-$106°40'$ East Longitude, about 45-47 km to the north of Jakarta in Jakarta Bay. It includes Pulau Panggang Village and Pulau Kelapa Village, Pulau Seribu Regency, North Jakarta Municipality and Daerah Khusus Ibukota Jakarta Province.

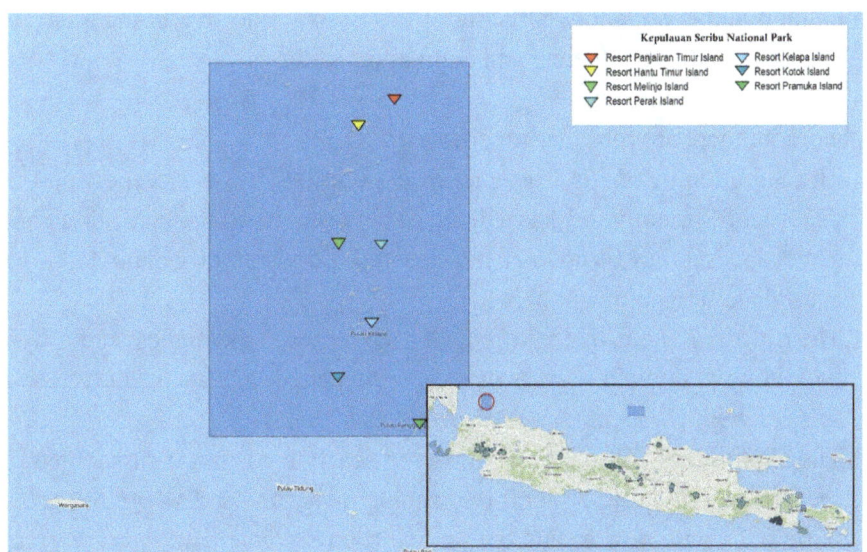

The archipelago that makes up Kepulauan Seribu consists of 106 islands spread over 80 km from north to south. The size of islands varies between 0.5 ha to more than 37 ha. The National Park covers 108,000 ha of this archipelago, of which 78 islands make up 526 ha of land. It is divided into four management zones:

- The Core Zone. There are three parts to the core zone, which is devoted to the preservation of species and ecosystem processes and the study of these species and ecosystems. Core zone I, which is located on Gosong Rengat Island, is devoted to the conservation of scaly turtle habitat. Core zone II, which is located on Penjaliran Timur, Penjaliran Barat, Peteloran Timur and Peteloran Barat islands as well as the near water areas of these islands is dedicated to the conservation of mangrove ecosystems and habitat for scaly turtles especially on Peteloran Barat and Peteloran Timur Islands. Core zone III, which covers Kayu Angin Bira Island, as well as the marine areas that surround them, is for the conservation of coral reef ecosystems, and research on these ecosystems.
- The Protected Zone. This covers core zones I and II and protects the habitat and food sources of protected species and the scaly turtle breeding grounds.
- The Utilization Zone. This zone is used for recreation and tourism development.
- The Traditional Utilization Zone. This zone is used to support the social, cultural and economic well being of the people who live in the area, for example, traditional fishing and cultivation methods.

In addition, there are four maritime zones:
- The Sublittoral Zone is from the lowest tide level to the deepest part of the beach. Normally this has a depth of between 20-40 meters, although at some places it may be up to 70 meters deep. Corals of the genera *Acropora* and *Porites* are dominant here
- The Littoral Zone is that part of the beach between the highest and lowest tide limits. It contains corals and when inundated, fish such as *Mycedium* sp., *Echinophyllia* sp., *Oxypora* sp. and *Pachyseris* sp.
- The Supralittoral Zone is that part of the beach that is never submerged by sea water. It consists of coral reef rubble or beach sand where the green turtle (*Chelonia mydas*) and the scaly turtle (*Eretmochelys imbricata*) lay

their eggs. The dominant plants are mangroves, which also provide habitat for water birds.
- The Land Zone is that part of the island above the Supralittoral Zone. There are usually a lot of coconut palms, *Cocos nucifera*, to be found there.

Climate and Topography

The average rainfall is more than 3,000 mm/year. Daily temperatures vary between 21.60^0-32.30^0C (about 27^0C on average). The change from high to low tide is around 1.5-2 meters depending on the season. The relative humidity varies between 67%-98% (around 80% on average). There are two main wheather patterns. The west/southwest season occurs from December to February, with rainfall between 500-1000mm per month. From November to April there are high seas with big waves and the force of the sea current is around 20-40 cm/second. The other weather pattern is from May to September when conditions are sunny, normally it is quite calm, and the water becomes clearer. These are the best months to visit this park.

Altitude varies between 0 and 7 m above sea level. There are shallow sea waters with coral islands and sandbars, reef flats, and reef slopes. On several island clusters there are drop offs, for example, Congkok, Sempit, Karang Bongko, Opak Kecil, Kotak Besar and Bira Kecil Islands. Several islands are almost at the same level as the sea, for example, Air Kecil, Payung Kecil, Ubi Besar and Gosong Islands.

History

On the 21st of July 1982, according to Decisison Letter of the Agriculture Minister Number 527/Kpts/Um/7/1982, some parts of Kepulauan Seribu were designated for Marine Preservation, coverimg an area of 108,000 ha

On the 14th of Oktober 1982, according to Decision Letter of the Agriculture Minister Number 736/Mentan/X/1982, the designation was changed to Kepulauan Seribu Marine National Park

On the 21st of March 1995, Decision Letter of the Forestry Minister Number 62/Kpts-II/95, was released to strengthen the status of Kepulauan Seribu National Marine Park.

Biodiversity and Ecosystems

Kepulauan Seribu Marine National Park, overall, is a tropical maritime sea water ecosystem. Because it consists of islands and clusters with their coral reefs as well as varied sea depths, it can be divided into the following local ecosystems:

The coastland ecosystem on the islands, occupying a small part of the park, which has terrestrial plants and animals

The mangrove forest ecosystem, on those parts of the coast that support mangroves

The coral reef ecosystem, which is the largest of the three. It is more varied and productive compared to the other two.

The plants that can be found in the park include coconut, *Cocos nucifera*, sea pandan, *Pandanus tectorius*, sea pine, *Casuarina equisetifolia*, noni, *Morinda citrifolia*, breadfruit, *Artocarpus atilis*, kecundang, *Cerbera adollam*, nyamplung, *Calophyllum inophyllum*, bogem, *Sonneratia sp.*, other mangroves such as *Rhizophora sp., Avicennia sp., Bruguiera eriopetala,* sea hibiscus, *Hibiscus tiliaceus*, babakoan, or crown flower, *Calotropis gigantea*, butun, *Barringtonia asiatica*, ketapang, *Terminalia catappa* and *Ipomoea sp.* The sea flora that are normally found are marine algae clusters such as *pRhodophyta* spp, *Chlorophyta* spp and *Phaeophyta* spp as well as seaweed clusters such as *Halimeda padina, Thalassia* spp, *Caulerpa* spp, *Gelidium* spp, *Sargassum* spp, *Fucus* spp and *Chondrus* spp.

There are about 257 documented species of coral. They include reef edge and sandbar species as well as several species of protected mollusks like giant clam, *Tridacna gigas*, goat head, batu laga, bahar root and susu bundar shellfish. There are about 113 species of fish and 78 other invertebrate species which are known to be associated with seagrass.

A notable bird species found in Kepulauan Seribu National Marine Park is the bondol eagle, *Haliastur indus*, which has become the zoological emblem of Jakarta. There are 17 other terrestrial and coastal species of bird in the park mostly found on Rambut island. Rambut island has been called "birds heaven" because at least 20,000 birds can be found there from as many as 22-40 species of water birds as well as the terrestrial species. During migrantion events there can be up to 50,000 individuals on the island. Some species that are easily seen are purple heron, *Ardea purpurea*, grey heron, , *Ardea cinerea*, great egret, *Egretta alba*, little egret, *Egretta garzetta*, eastern reef egret,

Egretta sacra, milky stork, *Mycteria cinerea*, glossy ibis, *Plegadis falcinellus*, oriental darter, *Anhinga melanogaster* and cattle egret, *Bubulcus ibis*.

The island also supports a big heronry, which can be viewed from an observation tower. The heronry includes a few pairs of breeding Milky Stork. There are usually Glossy Ibis. As the island is relatively undisturbed, it can also be a good place to see the Javan Myna, *Acridotheres javanicus,* a species becoming rare on the mainland. In May 2014 there was a Nicobar pigeon, *Caloenas nicobarica*, on the trail to the tower, so they probably visit the island irregularly.

The lagoon side of the reef is characterized by both small and massive branch corals such as *Acropora sp., Porites andrewsi* and *Porites lutea*. The reef crest is characterized by tube corals as well as the small and massive branch corals.

Local Communities and Culture

There are approximately 10,000 people in three villages living in the park and around 7,000 work as traditional fishermen. The people in Kepulauan Seribu consume lamun fruit of the seagrass, *Enhalus acoroides*, to supplement their diet. The main economic activities are fishing (include collecting coral reef creatures), and coconut farming, Tourism in this area has developed into a significant industry.

Pulau Kelapa and Pulau Panggang are the two main inhabited islands in the archipelago. Others, which are unhabited, may be occupied by fishermen during certain seasons, or rented or even owned by private individuals for tourism development. Twenty-three islands are privately owned and eight, including Putri, Melintang and Bidadari Islands, have been developed for tourism with cottages, bungalows, bars, restaurants, shops and camping grounds.

Tourism

This National Park may be the easiest to reach in Indonesia as it is located close to Jakarta. Popular destinations are Ayer Besar, Bidadari, West and East Antuk, Pelangi, Putri, Laki, Sepa Barat, Middle and East Kotok Besar and Rambut Islands. Almost 1.3 million tourists visited the park in 2015. Revenue from tourism in 2015 amounted to 5% of the revenue of the entire

Kepulauan Seribu regency. Usual tourist activities are diving, swimming, snorkeling, kayaking, photography and bird watching. Several interesting locations to visiting are:
- Jukung, Panjang, Macan, Genteng, Semut, Kelor and Petondan Islands where coral reefs are still in good condition, for diving and snorkeling
- Jukung, Petondan and Belanda Islands for recreational fishing
- Kelor, Onrust and Kayangan Islands have the remains of fortresses from the era of occupation by the Netherlands

Access and Transportation

Kepulauan Seribu Marine National Park can be reached from Jakarta in several ways:
- From Ancol to Bidadari Island (15 km) by motorboat takes about 30 minutes. There are public services twice a day, and four times a day on Saturday and Sunday
- From Donggala (Tanjung Priuk harbour) to Panggang Island, public boats are available twice a week and take about 5-7 hours. Every Saturday there is a jet foil, which takes less than an hour.
- There is a daily passenger service from Marina Ancol to Putri Island. The service to tourism sites is on Saturday and Sunday. It takes around 3-4 hours. The islands can also be reached by charter boat, which takes about 2-2.5 hours from this marina
- There is an airstrip on Panjang Island. Chartered aircraft such as Cessna are usually available from Monday to Thursday but twice each day on Saturday and Sunday, which takes about 25 minutes
- The best birding in the park is at Pulau Rambut, a bird sanctuary. Assuming you hire a boat to get to the park, along the way you can direct the driver to pass close to the many Buginese fish traps that can be seen on the 30 min journey. The wooden poles of these traps can support large flocks of Christmas Island frigatebird, *Fregata andrewsi*, some times as many as hundreds, and smaller numbers of Lesser frigatebird. *Fregata ariel*. If the birds are there you can often approach very closely. It is also worth looking out for Aleutian tern, *Onychoprion aleutica* around here, as they have been seen in Jakarta Bay.
- On Laki, Putri, Ayer, Pelangi, Kotok, Matahari, Sepa and Bidadari Islands there are hotel facilities as well as other types of accommodation.

The National Parks of Indonesia

- The Information Centre on Pramuka Island, can be reached within one hour by speedboat
- Other islands in the core zone can be reached from Pramuka Island by charter boat

Facilities

Facilities in Kepulauan Seribu Marine National Park include: the park ranger office, Tourist information Center, Security Post, Cottage, Meeting Rooms, Bungalows, a Camping Ground, Bars, Restaurants, Helipad, Shops, Motor Boats, Inflatable Boats, Life jackets and the Diving Facilities in the Panggang, Laki, Pelangi, Kotok, Matahari, Kelapa, Bidadari, Putri, Ayer, and Pramuka islands.

Park Office

Balai Taman Nasional Laut Kepulauan Seribu
Jl. Salemba Raya No. 9 Lt. III Jakarta Pusat 10440
Telp. (062-21) 3915773, 3103574
Fax. (062-21) 3915773
Call Center: 082149523053
Instagram: @ btn_kep_seribu
Email: tnkls@indo.net.id and informasi@tnlkkeulauanseribu.net

15. UJUNG KULON NATIONAL PARK

Geographic Location and Size

Ujung Kulon National Park is on the western end of Java island in Pandeglang Regency, Banten Province, about 78 km to the southwest of Pandeglang city and 153 km to the west of Jakarta city. It is located at $6^0 45'$ South Latitude and $102^0 00'$-$105^0 20'$ East Longitude. The park is bordered by the Sunda Strait and the Indian Ocean, except on the east side, where it is bordered by Mount Honje (620 m).

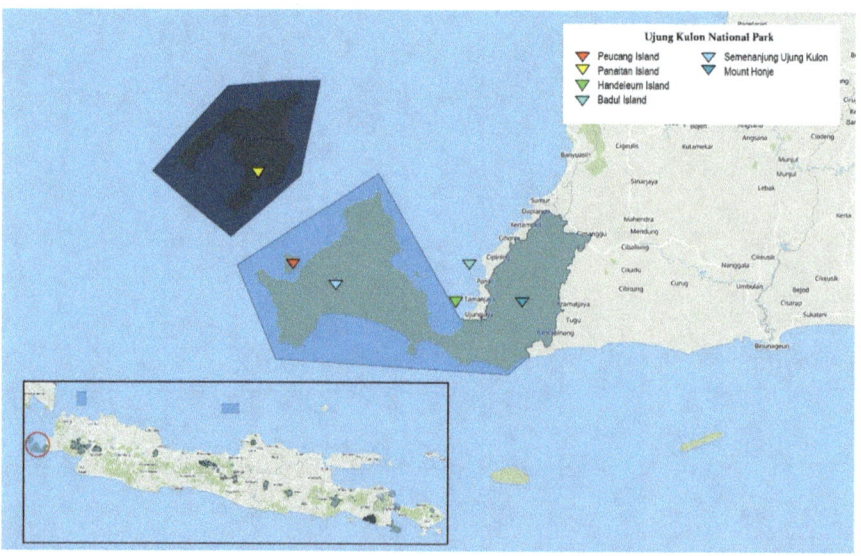

The park is 122,956 ha in area, with a land area of 78,619 ha, consisting of of Semenanjung Ujung Kulon (39,120 ha), Gunung Honje (19,489 ha) and the islands of Panaitan (17,500 ha), Peucang (472 ha), and Handeuleum (220 ha). The surrounding marine area covers 44,337 ha.

Climate and Topography

The rainfall averages 3,140 mm/year (recorded at Tanjung Layar). Temperature ranges between 20^0 and 30^0C and humididty ranges between 80% and 90%. Given its location between the Indian Ocean on the west and south and the Sunda Strait on the north, the park is affected by the strong seasonal winds from the west. The wettest season is between October and April when a seasonal northwest wind brings rain. During this season average rainfall is around 200 mm/month, rising to more than 400 mm in December and January. The dry period is May-September, when rainfall normally exceeds 100 mm/month. During the drier season the wind is from the north.

The topography is varied, ranging from the waters of Sunda Strait to the peak of Mount Honje, at 620 m asl. Ujung Kulon Peninsula, which is shaped like a triangle, has low hills with an altitude of around 50 m in the middle part. The High Plain of Telanca (100-140 m asl) is the area where streams rise that flow west. On the western end of the Peninsula are Mount Guha Bendang, 550 m asl and Mount Payung, 480 m asl.

History

In 1892, SH. Koorders from the Netherlands and the founder of the "Association of the Netherlands East Indies Nature Preservation" introduced this area to the world in many of his papers.

In 1910 the Javan rhinoceros became totally protected

In 1912 following on from the above, Ujung Kulon peninsula was protected and hunting was restricted

From November 1921 until 1937 Ujung Kulon was designated as a Preservation area

From 1937 until 1958 the area was as a Wildlife Reserve, and included Handeuleum, Peucang and Panaitan Islands

In 1958 the status changed to Wildlife Reserve that also included some parts of Mount Honje

- In 1980 it was declared Indonesia's first National Park with an area of 120,551 ha
- On the 1st of February 1992, Ujung Kulon National Park was listed as a World Heritage Site by UNESCO
- On the 26th of February 1992, Decision Letter of the Forestry Minister Number 284/Kpts-II/92, was released to strengthen the status of Ujung Kulon National Park, and the area increased to 122,956 ha.

Biodiversity and Ecosystems

Many different ecosystems are represented in Ujung Kulon National Park including mangrove forest, beach forest, freshwater swamp forest, grass meadow, low plain tropical forest, primary rain forest, secondary forest/shrubland, low mountain tropical forest and coral reef. There are also unique habitats such as the sand hills on the south coast of Ujung Kulon Peninsula and tropical beach lowland forest on Peucang Island. Ujung Kulon contains the largest tract of natural forest remaining in Java.

Mangrove forest is the dominant forest on Panaitan Island and the mangrove forest on Hamdeuleum is still in good condition. These forests are dominated by the genera *Avicennia*, *Sonneratia* and *Rhizophora*.

Barringtonia and *Ipomoea* are the dominant genera in the two beach communities. The *Baringtonia* community is characterized by species such as nyamplung, *Calophyllum inophyllum*, butun, *Barringtonia asiatica*, Chinese kampis, or lantern tree, *Hernandia peltata*, ketapang, *Terminalia catappa* and cangkil, *Pongamia pinnata*. The *Pes-caprae* community is the pioneer vegetation, normally found along the coast near the high tide mark. It is characterized by katang-katang, *Ipomoea pes-caprae*, kutut tiara, *Spinifex littoreus* and sea bean, *Canavalia maritima*, as well as small trees such as nyamplung and ketapang.

A narrow seasonal freshwater swamp forest is on the north of Ujung Kulon Peninsula near Alang-alang Bay, Nyawaan area, Nyiur Jamang and Sungai Cihandeuleum. It is characterized by cattail, *Typha angustifolia*, sedges such as *Cyperus pilosus*, and lampeni, *Ardisia humilis*.

Rainforest covers most of Ujung Kulon peninsula, Panaitan Island, Peucang Island and Mount Honje. On Ujung Kulon Peninsula and Mount Honje the rainforest is characterized by different species of palms, especially langkap, *Arenga obtusifolia*. Langkap is always found in pure stands on lower areas, especially in the northwest and northeast. A stand is normally broad and forms a tight crown about 10-15 meters in height. Other palms include nibung, *Oncosperma tigillarium*, aren, *Arenga pinnata*, sayar, *Caryota mitis* and salak, *Salacca edulis*, a which produces a popular edible fruit. At higher altitudes bing-bin, or ivory cane palm, *Pinanga coronate* is normally found. On the higher slopes, trees such as saninten, *Castanopsis* sp. and other species of the *Fagaceae* family can be found. Due to their high humidity, east facing slopes have dense forests of species such as janitri, *Elaeocarpus sphaericus*, a conifer, cangkudu badak, *Podocarpus nerifolius*, palahlar, *Dipterocarpus hasseltii*, kipela, *Aphana misxis* and *Eurya* sp. On tree branches and on the soil thick moss grows and there are many epiphytes such as the orchid, *Frycinetia* sp., and bird's nest fern, *Asplenium nidus*.

On Panaitan Island the ecosystems consist of dry beach forest, mangrove forest and lowland rainforest containing many species of palm trees. There are also deciduous plants such as munung, *Pterocymbium javanicum*, kibonteng, *Urandia secundifolia* and bungur, *Lagerstroemia speciosa*. On Peucang Island there is a mixed forest with species such as *Maranthes corymbosa*, bungur, *Bombax valetonii*, and kepuh, *Sterculia macrophylla*. The grass meadows in Cidaon and Ciujungkulon are grazed by wild cattle.

Ujung Kulon National Park is famous worldwide as the last remaining stronhold of the Javan Rhinoceros, *Rhinoceros sondaicus*. There are only about 50-60 animals, and it is the only place in the world where it has been able to breed naturally in recent decades. Besides the rhino there are about 30 species of other mammals. Examples are banteng, *Bos javanicus*, Timor deer, *Cervus timorensis*, roe deer, *Muntiacus muntjak*, and otter, *Lutra lutra*. Many different cats are also found in the park including fishing cat, *Prionailurus viverrinus*, marbled cat, *Pardofelis marmorota*, Javan leopard, *Panthera pardus melas*, and Javan leopard cat, *Prionailurus bengalensis javanensis*. Primates found in the park are Javan gibbon, *Hylobates moloch*, Javan surili, *Presbytis comata*, West Javan langur, *Trachypithecus mauritius* and Javan coucang, *Nyticebus javanicus*.

Approximately 250 species of bird have been recorded in the park, Including wreathed hornbill, *Aceros undulatus*, Rhinoceros hornbill, *Buceros rhinoceros*, Nicobar pigeon, *Caloenas nicobarica*, beach stone-curlew, *Esacus magnirostris*, black headed fruit dove, *Ptilinopus melanospila*, crested serpent eagle, *Spilornis cheela* and grey cheeked green pigeon *Treron griseicauda*.

Java endemic birds that can be found in the park are Javan eagle, *Spizaetus bartelsi*, Javan coucal, *Centropus nigrorufus*, Javan sunbird, *Aethopyga mystacalis* and Javan Kingfisher, *Halcyon cyanoventris*.

Other animals include true frogs, e.g. *Rana chalconata*, puru frog, *Bufo biporcatus*, Javan toad, *Limnonectes macrodon*, estuarine crocodile, *Crocodylus porosus*, pythons, *Python reticulatus* and *Python molurus*, gharial, *Tomistoma schlegelii*, freshwater turtle, *Siebenrockiella crassicollis* and Asian water monitor lizard, *Varanus salvator*. Marine turtles found in the waters of the park are green turtle, *Chelonia mydas*, leatherback turtle, *Dermochelys coriacea* and hawksbill turtle, *Eretmochelys imbricata*.

Local Communities and Culture

The local communities around Ujung Kulon National Park consist mainly of Sunda people but there are others such as Java, Madura and Bugis. The most common handicraft is Javan Rhino statues. There are 13 villages on the east side of the park near Mount Honje, which are within Sumar and Cimanggu Regencies.

These communities have been there, interacting with the forests in their environment, for many generations. They utilize the forests to meet their daily needs, from household goods to spiritual needs. Household goods include food (hunting for deer, fishing, rootcrops, fruit picking), firewood, construction and handicraft material, medicine, farming plots, settlements, and cattle pastures. The locals work as fishermen, coconut pickers and dry land farmers. The shape of their houses and their farming practices are still traditional. In addition, there is the macabre "debus" magic art demonstrating the invulnerability of its adherents, and several other arts such as "syaman and bedug", the magic dancing with drums

Tourism

Exisiting and potential tourism attractions include:
- Taman Jaya, which has a hot water spring, a beach, opportunities for adventures in nature, a botanical garden and opportunities for wildlife and nature photography
- Handeuleum Island has mangrove forest and wildlife, which can be seen from a kayak paddling along the river. Hiking, photography, and camping

are also common activities. At Cigenter you will find the Javanese rhinoceros, a waterfall and a fast-running river.
- Peucang Island has beaches for swimming and diving in the sea, unique and beautiful natural views, kayaking and camping. There is also a restaurant and hotel. There is also a 600 m long beach called the Golden Beach, consisting of white sand, and clear and calm blue sea. Approximately 1 km to the south there are coral reefs with a large variety of various species of beautiful and extraordinary corals and fishes.
- Panaitan Island has a cultural attraction in the form of statues of the Hindu gods, Ganesha and Shiwa on top of Mt. Raksa (320 m). This Island provides an astonishing attraction in the form of its unique seagarden where the Sunda Strait meets the Indian Ocean. Several other locations on Panaitan Island are worth visiting, for example, Semadang, to enjoy the beautiful corals and sloping beach while snorkeling and swimming, Lentah, for scubadiving, coral reef and tropical fishes, Bajo, to enjoy the beautiful sandy beach scenery and white sand, and to surf and the Kasuaris Swa, a white sandy beach lined with coconut trees, which is popular with surfers.
- Layar Cape has a Dutch colonial lighthouse that is actually still operating. People go to Layar Cape to meditate in the Sanghian caves.
- Citerjun has a 2.5 meter high waterfall. The water is fresh and abundant and flows towards the beach where it meets and mixes with the salt sea water.
- Ciramea beach is a turtle nesting beach where, at the right time of year, turtles can be seen laying eggs.
- Visitors who require tour guides can contact the designated National Park Management Office

The best time to visit Ujung Kulon National Park is between the months of April and October, which is the period where there are no high seas. Between the months of December and March the ocean waves are quite dangerous and it is not a pleasant time to visit.

Access and Transportation

Ujong Kulon National Park Office and Information Center is at Jl. Perintis Kemerdekaan No. 51, Labuan, Pandeglan SubRegency, Banten. The park can be reached as follows:
- From Jakarta through Tangerang–Serang–Pandeglang–Labuan, a distance of approximately 153 Km. It will take 3 to 4 hours using public transport.
- From Bogor through Rangkas Bitung–Pandeglang–Labuan is a distance of approx. 160 Km. It will take 3.5 to 4 hours using public transport.

Taman Jaya. This region is at the entrance to Ujung Kulon National Park and is the transit terminal leading to the islands in the Tamanjaya area. It can be reached by land transportation in the form of a minibus leaving from the Labuan terminal to Sumur, and then continuing on by motorbike to Tamanjaya ($\pm$ 19 km). The traditional route is from Labuan harbor using a boat from Tamanjaya. Other tourism sites can be reached by following an existing patrol road.

Handeuleum Island is located north of Ujung Kulon. It can be reached by rental speedboat from Tamanjaya harbor, which is 9 km and will take approximately 45 minutes, or from Sumur, which is approximately 20 km and will take 1.5 hours.

Peucang Island can be reached by:
- Renting a speedboat from the Tamanjaya harbor, which is approximately 38 km and will take 3.5 hrs. The boat leaves Labuan twice a week on Monday and Friday and returns on Thursday and Sunday. This boat can be booked through the National Park Office or The Wana Wisata Alam Hayati Office at Labuan.
- by following the patrol road from Karangranjang to Tamanjaya (5.5 hrs), then by foot following the hiking path at pantai selatan. There is a crossroads at Cibunar. The north branch will lead to Ciadon and Peucang Island, while the other crosses Mt. Payung to Sanghiang Sirah Bay, which is a holy place.
- From Sumur a rental speedboat will cover the approximately 43 km in 2.5 hours
- From Carita, the fastboat Wanawisata III will take 2.5 hours. An ordinary boat it will take 6 hours.

Panaitan Island can be reached using a rental boat from Peucang Island harbor. It will take about 30 minutes. Mt. Raksa peak can be reached on foot by following a hicking track in the Citambuyung area.
- The Ujung Kulon Peninsula contains several tourism sites such as:
- Cidaon Game Reserve. This area consists of cattle grazing pasture. It can be reached from Peucang Island by boat followed by a walk of 4kms.
- Layar Bay (Tanjung Layar). This area has the Sanghiang Sira cave and the spectacular view of the west coast of the peninsula. It can be reached from Tamanjaya by following a footpath or from Peucang Island by using a motorboat.
- Cigenter can be reached by using a motorboat from either Handeuleum Island or Peucang Island.

Facilities

- In general, accommodation is available mainly on the north coast at Tamanjaya, Handeuleum Island, and Peucang Island. In the villages of Cegog and Rancecet, located in the south, there is accommodation owned and managed by the local community. Both areas can be reached on foot from Tamanjaya.
- At Labuan there is the Ujung Kulon National Park office, an information center, souvenir shops, minibus terminal located in front of the local Police Office, and motorboats and hotels.
- At Tamanjaya there are camping sites, restaurants, accommodation, harbor and motorboats, hiking paths and patrol paths, watch towers and posts and the Sundajaya Homestay.
- At Citalang and Nyawaan there are hiking paths and camping sites.
- There are camping sites provided at Citadahan and Karangranjang, located near a hiking path.
- On Handeuleum Island there are camping sites, restaurants, accommodation a harbor and motorboats, watch posts and towers and hiking paths.
- On Peucang Island there is an information center, camping site, restauarants, accommodations a harbor and hiking paths.
- On Panaitan Island there is a harbor and hiking paths.
- Other facilities incude watch posts, observation Towers Cafeterias, Shelters, harbors at Cigenter and Ciujungkulon, 20 patrol roads and souvenir shops at Carita beach.

- A visitor management facility in Ujung Kulon National Park is run by the ana wisata Alam Hayat (a nature tourism company). This company provides, amongst other things, motorboat rentals to reach the northern part of Ujung Kulon from Labuan.

Park Office

Balai Taman Nasional Ujung Kulon
Jl. Perintis Kemerdekaan No.51 Labuan, Serang, Pandeglang, Banten
Telp. (+62-253) 801731, 804681
Fax. (+62-253) 804651
Call Center: 08111238884
Instagram: @ btn_ujungkulon
Email: info@ujungkulon.org or balai_tnuk@yahoo.com
Twiter: @ujungkulonNP

16. GUNUNG GEDE PANGRANGO NATIONAL PARK

Geographical Location and Size

Gunung Gede Pangrango National Park (Taman Nasional Gunung Gede Pangrango, or TNGGP) is located between $6°41'$ and $6°51'$ south latitude and $106°51'$ and $107°02'$ east longitude. It is in West Java Province and covers parts of three Regencys, namely:
- Bogor Regency, with 25% of TNGGP covering an area of approximately 4,514 ha.
- Sukabumi Regency, with 30% covering an area of approximately 6,782 ha.
- Cianjur Regency, with 45% covering an area of approximately 3,899 ha.

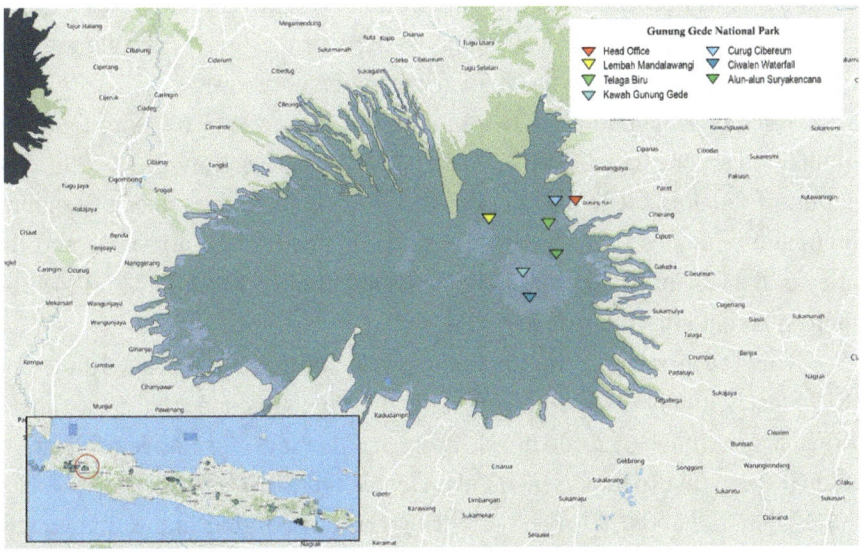

To the east and north are the Regencies of Cianjur and Bogor. Sukabumi Regency is to the west and south. The landscape outside the park consists of production forests, tea plantations and land owned by local communities.

The total area of Gunung Gede-Pangrango National Park is 15,195 ha, covering Cibodas Nature Reserve, Cimungkat Nature Reserve, Gunung Gede Pangrango, Situ Mountain Nature Park and the forest area on the slopes of Mount Gede-Pangrango.

Climate and Topography

The average rainfall ranges between 3,000 and 4,200 mm/ year. The rainy season is from October to May with an average rainfall of about 200 mm/month, reaching a peak from December to March when rainfall exceeds 400 mm/month.

The drier season occurs from June to September when average rainfall is less than 100 mm/month. Although the humidity is generally high, in the dry season it varies from 30% at night up to 90% in the afternoon.

In the daytime the average temperature in Cibodas is about 18^0C, at the peak of Mt. Gede and Mt. Pangrango it is about 10^0C, but it can reach to between $0-5^0C$, and is often covered with thick fog.

In general, the wind is a monsoon wind that changes direction according to the season. In the rainy season, especially between the months from December to March, the wind blowing from the southwest can be very strong sometimes resulting in the destruction of parts of the forest. Throughout the dry season, the wind blows more gently from the northeast.

TNGGP is named after two mountains located side by side, Gunung, mountain in Indonesian, Gede, 2,958 m and Gunung Pangrango, 3,019 m. The summit of Gunung Gede has a crater that is older than the crater of Gunung Pangrango with steep rock walls. These walls open to the northeast and the valley of the Cibatu River. Mgunung Gede first erupted around 1747-1748 and the most recent eruption was in 1948.

The topography is mountainous with altitudes between 1,000 and 3,019m asl. Gunung Gede and Gunung Pangrango are connected by a plateau located at an altitude of 2,400 m asl. There are also several small lakes, namely Lake Gayonggong (near Cisaat), Lake Denok and Situ Gunung.

There are about 50 rivers and tributaries discorging in a radial pattern from the park. Major rivers include the Cimandiri River, which flows to the south and into the sea at Pelabuhan Ratu, the Angke River, which flows into the Java Sea and the rivers Cikundul and Cianjur Leutik flowing eastward to join the Citarum river, which in turn flows north towards Jakarta.

History

In 1819 C.G.C. Reinwardt was the first European to climb Mt. Gede. He was followed by many others including A.R. Wallace in 1861. C.G.G.J. van Steenis made a collection of plants, which formed the basis for the The Mountain Flora of Java, released in 1972.

In 1830 the Cibodas Botanical Garden was established on the slopes of Gunung Gede, which is where the first kina, or quinine, plants, Cinchona spp. And other exotic species of plants were first cultivated in Indonesia.

In 1889 the Cibodas Botanical Garden was expanded to the borders of the primary forest, located above the Garden.

In 1919 Cimungkat Nature Reserve, with an area of 56 ha was established on the southern slopes of Gunung Pangrango.

In 1925 the Cibodas Botanical Garden was expanded and re-classified as the Cibodas – Gunung Gede Nature Reserve, with an area of 1,040 ha.

In 1975 a Tourism Park was established in the Gunung Situ region with an area of 120 ha.

On the 6th March 1980, the Minister of Agriculture declared that the Cimungkat Nature Reserve, Cibodas Nature Reserve, Gunung Gede-Pangrango Nature Reserve and the Tourism Park at Gunung Situ and the surrounding natural forest on the slopes of Gunung Gede and Gunung Pangrango be combined to form Gunung Gede Pangrango National Park, with an area of 15,000 ha. This was done simultaneously with the establishment of four other Indonesian national parks, to coincide with the launching of the World Conservation Strategy.

In 1982 the Minister of Agriculture issued statement No. 736/ Mentan/1982, declaring that the area of the Gunung Gede Pangrango National Park was 15,196 ha.

In 1991 Gunung Gede Pangrango National Park was declared a Biosphere Reserve by UNESCO.

Gunung Gede Pangrango National Park is also a sister park with Malaysian parks.

Biodiversity and Ecosystems

Most of the park consists of forested mountains, the forests of which are in relatively good undisturbed condition. This tropical rainforest ecosystem can be divided into three zones based on elevation:

The sub-montane zone (1000-1500 m). The forest in this zone is characterized by high species diversity. There are many big trees such as Rasamala, *Altingia excelsa*, that can reach a height of 60m, *Castanopsis argentea*, *Antidesma tetradrum*, *Litsea sp.*, and shrubs such as *Ardisia fulginosa* and *Dichroa febrifuga*. The dominant tree species in general originate from the families Fagaceae and Lauraceae. There is a dense understory, which includes many epiphytes, ferns such as *Asplenium nidus* that can reach 2 m in height, about 208 species of orchids and the red moss, *Sphagnum gedeanum*.

The montane zone (1500-2400 m). A prominent tree species here is puspa, also known as needlewood tree, *Schima wallichii*. Conifers such as *Dacrycarpus imbricatus* and *Podocarpus neriifolius* are found at these highr altitudes, with diversity generally decreasing with altitude.

The subalpine zone (over 2,400 m). The forest in this zone has a lower canopy. The dominant species are *Rhododendron retusum*, *Rhododendron javanicum*, *Myrsine avenis*, *Selliguea feei* and cantigi, *Vaccinium varingiaefolium*. Cantigi can form monocultures, as it does in the crater areas. Typical plants found on the summits are low-growing perennials such as edelweis flower, *Anaphalis javanica*. Other habitats include lakes, swamps, subalpine pastures, volcanic craters and lowland forests in the south-west.

Of the more than 1,000 species of plants recorded in the park, notable specimens not already listed above include ratan, *Calamus sp.*, Hasselt's orchid, *Dendrobium hasseltii*, kondang, or veriagated fig, *Ficus variegata* and kantong semar, a carnivorous pitcher plant, *Nepenthes gymnamphora*. Plants endemic to Java include Javan edelweis, *Anaphalis javanica*, lumut merah, *Sphagnum gedeanum*, *Dioscorea blumei*, *Dioacorea platycarpa*, *Amomum pseudofoetens*, and several species of orchids such as *Corybas praetermissus*, *Malaxis sagittata*, *Stigmatodactylus javanicus*, and *Liparis mucronatus*.

There is a great diversity of wildlife found in TNGGP with some unique species, such as the worm cacing sonari, *Polypheretima elangata*, which can reach 60 cm in length and emits a high-pitched sound.

250 of the 450 species of birds found on Java island, inhabit the park, of which 25 of are Javan endemics. Two Javanese eagles, Javan eagle, *Spizaetus*

bartlesi and Javan tesia, *Tesia superciliaris*, are endemic birds only found at a few locations on the island of Java, one of which is in this national park. Other notable birds include the crested serpent eagle, *Spilornis cheela*, Sumatran green pigeon, *Treron oxyura* and the spotted wood owl, *Strix seloputo*. The park is also an important mammal habitat. More than 100 species of mammals are found here and several are Javan endemics. These include Javan gibbon, *Hylobates moloch*, Javan surili, *Presbytis comata*, Sunda leaf monkey, *Trachypithtecus mauritius*, Javan coucang, *Nycticebus javanicus* and pangolin, *Manis javanica*. Other mammals include a flying squirrel, *Hylopetes bartelsi*, Javan porcupine, *Hystrix javanica*, Javan leopard, *Panthera pardus melas* and Javan leopard cat, *Prionailurus bengalensis javanensis*.

The herpetofauna includes a horned frog, *Megophyrs montana*, a flying frog, *Rhacophorus reinwardti*, Javan flying frog, *Rhacophorus javanus*, bleeding toad, *Leptophryne cruentata*, a blind snake, *Calamaria linnaei*, green viper snake, *Trimeresurus puniceus* and many others.

Local Communities and Culture

Villages surrounding the park include Pangrango Gede village, Cikahuripan village, Sukamaju village and Sukamanis village. Rice is grown where possible, but most of the land used for cultivation in this area is sloping, some of it steeply so. Crops include vegetables and perennials such as calliandra, bamboo, jeunjing and aren. The surrounding communities rely heavily on what they can obtain from forest products such as wood, bamboo, rattan and others. Agroforestry systems should be developed in the buffer zone around the park because it would have a positive impact in preserving the forest of the park as well as provide communities with incomes.

Tourism

TNGGP is one tourist attraction among others nearby, such as the Puncak and Taman Safari Indonesia. In general the tourism attraction of this area is mountain views of nature with fresh, clear and brisk air. The park is the only conservation area that limits the number of visitors. Several places frequently visited are:

Telaga Biru, ± 5 ha, altitude 1,575 m asl, 1.5 km (15 minutes walk) from the Cibodas entrance. The water is blue due to a blue alga.

Gayonggong Swamp formed in an extinct volcanic crater.
- The Cibeureum waterfall, altitude 1,625 m asl, 2.8 km (1 hr walk) from the Cibodas entrance.
- Hot Water Springs, altitude 2,150 m asl, approximately 5.3 km (2 hrs walk) from the Cibodas entrance. The spring is located next to the hiking trail, so it is a perfect place to rest. The water temerature can reach 75^0C and contains a high amount of sulfur.
- Puncak and the Crater of Gunung Gede, altitude 2,958 m asl, 9.7 km (5hrs walk) from the Cibodas entrance. From here there are breathtaking sunrises and sunsets, and views over Cianjur, Sukabumi and Bogor city.
- Gunung Pangrango Peak, altitude 3,019 m asl, approximately 3 km (3 hrs walk) from Kandang Badak or 11 km (6.5 hrs walk) from the Cibodas entrance.
- The Cibeureum Selabintana Waterfall, altitude 1,350 m asl, 2.4 km (45 minutes walk) from the Selabintana entrance.
- Sawer Waterfall, altitude 1,200 m asl, 1.85 km (20 minutes walk) from the Situgunung entrance.
- Kandang Batu, altitude 2,200 m asl, 5.6 km (2.5 hrs walk) from the Cibodas entrance.
- Kandang Badak, altitude 2,400 m asl, 7.8 km (3.5 hrs walk) from the Cibodas entrance.
- Alun-Alun Suryakencana, altitude 2,750 m asl, 11.8 km (6 hrs walk) from the Cibodas entrance.
- Selabintana, where there is a waterfall and camping site (capacity 150 people).
- Mt. Situ Recreation Park, where there is a lake, waterfall, and camping site.
- Mt. Putri, camping site.
- Cibodas Botanical Garden.
- Badogol Conservation Education Area is located near the Hotel Lido, Bogor. It is an education and student facility, which can accommodate 40 students. There is a canopy bridge between trees.
- Tropical Forest Research Center, which is a collaboration between Conservation International Indonesia, the University of Indonesia and the National Park. There are crisscrossed trails to study animals and also permanent plots to study plant ecology. A research center with 3 bunk beds is availble for students and researchers.

- The Javan gibbon Rehabilitation Center and animal hospital is located near the south gate in the Badogol area of the TNGGP. Tourists can only see this rehabilitation center from a distance due to strict regulations limiting access.

Tourists and Visitors who need a guide can contact the National Park.

It is prohibited to bring electronic equipment, firearma, knives, and detergent, or to cut plants or to move plants or wildlife to another location.

It is prohibited to throw seeds in the area or to litter.

During the dry season visitors have to be careful, due to the plants being very dry and can easily catch fire.

Climbers of the Gunung Gede-Pangrango peak, are recommended to bring warm jackets, flashlights, camping matress and sleeping bags, tent, simple cooking utensils, First Aid kit, and should have a permit to climb the mountain, when:
- Leaving the Cibodas-Cibereum trail to go to the hotsprings;
- Going down the Selabintana trail to the Cibeureum waterfall
- Entering the National Park from the Mt. Putri entrance at the back of the Bobojong camping site.

Visitor service and climbing permits can be obtained from the TNGGP Office, at the following times:
- Monday to Thursday : $07.^{30}$ - $14.^{30}$
- Friday : $07.^{30}$ - $11.^{00}$
- Saturday : $07.^{30}$ - $13.^{30}$
- Sunday : $07.^{30}$ - $14.^{00}$

TNGGP provides slide shows at its Wisma Cinta Alam to visitor groups on request.

Access and Transportation

The Park receives many visitors, especially on weekends. There are four entrances, two in Cianjur Regency at Cibodas and Mt. Putri, and two in Sukabumi Regency at Selabintana and Mt. Situ Tourism Park.
- Cibodas is the main entrance and is where the National Park office is located. It can be reached by car from Jakarta (100 km) duration 2.5 hrs,

from Bogor through Ciawi (36 km) duration 1 hr., or from Bandung (89 km) through Cianjur and Cipanas, which will take 2 hrs.
- Gunung Putri is located near Cibodas and can be reached through Cipanas and Pacet.
- Selabintana, can be reached by car from Bogor (60 km) duration 1.5 hr, and from Bandung (90 km) duration 2 hrs. From the center of Sukabumi city there is public transportation to Selabintana, which stops at Pondok Halimun. From there it is possible to rent a car or motorbike, or walk the 6 km.
- Taman Wisata Situ Gunung is 15 km from Selabintana. It can be reached by public transport from Bogor or Sukabumi and then on to the destination.

Facilities

This park is located near many medium sized cities such as Bogor, Cianjur and Sukabumi, and close to the large cities of Bandung (Capital city of West Java) and Jakarta (Capital City of Indonesia). In Bogor there are many facilities including hotels. Along Puncak road connecting Bogor and Cianjur cities, there are thousands of villas and home stays. Even the headquarter of the national park office is located on this road, approximately 30 km from Bogor. From Bogor to Sukabumi, there are also hotels in Lido, one with a golf course and there are other tourist attractions such as a Conservation Education Center.
- Cibodas provides guesthouses, Wisma Cinta Alam, Information Center, Gunung Gede-Pangrango National Park management office, Youth Cottage Cibodas, Cibodas Botanical Garden, hotels, housing and home stay, tour guides, shelters and toilets.
- Camping is provided at at very attractive locations ranging from simple, such as at Kandang Batu, Kandang Badak and Alun-Alun Suryakencana at the peak of Gunung Gede to more salubrious, such as the campgrounds of Selabintana and Gunung Putri.
- The Bodogol Nature Conservation Education Center includes a camping site where visitors can learn the ways of good camping and how to reduce impacts. There is also a 100 m canopy walkway.
- Gunung Situ Park is provided with a Wisma Cinta Alam, Information Center, pathways with signs, shelters, bathrooms as well as other accommodation facilities.

Park Office

Balai Besar Taman Nasional Gunung Gede Pangrango
Jl. Raya Cibodas, Cipanas PO.Box 3 Cipanas,Cianjur, Jabar 43253
Telp: (+62-263) 512776, Fax: +62-263-519415
Call Center: 08119155815
Instagram: @ bbtn_gn_gedepangrango
e-mail   : info@gedepangrango.org and info@gedepangrango.org

17. GUNUNG HALIMUN SALAK NATIONAL PARK

Geographical location and Size

Gunung Halimun Salak National Park, Taman Nasional Gunung Halimun Salak, or TNGHS, is located between $6°37'$ and $6°53'$ south latitude and $106°21'$ and $106°38'$ east longitude. It is about 100 km south-west of Jakarta, 20 km southwest of Bogor and 10 km north of Pelabuhan Ratu. The park is located in West Java Province and covers parts of three Regencies

- Bogor Regency consisting of 5 districts and 13 villages.
- Sukabumi Regency consisting of 3 districts and 18 villages.
- Lebak Regency consisting of 4 districts and 19 villages.
- The total area of tnghs is 113.357 ha.

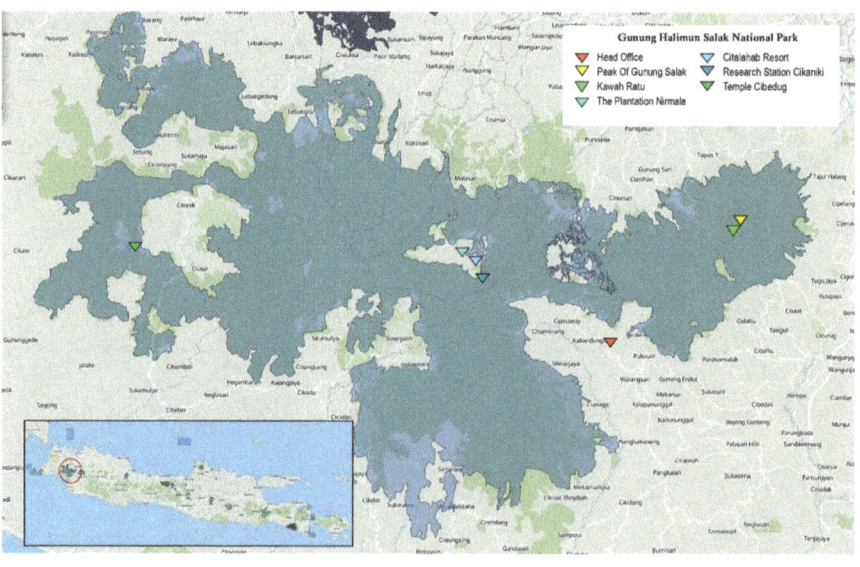

Climate and Topography

Average annual rainfall is between 4,000 and 5,000 mm. Daily temperatures rang between 18 and 26°C, with a humidity of about 85%. The rainy season occurs in the months from October to April, and the dry season in the months from May to September.

Topography is in general mountainous with an altitude of between 500 and 2,211 m above sea level, with slopes between 25% and 65%.

Gunung Halimun Salak National Park is a mountainous region that is geologically old with volcanic rocks such as breccias, basalts and andesite lava. There are more than 10 mountains, among them Mt. Halimun (1,929 m), Mt Salak (2,211 m), Mt. Sanggabuana (1,919 m), Mt. Halimun South (1,744 m), Mt. Botol (1,785 m), Mt. Amdan (1,463 m) and Mt. Kendeng (1,764 m), as well as several other peaks between 800 and 1200 m above sea level.

More than 50 rivers disgorge from the park including the Ciberang/ Ciujung, Cidurian, Cisadane, and Cimadur Rivers.

History

Between 1924 and 1934 under the Dutch colonial government there was a Protected Forest with an area of 39,941 ha.

On the 11th of January 1979, based on the Decree of the Minister of Agriculture No. 40/Kpts/Um/I/1979, the status of the region was altered to a Nature Sanctuary with an area of 41,710 ha.

On February the 26th, 1992, based on the Minister of Forestry Decree No. 282/Kpts-II/92, the status of the area was transformed into Mt. Halimun National Park with an area of 40,000 ha. Then in 2003, based on ministerial decree 175/Kpts-II/2003, the area of Mt. Halimun National Park expanded to include forest areas of Mount Salak, Mount Endut and some other surrounding forest areas, making the total area of the park 113,357 ha.

Biodiversity and Ecosystems

The ecosystems of Gunung Halimun Salak National Park are diverse. This area serves as a life support system, particularly its climate and hydrological functions, for the Bogor, Lebak and Sukabumi Regencies and the city of Jakarta. It also supports science, education and training, as well as nature tourism.

There are about 1,000 species of plants including rare orchids and 17 species of *Ficus*. The ecosystems in this region can be distinguished based on altitude, as follows:

Highland rainforest (500-1000 m asl). In this zone there has been a lot of damage. The forest has become a secondary forest with rampant undergrowth and tree pioneers like Kareumbi, *Omalanthus populneus*, cangcaratan, *Nauclea lanceolata* and manggong, *Macaranga rhizoides*. Also found here are several commercial species such as suren, or Chinese cedar, *Toona sinensis*, rasamala, *Altingia excelsa*, riung anak, *Castanopsis javanica*, keruing, *Dipterocarpus spp*. and puspa, *Schima walichii*.

Sub-montane forest (1,000-1,500 m asl). These forests have high species diversity. The dominant species here include rasamala, *Altingia excelsa*, puspa, *Schima walichii*, pasang, *Lithocarpus sp*., suren, *Toona sinensis*, jamuju, *Dacrycarpus imbricatus*, baros, *Magnolia blumei*, waru sintok, *Cinnamomum sintok*, kiputri, *Podocarpus neriifolius*, *Antidesma montanum*, *Eurya acuminata* and *Evodia aromatica* with a sparse undergrowth. Also found here are 75 species of orchids, such as *Bulbophylum binnendykii*, *Bulbophylum angustifolium*, *Bulbophylum scottifolium*, *Bulbophylum violaceum*, *Coelogyne correa*, *Cymbidium sundaicum* and *Dendrochilum raciborsckii*, and various kind of epiphytes and climbing plants such as rattan, *Calamus sp.*

The montane forest zone (above 1,500 m asl) is dominated by species of the Fagaceae family such as the pasang, *Quercus sp*., jamuju, *Dacrycarpus imbricatus* and the kiputri, *Podocarpus neriifolius*.

Historically, the Javan tiger, *Panthera tigris sondaica* and the Javan rhino, *Rhinoceros sondaicus* occurred in the area covered by this park but are now locally extinct. Prominent animal species include the wild dog or ajag, *Cuon alpinus*, pangolin, *Manis javanica*, leaf monkey or lutung, *Tracypithecus mauritius*, wild boar, *Sus scrofa*, Javan mouse-deer or Kanchil, *Tragulus javanicus*, and skunk, *Mydaus javanensis*. The park contains three species of primates endemic to Java, Javan gibbon, *Hylobates moloch*, Javan surili, *Presbytis comata*, and Javan coucang *Nycticebus javanicus*. Tthe Javan gibbon population is the largest in the world, at more than 1,600 amimals. (Supriatna et al 2003). Many different cats can be found including Javan leopard cat,

Prionailurus bengalensis javanensis, fishing cat, *Prionailurus viverrinus*, and marbled cat, *Pardofelis marmorota*.

There are more than 250 species of birds of which 30 are endemic. They include crested goshawk or Elang-alap jambul, *Accipiter trivirgatus*, white-flanked sunbird, *Aethopyga eximia*, barred eagle owl, *Bubo sumatranus*, rhinoceros hornbill, *Buceros rhinoceros*, red jungle fowl, *Gallus gallus* and Javan trogon or luntur gunung, *Harpactes reinwardtii*. Javan endemic birds that occur in the park include grey-cheeked tit-babbler, or Java ciung-air, *Macronous flavicollis*, Javan eagle, *Spizaetus bartelsi*, chestnut-bellied partridge, *Arborophila javanica*, Javan Scops owl or Celepuk, *Otus angelinae*, Javan cochoa, or Ciung mungkal, *Cochoa azurea*, Javan fulvetta, or Wergan, *Alcippe pyrrhoptera*, Javan tesia, *Tesia superciliaris*, pygmy titi or Javan cerecet, *Psaltria exilis*, Javan grey-throated whiteye, or opior, *Lophozosterops javanicus* and flame-fronted barbet, or Takur tohtor, *Megalaima armillaris*.

Local Communities and Culture

The communities residing in the surrounding area consist mostly of the Sundanese tribe, especially from the Kasepuhan Citorek and Cicemet communities, who still adhere to their cultural traditions. In this society there are several traditional ceremonies, which include:
- Nandur, which occurs when the rice harvest is about to start.
- Meupeuk pare berkah, which occurs when the padi is starting to bear rice.
- Nganyaaran, which occurs when they are going to store the harvested rice in designated storage sheds.
- Seren Tahun, is practiced by the Kasepuhan Banten Kidul community, around the month July, as a sign that the farming period of the past year is over.
- Ngaruwah, thanksgiving during the Javanese month of Ruwah

The Kasepuhan communities reside in the villages of Cicarucub, Cisungsang, Bayah and Kidul in South Banten. In the west there is the Badui tribe, indigenous people of West Java. The Badui still uphold their traditional way of life, which, it can be fairly said, has not been influenced by other cultures.

The Kasepuhan community has a unique pattern of forest management, similar to a modern zoning concept. The forest is divided into four zones, namely Leuweung Kolot (not to be disturbed), Leuweung Titipan (must have permission from the Girang Elders; the traditional leaders), Leuweung Sirah Cai (forest reserved as a source of water) and Leuweung bukaan (can be used) (Harada et al., 2001). Forest products are used extensively by the communities as a source of construction timbers and household appliances, firewood, ferns, ornamental plants, rattan, plant foods, medicinal plants and herbs used in traditional ceremonies.

Tourism

- Waterfalls. In the National Park there are about eight waterfalls which have development potential as follows:
  - In Sukabumi Regency the waterfalls Curug Cimantaja and Cipamulaan in Cikiray, Cikidang, as well as the waterfalls Citangkolo and Ciraksamala around Mekarjaya
  - In Bogor Regency the waterfalls Curug Piit and Cihanjawar near the tea plantation Nirmala Agung, and the waterfall Curug Ciberang at the Village Cisarua, Cigudeg.
- Candi Cibedug. Here there are small temples from the Megalithic era, located in the southwest. They can be reached on foot, which is 8km and will take 2 hours from the village of Citorek, crossing small lakes with beautiful scenery.
- The mountain peaks in this region can be climbed by following paths and trails, but the conditions are often quite difficult. Examples are the trails in the Cisangku area to the summit of Mount Kancana and the trails to the summit of Mount Halimun. Mount Salak (2,211 m above sea level) can be reached from several roads.
- Tea Plantation Nirmala. Located in the east near the main entrance at Cipeuteuy.

Citarik river, for rafting activities

Visitors who require tour guides can contact the National Park Management office.

Because Mt. Halimun is a fairly remote forest area, each visitor must submit a copy of an ID card or other valid identification and also a statement of approval from parents for those visitors under the age of 17.

Each visitor is required to bring a letter of admission (Entrance Permit), sufficient food and water or other drinks, check their luggage at the checkpoint, walk on designated tracks and stop at designated rest points, carry any rubbish back with them and to report to the officer when leaving the park.

Do not bring pets, weapons or hunting equipment, musical instruments. Do not damage sites in any way and do not make a bonfire with wood, twigs or the like in the forest.

When entering Halimun Salak National Park you must register at the National Park Office (Kabandungan) which can be done on the following days:
- Monday - Friday : $07.^{30}$-$14.^{30}$
- Saturday : $07.^{30}$-$13.^{30}$

The best time to visit is during the summer (dry season), which is from about the months of May to October. During the rainy season the terrain is very difficult to access.

Access and Transportation

Halimun Salak National Park can be reached from three directions, from Bogor, from Rangkasbitung (Lebak) and from Sukabumi. The condition of the roads through villages around the park's boundaries is relatively good. The routes are as follows:

From Bogor Regency:
- Bogor to Leuwiliang (20 km) by public transportation will take 30 minutes.
- Leuwiliang to Nanggung (15 km) by public transportation will take 20 minutes.
- Nanggung - Cisangku (15 km) by motorbike (ojek) will take 1 hour.

From Lebak Regency:
- Rangkasbitung (Lebak) to Bayah (150 km) by public transportation will take 2 hours.
- Bayah to Ciparay (36 km) by bus or motorbike (ojeg) will take 2 hours.

From Sukabumi Regency:

- Sukabumi to Parungkuda (20 km) by bus will take 30 minutes.
- Parungkuda to Cipeuteuy (30 km) by public transportation will take 1 hour.

Facilities

Camping sites can be found in the villages of Cikaniki and Citalahab, located on the road between the Nirmala Tea Plantation and Kabandungan village.

A Canopy Trail of 110 m is located in the village of Cikaniki.

Facilities present in Gunung Halimun National Park include the Pondok Kerja (work cabin), Watch Posts, a Guest Inn, Information Center and Hiking trails.

Park Office

Balai Taman Nasional Gunung Halimun
Jl. Raya Cipanas, Kabandungan Kotak Pos 2 Parung Kuda Sukabumi 43168
Telp./ Fax. (062-266) 62125657
Call Center: 08572188866
Instagram: btn_gn_halimunsalak
Email: tnhalimunsalak@gmail.com or tnhalimunsalak@yahoo.com

18. KARIMUN JAWA ISLANDS NATIONAL PARK

Geographical Location and Size

The Marine National Park, Karimun Jawa, is located between $5^0 40'$-$5^0 71'$ South Latitude and $110^0 04'$-$110^0 41'$ East Longitude in the Java Sea about 113 kms to the north of Central Java. Administratively it falls under the Karimun Jawa District, Jepara Regency, Central Java Province.

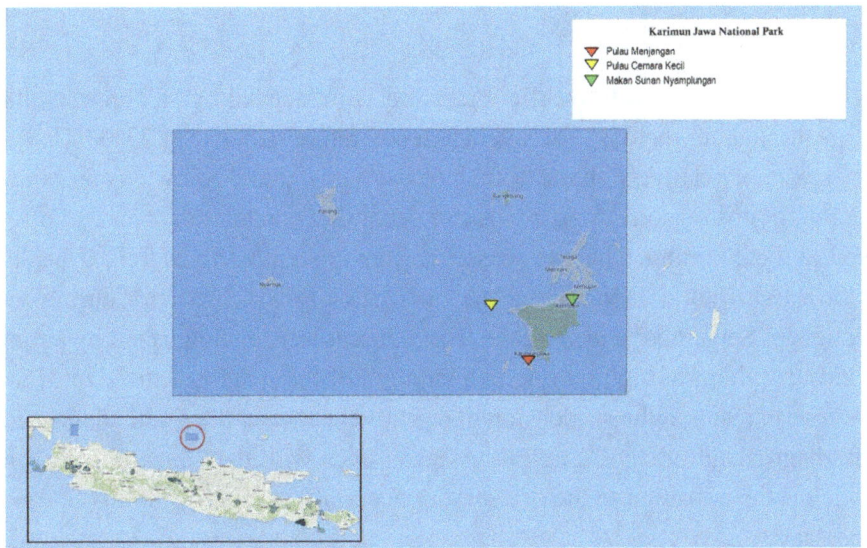

The area of this park is 111, 625 ha, consisting of 27 islands covering an area of 7,033 ha and the surrounding waters covering 104 592 ha. The park is divided into several zones as follows:

- The Core Zone is reserved for the conservation of genetic resources and the protection of ecological processes It includes Geleang Island and Bird Island and surrounding seas. The core zone is closed to any form of visit, except for scientific research.
- The Protected/Rimba Zone is similar to the core zone, but limited tourist activities can take place It includes the islands of Krakal Besar, Krakal Kecil, Menyawakan, Cemara Besar, Cemara Kecil, Bengkoang, Kamujan and part of Karimun Jawa island and the surrounding seas.
- The Utilization Zone is where activities that support the development of National Parks such as tourism can take place. It includes the islands of Menjangan Besar, Menjangan Kecil, Kumbang, Kembar, Karang Katang, Karang Besi and parts of Parang Island, Karimun Jawa island and Kemujan island and the surrounding seas.
- The Buffer Zone is an area that can be utilized by the local people in a traditional manner and is the dwelling place of the local population, including Mosquito Island, Genting Island, Parang Island, parts of the Kemujan Island and Karimun Jawa island, as well as other small islands in the surrounding waters.

Climate and Topography

This area has a tropical climate and is influenced by sea breezes that blow throughout the day. The average temperature is between 23^0-32^0C. The dry season is relatively short, from June to August and is known as the timur season. In this season strong winds can occur that contain moist air and often lead to local rainfall. The wet season is from November to March. It begins with a transition season, called the barat season, which occurs during the months of September and October. The wet season is distinguished by strong winds, resulting in high waves. The strength of the wind and the size of the waves can result in the suspension of sea travel between the small islands and the main island of Java. The rainy season ends with the pancaroba season, which is the transition to the dry season, occuring during the months of April and May.

The park consists of sea waters and island clusters that reach altitudes of between 65-500m asl. The islands are generally lowlands with white sandy beaches and lots of coconut trees. Karimun Jawa island, is the largest in the National, which takes its name from this island.

History

On the 26th of October 1982, a Letter of Recommendation to the Minister of Forestry was issued by the Governor of Central Java, No. 556/21378, which proposed to establish Karimun Jawa Island as a Marine National Park.

On the 14th of November 1983, Letter No. 2865/VI-Sek/83 was issued, which also proposed to the Minister of Forestry that the Karimun Jawa Island area be established as a Marine National Park.

On the 9th of April 1986, decree No. 123/Kpt-II/1986 was issued by the Ministry of Forestry stating that the Karimun Jawa islands and the surrounding waters were designated as a Laboratory and Marine Nature Reserve with an area of 111,625 ha.

On the 29th of February 1988, based on the Decree of the Minister of Forestry No. 161/Kpts-II/88, the Karimun Jawa Islands and surrounding waters were declared to be the Karimun Jawa Islands Marine National Park, with an area of 111,625 ha.

Biodiversity and Ecosystems

The Park's ecosystems are mainland ecosystems such as lowland forests, beach and coast forests and mangrove forests, and marine ecosystems, such as the reefs, and algae and seagrass beds.

Typical land plants include the Wijayakusuma, *Pisonia grandis*, Gondorio, *Bouea macrophylla*, sentigijambon, *Acmena acuminatissima*, uyah-uyahan, *Procris penduculata*, coconut, *Cocos nucifera*, nutmeg, *Myristica sp.*, dewadaru, *Cristocalyx macrophylla*, a characteristic Karimun Jawa species, mangrove, *Avicennia* sp., butun, *Barringtonia asiatica*, *Bruguiera sp.*, *Casuarina fistula*, sea hibiscus, *Hibiscus tiliaceus*, casurarina, *Casuarina equisetifolia*, ketapang, *Terminalia catappa* and another mangrove, *Rhizophora* sp.. In the waters there are marine plants such as seaweed and seagrass. There is one species endemic to Karimun Jawa, *Terminalia kangeanensis*. This is a flowering plant in the order Myrtales, which is found in mangrove habitat. According to the IUCN red data list, it is vulnerable.

Marine species can be found to a depth of 55 meters. The corals represent 33 genera from 12 families. There are sponges, soft corals and bahar root, *Anthipatharia*. Apart from corals and sponges, other animals include red clam,

Tubifora musica, giant clam, *Tridacna gigas*, 242 species of reef fish, green turtle, *Chelonia mydas*, hawksbill turtle, *Eretmochelys imbricata*, leatherback turtle, *Dermochelys coriacea*, dolphins and sharks.

Species of land animals include the Javan rusa deer, *Cervus timorensis*, barking deer, *Muntiacus muntjak*, pangolin, *Manis javanica*, porcupine, *Hystrix brachyura*, long-tailed macaque, *Macaca fascicularis*, Javan leopard cat, *Prionailurus bengalensis javanensis*, and monitor lizard, *Varanus salvator*.

There are about 50 bird species, including Nicobar pigeon, *Caloenas nicobarica*, tongtong, or lesser adjutant stork, *Leptoptilos javanicus*, white bellied sea eagle, *Heliaeetus leucogaster*, small Pecuk Padi, or little cormorant, *Phalacrocorax niger* and trinil, or common sandpiper, *Tringa hypoleucos*.

Local Communities and Culture

Only five islands of the 27 island clusters in the National Park area are inhabited. They are: Karimun Jawa Island, Kemujan Island, Parang Island, Nyamuk Island and Genting Island.

Tourism

Situated between western Indonesia and eastern Indonesia, this park is a strategic marine tourism site. Visitors and tourist coming to this area can enjoy the beauty of the sea, swimming, snorkeling, diving, water skiing, sailing, sunbathing and fishing. Following are some tourism highlights:
- The waters do not have strong currents and the water is clear and clean. There is a diversity of corals with a variety of clourful fish, as well as a beaches with beautiful views.
- There is a waterfall and a fort that was built during the Portugese era in Indonesia. There is also a historical monument, the grave of Sunan Nyamplungan, as well as the site of the first anchorage of a Dutch colonial ship at Karimun Jawa (greensand).
- Cilik and Tengah islands are the best for snorkeling.
- Ujung Gelam has a very quiet and tranquil beach.
- The best time to visit is in the dry season from May to September.

Access and Transportation

Karimun Jawa Marine National Park can be reached by using air or sea transportation, as follows:
- Air transportation from Ahmad Yani airport in Semarang to Dewodaru airport on Kemujan Island operates once a week.
- Sea transportation from Kartini harbour in Jepara to Karimun Jawa Island (100 km) using the Dewodaru and Tongkol motorboat that operates each Monday and Thursday, except when there is a special occasion, or from Semarang harbor (113 km).

Facilities

There are many guest houses, home stays and hotels on Tengah and Menyawakan islands. Bikes can be rented at special bike rentals, and boats are available to go diving and snorkeling.

Supporting facilities include the Coral Research Center on Mayangan Island, guest houses, camping sites, footpaths and tracks, radio communication equipment, fresh water sources, and Watch Posts.

Park Office

Balai Taman Nasional Karimun Jawa
Jl. Sinar Waluyo Raya Nomor 248
Semarang, Jawa Tengah - 50273
Telp: (024) 6735419, 76738248
Call Center: 0811 2799 111
Instagram: @ btn_karimunjawa
Email: info@karimunjawanationalpark.org

19. GUNUNG CIREMAI NATIONAL PARK

Geographic Location and Size

Gunung Ciremai National Park (TNGC) is located between 108 °21'35 " and 108 ° 28'00" east longitude and 6° 50'25" and 6 ° 58'26" south latitude. Administratively it is part of two Regencies, Kuningan Regency and Majalengka Regency. It is located about 50 km south of the city of Cirebon, West Java.

The size of Gunung Ciremai National Park is 15,859.17 ha, consisting of 8,931.27 ha within Kuningan Regency, and 6,927.9 ha within Majalengka Regency. The park borders the Cirebon Regency in the north.

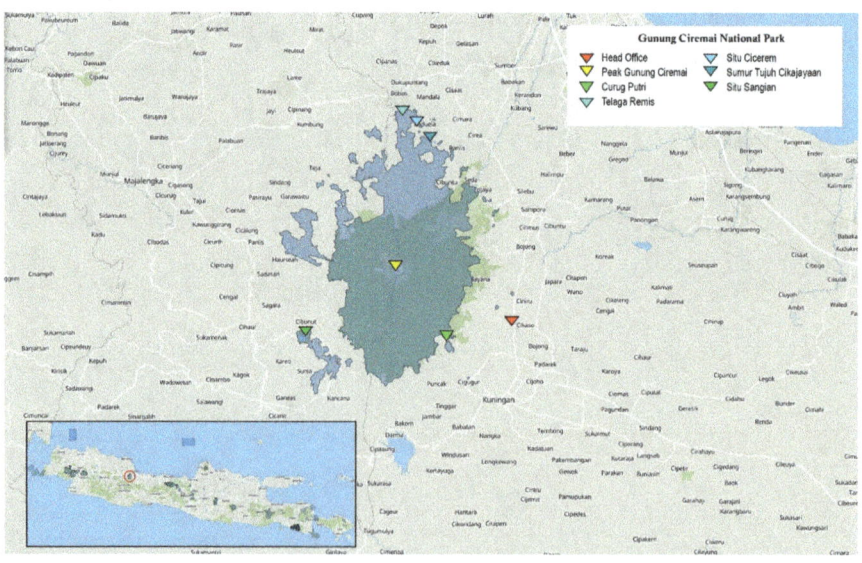

Climate and Topography

The topography is mostly hilly (64%) and steeply mountainous (22%) with Ciremai peak, an active volcano, reaching 3,087 m asl making it the highest mountain in West Java.

History

Gunung Ciremai was designated as a national park by Minister of Forestry Decree No. 424/Menhut-II/2004, dated October 19th, 2004. This decree designated an area of 15,500 hectares located in the Regencies of Kuningan and Majalengka, West Java Province, as a National Park. The establishment of the Ciremai Mountain forest area into a national park was proposed by the Kuningan Regency government through letter No. 522/1480/Dishutbun dated July 26, 2004, regarding the "Proposal Ciremai Mountain Forest Area as a Conservation Area" and the Majalengka Regency government through letter No. 522/2394/Forestry dated August 13, 2004, concerning the "Proposal Mount Ciremai as a Conservation Area".

Biodiversity and Ecosystems

The vegetation in Gunung Ciremai National Park can be classified into three main zones based on elevation.
- Tropical lowland and sub montain zone, between 0-1,000 m asl. With a hilly subzone between 500-1,000 m.
- Tropical Montain Zone, between 1,000-2,400 m asl. With a sub-montain zone between 1,000-1,500 m.
- Sub-alpine zone, above 2,400 m.

There is wet montain forest in the south (Cigugur and surroundings) and drier montain forest in the north in the Setianegara area and surroundings.

The wet montain forest from Cigugur north to the peak of Mt. Ciremai is rich in tree species. The list includes saninten, *Castanopsis argentea, C. javanica, C. tungurrut,* pasang, *Lithocarpus elegansand L. sundaicus,* jenitri, *Elaeocarpus obtusus, E. petiolatus, E. stipularis,* mara, *Macaranga denticulata,* kareumbi, *Omalanthus populneus.* aneka jirak, *Symplocos fasciculata, S. spicata, S. sessilifolia, S. theaefolia,* various species of ara, or fig, including *Ficus padana* and *F. racemosa*, puspa, *Schima wallichii* and ki sapu, *Eurya acuminata*.

The drier forst around Setianegara, is dominated by various species of huru or medang, *Litsea* spp., saninten, *Castanopsis. argentea* and *C. javanica*, mara, *Macaranga tanarius*, mareme, *Glochidion* sp., bingbin, *Pinanga_ javana*, and pandan gunung, *Pandanus* sp. The sub-alpine zone is dominated by the conifer jamuju, *Dacrycarpus imbricatus*, where it forms a distinct ecosystem of its own.

The park has been designated by BirdLife International as an Important and Endemic Bird Area. Several species are categorized as vulnerable by IUCN, such as Javan owl, *Otus angelinae* and Javan Ciung-Mungkal, *Cochoa azurea*. Also, at least 18 species are recorded as restricted area birds, such as the puyuh-gonggong jawa, or chestnut bellied partridge, *Arborophila javanica*, walik kepala-ungu, or pink headed fruit dove, *Ptilinopus porphyreus*, takur bututut, or brown-throated barbet, *Psilopogon corvinus*, berkecet biru-tua, or Javan blue robin, *Cinclidium diana*, poksai kuda, or rufous-fronted laughingthrush, *Garrulax rufifrons*, cica matahari, or spotted crocias, *Laniellus albonotatus*, opior jawa, or Java grey-throated white-eye, *Lophozosterops javanicus*, and the kenari melayu, or mountain serin, *Serinus estherae*.

All Javan mammals except the Javan rhino and Javan Banteng are found in the park. There are many endangered mammals such as Javan gibbon, *Hylobates moloch*, grizzled langur, *Presbytis comata*, West Javan langur, *Trachypithecus mauritius*, Javan coucang, *Nycticebus javanicus* and long-tailed macaque, *Macaca fascicularis*. No in-depth research has been carried out to determine the population sizes of these primates so their status remains uncertain. Many different cats are found in the park such as Javan leopard, *Panthera pardus melas* and Javan leopard cat, *Prionailurus bengalensis javanensis*.

Tourism

The nature tourism sites and attractions present in and around Gunung Ciremai National Park are diverse and widespread. They include
- Situ Sangiang (Sangiang Lake) is located to the south of Talaga city at a height of between 600-800 m asl. The Tourism activities that are available here include nature cross country walking, boating, fishing and camping. Situ Sangiang is close to the village of Wates, Majalengka Regency. Situ

Sangiang is approximately 26 km from the center of Majalengka city by rental car.
- Climbing Ciremai Mountain. Groups of hikers, including parties of students, often climb Mt. Ciremai although care is needed because it takes at least 10 hours. There are three places to start the climb from, Linggar Jati, Linggasana and Majalengka. From Linggar Jati, there are 12 shelters starting from Cibunar village. This route is the most popular because of the breath-taking views toward the cities of Kuningan to the south and Cirebon to the north .

Access and Transportation

From Jakarta there is a train from Gambir station to Cirebon, which takes three hours. There are also regular buses and rental cars can be hired to drive to central Java on a good toll road for 3-4 hours. From Cirebon, there are cars available or regular busses going to tourism locations such as Situ Sangiang (lake), which takes 1-2 hours. There are also busses from Jakarta to Kuningan City, which stop at Linggar Jati or Cilimus.

Facilities

In Cirebon and small cities around the park there are many good hotels. There are also many small hotels and villas that can be rented at Cilimus and Linggar Jati such as Horizon Grage that has a natural hotspring inside the rooms (Kuningan Regency)

Park Office

Balai Taman Nasional Gunung Ciremai
Jl. Raya Kuningan - Cirebon Km. 9 No. 1
Manis Lor, Jalaksana, Kuningan – 45556
Jawa Barat
Telp. And FAX +62-232) 613152
Call Center: 08112187411
Instagram: @ gunung_ciremai
Email: bTaman Nasional_gciremai@ymail.com

20. GUNUNG MERAPI NATIONAL PARK

Geographic Location and Size

Gunung Merapi National Park lies in central Java. Administratively it falls within two provinces, Central Java and Yogyakarta. Its location is between 07°22`33" and 07°52`30" South latitude and 110°15`00" and 110°37`30" East longitude. Its total area is approximately 6,410 ha, with 5,126.01 ha in Central Java and 1,283.99 ha in Special Region Yogyakarta.

Gunung Merapi National Park covers parts of several regencies, Magelang, Boyolali and Klaten in Central Java, and Sleman in Yogyakarta.

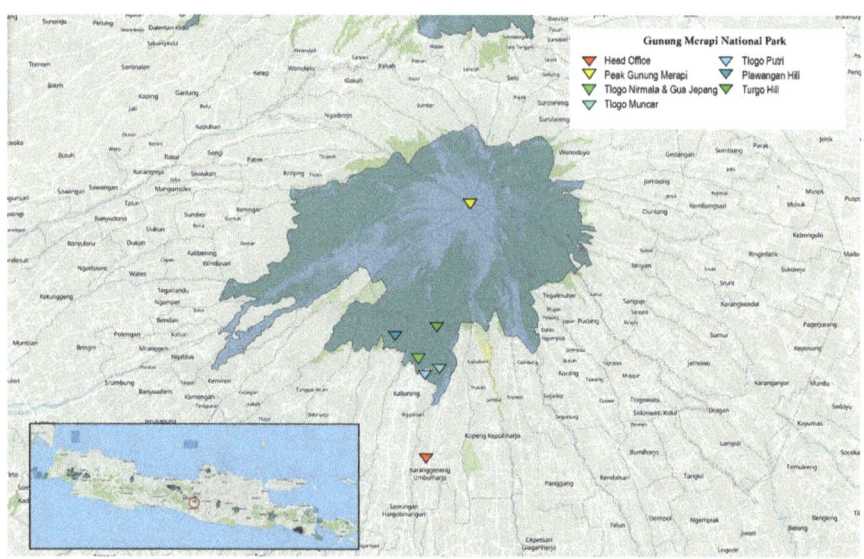

Climate and Topography

The Park has a tropical monsoon climate, which is characterized by high rainfall during the wet season from November to April, then changes to the drier season from May to October. Mean annual rainfall is between 2,500 and 3,000 mm. Rain variation across the slopes of the park is influenced by orographic effects. Like other tropical monsoon areas, the variation in air temperature and humidity is not high. The temperature lies around 20° to 33° C and air humidity varies between 80% and 99%. Elevation is between 600 and 2,968 m asl. Topography ranges from slightly sloped and hilly to steeply sloped and mountainous. In the north lies a highland area between Mt Merapi and Mt Merbabu, around Selo District in Boyolali Regency. In the south, Merapi's slope continues to flatten out all the way to the shore of the Indian Ocean, crossing through the Yogyakarta region. Right before the foot of the mountain lie two hills, Turgo Hill and Plawangan Hill, which are part of the Kaliurang tourism area.

History

The forests of Merapi Mountain have been in a protected area since 1931, when they were set aside to protect the water sources of Sleman, Yogyakarta, Klaten, Boyolali, and Magelang regencies and their rivers and life support systems. It continues to perform those functions today.

Merapi Mountain was established as a national park through the Decree of the Minister of Forestry Number 134/Menhut-II/2004, on the 4th of May 2004. Gunung Merapi National Park was officially declared on October 17th by President of Indonesia, Megawati Soekarnoputri.

Biodiversity and Ecosystems

The natural ecosystem of the park is a tropical mountain forest that is influenced by volcanic activity. This ecosystem can be divided into
- Low Mountain Forest (1200-1800 m asl), dominated by trees reaching up to 60 meters. Many ferns grow attached to trees in heights up to 15 meters. These species are common components in mountain areas, especially in low mountain areas (Whitten et al., 1999).
- High Mountain Forest (1800-3000 m asl), which is characterized by the many mosses that cover the ground and every tree branch, 2 to 3 meters above ground level.

- Field Ecosystem near villages. This man-made ecosystem is dominated by crops such as cassava, red pepper and corn.

The lower mountain vegetation contains at least 154 plant species including Sumatran pine, *Pinus merkusii*, introduced black wattle, *Acacia decurrens*, puspa, *Schima wallichii*, and bambu apus, *Gigantochloa apus*. There are also endemic Javan plants including saninten, *Castanopsis argentea*. The area also has a diverse fungal and fern flora. There are at least 43 fern species found in the park, such as *Adiantum cuneatum, Blechnum patersonii, Cyathea contaminans, and Selaginella wildenowii.*

The mammals include Javan leopard, *Panthera pardus melas*, Javan leopard cat, *Prionailurus bengalesis javanensis*, civet, *Paradoxurus hermaproditus*, squirrels, *Lariscus insignis* and *Colosciurus notatusi*, long-tailed macaque, *Macaca fascicularis*, ebony leaf monkey, *Trachypithecus auratus*, wild pig, *Sus scrofa, S. vittatus*, barking deer, *Muntiacus muntjak*, and Timor deer, *Cervus timorensis*. There are at least 99 species of birds, some of which are Javan endemics such as Javan eagle, *Spizaetus bartelsi*, Javan lonchura, *Lonchura leucogastroides*, Javan honeyeater, *Aethopyga mystacalis*, white-flanked sunbird, *Aethopyga eximia*, blood-breasted flowerpecker, *Dicaeum sanguinolentum*, mountain kingfisher, *Halcyon cyanoventris* and serindit jawa, or yellow-throated hanging parrot, *Loriculus pusillus*. Other bird species commonly seen in the park include elang hitam or black eagle, *Ictinaetus malayensis*, jalak suren or pied mynah, *Strurnus contra*, red-breasted parakeet, *Psittacula alexandri*, oriental hobby, *Falco severus*, crested serpent eagle, *Spilornis cheela* and walet gunung or volcano swiftlet, *Collocalia volcanorum*.

Tourism

Mount Merapi is the most active volcano in Indonesia that surrounded by many big cities, Yogyakarta, Sleman, Muntilan, Boyolali, Magelang, Klaten and others. It is stratovolcano that has erupted since 1548. It erupts almost every 5-10 years killing many people with pyroclastic flows down to the villages. One of the largest eruptions was in 2010 when more than 350 people were killed and thousands were forced to flee their homes. Since then, many small eruptions have been seen such as in 2013, 2018 and 2020. The smaller eruptions can be seen from Yogyakarta at night and are therefore a tourist attraction.

Other tourist attractions in the park include
- Kaliurang (Sleman Regency, DIY) is a tourist attraction featuring natural resources typical of tropical areas, particularly rare flora and fauna, active volcano, natural scenery, and mountain panoramas around active volcanic areas.
- Selo (Boyolali Regency, Central Java) features natural scenery, especially mountain panoramas.
- Musuk - Cepogo (Magelang Regency, Central Java) features natural scenery, specifically valley and steep mountain slope panoramas.
- Ketep (Magelang Regency, Central Java) is a natural tourist attraction featuring volcanic area panoramas.
- Deles (Klaten Regency, Central Java) features rare fauna, active volcano, natural scenery, and mountain panoramas around active volcanic areas.
- Climbing the mountain to see the caldera is an attraction. A special guide is needed to climb this mountain, due to the many dangerous gasses originating from the volcano.

Access and Transportation

Hiking Access

Currently there are three hiking trails that reach this National Park: Kinahrejo/Kaliadem trail from the south side; Babadan trail on the western slope; and Selo/Plalangan trail from the north. Climbing from the north is recommended because it is less steep. A good time to climb is between June and August when the dry season has begun and the wind is not too strong. Because of the risk of eruption climbing this mountaion requires extra caution.

Kaliurang Access

Kaliurang can be reached by a Yogyakarta - Kaliurang bus, which will take approximately 30 minutes. Several facilities in Kaliurang include hostelry, viewing post, playground, Tlogo Putri swimming pool, tennis court, parking area, and bus terminal.

Facilities

The closest city to the park is Jogjakarta, one of the largest tourist destinations in Indonesia. Jogjakarta has many accommodation options from luxurious hotels to bed and breakfasts and home stays. Millions of tourists visit Jogjakarta primarily to see the magnificent World Heritage Site, the Budhist temple, Borobudur that was built in 800 AD and the many other large Hindu temples such as Prambanan and Roro Jonggrang.

Park Office

> Balai Taman Nasional Gunung Merapi
> Jl.Kaliurang km.22.6, Banteng, Hargo Binangun, Sleman, Yogyakarta 55171
> Telp: +62-274-4478664, FAX: +62-274-4478665
> Call Center: 0813 2769 1368
> Instagram: @ btn_gn_merapi
> Email: tngm_jogja@yahoo.com or tngmjogja@gmail.com

21. GUNUNG MERBABU NATIONAL PARK

Geographic Location and Size

Gunung Merbabu National Park covers an area of 5,725 ha and is in three Regencies, Boyolali, Magelang, and Semarang in Central Java Province. Merbabu Mountain has an elevation above 3,142 m asl with mountain slopes more than 40 degrees. Mt. Merbabu is a dormant stratovolcano. The name *Merbabu* can be loosely translated as 'Mountain of Ash' from the combined Javanese words; *Meru* meaning "mountain" and *awu* or *abu* meaning "ash". There are five craters in the park, Condrodimuko, Kombang, Kendang, Rebab, and Sambernyowo.

History

Gunung Merbabu National Park was established based on the Decree of the Minister of Forestry Number 135/Menhut-II/2004 on the 4th of May 2004, the same date as its neighbouring Gunung Merapi National Park.

Previously, it was Merbabu Mountain protected forest area and Tuk Songo Nature Park. On the 30th of December 2005, the park's management was handed to Central Java's Conservation Office (BKSDA). The Technical Implementation Unit (UPT) of Merbabu Mountain National Park Agency was established in June 2006, based on Regulation of the Minister of Forestry Number P29/Menhut-II/2006 about new National Park organization and working arrangements.

Climate and Topography

Merbabu Mountain has an elevation of 3,145 m asl at the Kenteng Songo peak. This mountain is known as an inactive volcano although it has 5 craters. It has two summits, Syarif Peak (3,119 m) and Kenteng Songo Peak (3,142 m). The park has a tropical monsoon climate, which is characterized by high rainfall during the wet season from November to April, which decreases during the drier season from May to October. Mean annual rainfall is between 2,500 and 3,000 mm. Temperatures range from around 20°C to 33°C.

Biodiversity and Ecosystems

The ecosystems of Merbabu Mountain can be divided into three types based on elevation.

- Low Mountain Forest (1,000-1,500 m asl) overgrown by secondary forest vegetation species with Sumatran pine, *Pinus merkusii*, and puspa, *Schima noronhae*.
- High Mountain Forest (1,500-2,400 m asl), overgrown by black wattle, *acacia decurrens*, puspa, *Schima noronhae*, sengan gunung, *Albizia falcatarie*, sowo, *Engelhardia serrata*, cemara gunung, *Casuarina montana*, and pasang, *Quercus sp.*
- Sub-Alpine Mountain Forest (2,400-3,142 m asl), overgrown by grass and Edelweiss Jawa, *Anaphalis javanica*.

Bird species in Gunung Merbabu National Park include javanese eagle, *Spizaetus bertelsi,* which is endemic to Java, black eagle, *Ictinaetus malayensis*, crested serpent eagle, *Spilornis cheela*, alap-alap sapi or spotted kestrel, *Falco moluccensis*, and Javan junglefowl, *Gallus varius.* There are also various mammals such as barking deer, *Muniacus muntjak*, porpcupine, *Hystrix javanica*, civet, *Paradoxurus hermaphroditus*, long-tailed macaque, *Macaca fascicularis*, ebony leaf monkey, *Trachipithecus auratus* and Javan leopard, *Panthera pardus melas.*

Local Communities and Culture

The local community livelihoods are dominated by vegetable farming, trading, and laboring. Tourism businesses in the form of cottage rentals, home stays, decorative plant stalls, renting binoculers, food stalls and small grocery shops are sideline jobs.

Merbabu Mountain's local community culture includes *ketoprak, campur sari, kuda lumping, tari soreng, turonggo seto, jatilan, budi tani, jelantur,* and other forms of dances. Merbabu Community Rembug Forum is a community organization in the Merbabu Mountain area for conveying community aspirations. The Government collaborates with this forum in the development and management of Gunung Merbabu National Park.

Tourism

- The Kopeng tourist area is located in Getasan District, Semarang Regency. It is easily accessible by sealed roads, located 14 km from Salatiga City and 27 km from Magelang. This attraction comprises mountain ecotourism and *umbul songo* springs, camping grounds, hiking trails and is supported by hotels and cottages, children's playground, vegetable market, and decorative plant stall.
- Selo Pass is in Selo District, Boyolali Regency is easy to reach and is located 20 km from Boyolali City and 14 km from the Ketep Pass tourist attraction. The attraction here is Merbabu and Merapi Mountain panoramas. Facilities include Joglo Mandala (a place used for making arts in the Selo area), souvenir stands, volcano theater, New Selo viewing post, Children's playground, bungalows, and home stays.

- Kedung Kayang Waterfall is located in Wonolelo Village, Sawangan SDistrict, Magelang Regency. It is easy to reach, being 4 km from Ketep and 10 km from Selo, on the Ketep Pass-Selo Pass route. This attraction is in the form of waterfalls that are 10 to 12 m in height and is supported by facilities such as a parking area and shelter.

Access and Transportation

Access to the park is easy because public transportation is available by bus, small bus and motor bike. To go to the hiking route Wekas, from Yogyakarta to Magelang take a bus or rent a car. From Magelang take a small bus or rental car or motor bike to Wekas. To go to Kopeng, take bus or rental car from Yogyakarta to Magelang. Then take another bus or rental car from Magelang to Salatiga and stop at Kopeng.

Hiking route

Base-camp Wekas to Post I

From base-camp Wekas the hiking route starts with village roads and later continues with tracks made from cement that are quite steep. This cement road ends at a cemetery. The route continues with an unsealed track that is very clear. Along the route to Post I the forest is not too dense. There are residences, fields and a pine forest.

Post 1 to Post 2

Post 1 is a flat plain that is large and open, but there is no water source at Post 1. It is only used as a rest stop by hikers. People rarely stay overnight there. The track from Post 1 to Post 2 is very clear and is quite steep. After post 1 there is a water pipe and the track route to Post 2 follows this pipe. Approaching Post 2, the footpath flattens and the summit ridge can be seen on the left side. It takes approximately 2.5 hours from Post 1 to Post 2.

Post 2 to the three-way junction from Kopeng

Post 2 is also a large plain, and in peak hiking season there are stalls set up by local residents. The scenery is very beautiful at this Post. If the weather is good, the antena summit, hellypad summit, and other ridges can be seen. To the left of the post lies ridges and deep valleys, and at the bottom of the valley there is a waterfall. There is a water source at this post that comes from perforated water pipes. Post 2 is often used as a place to sleep overnight by hikers. Not far from Post 2 up a steep track there is a three-way junction. Take the left route, which ascends uphill to a large rock, then turn left until the track meets with the track from the Kopeng direction. It will take approximately 2 hours from Post 2 to the three-way junction from Kopeng.

Three-way junction from Kopeng to Hellypad.

The convergence of the Wekas route with the route from Kopengis is marked by a border marker between 2 regencies. From here, Post 2 and the route towards the summit is clearly visible and not far from this three-way the hellypad is visible.

Hellypad

This location is an open plain capable of accommodating 4 to 5 tents and from here people can enjoy the scenery in every direction. Syarif and Keteng Songo summit are clearly visible. Down towards the crater, there is water, which sometimes tastes a little acidic because of the sulfur. The hellypad is a good place to camp because the view is very good. But the tent should be able to withstand strong winds because the site is very open.

Facilities

The park is close to the major city of Jogjakarta, one of the largest tourist destinations in Indonesia, and to Semarang, the large capital city of Central Java. In Jogjakarta and Semarang there are many hotels plus hundreds of bed and breakfasts and home stays. Other nearby cities are Magelang, Sleman, and Boyolali where the headquarters of the park is located.

Park Office

Balai Taman Nasional Gunung Merbabu
Jl. Merbabu No. 136 Boyolali - 57316, Central Java
Telp. (+62-276) 3293341, 3293347
Fax. (+62-276) 3293342
Call Center: 08112950970
Instagram: @ btn_gn_merbabu
Email: Taman Nasional_merbabu@yahoo.co.id

22. GUNUNG BROMO SEMERU TENGGER NATIONAL PARK

Geographical Location and Size

Bromo Tengger Semeru National Park is located between $7^0 54'$ and $8^0 13'$ South Latitude and $112^0 51'$ and $113^0 04'$ East Longitude. It is in East Java Province and covers parts of four Regencies: Malang, Pasuruan, Lumajang and Probolinggo. The park is about 17 km east of the city of Malang on a series of volcanoes that are part of the chain of mountains that stretch along the island of Java. In the northern part of the Bromo-Tengger Caldera region there is the beautiful and unique Tengger-Semeru caldera, with a diameter of 8 to 10 km, 5,290 ha in area with steep walls between 200 and 600 meters high.

Bromo Tengger Semeru National Park covers 50,276.20 ha, including Mt. Bromo-Tengger and Mt. Semeru, the highest volcanic mountain on the Island of Java.

Jatna Supriatna and Chris Margules

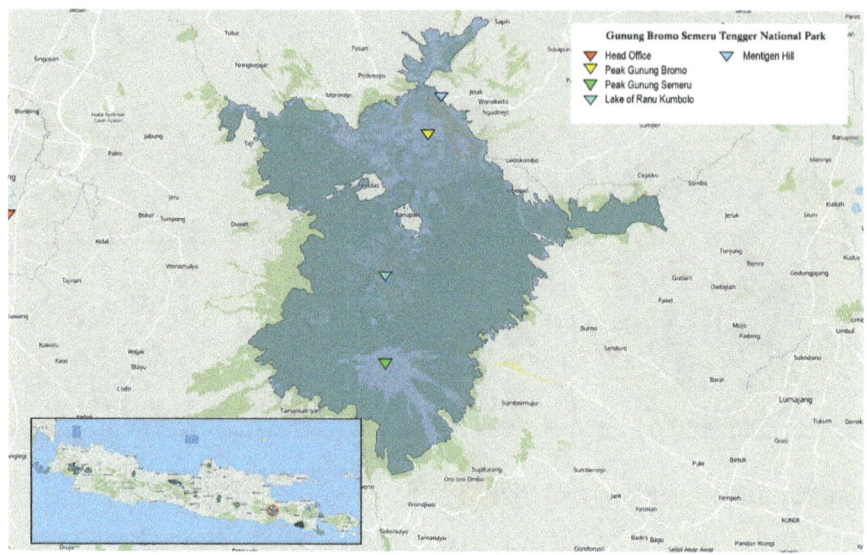

Climate and Topography

There are noticeable differences between the dry season and the wet season. The dry season occurs from May to November, and the wet season from November to March. Average rainfall is between 1,600 and 1,800 mm/year. The temperature varies from 3^0 to 18^0C and humidity varies between 42% to 45% in the dry season and 90% to 97% in the wet season. Strong winds (30-60 knots) often change the shape of the sand surface at the Caldera of Mount Bromo.

The topography of Bromo Tengger Semeru National Park is generally mountainous and hilly from undulating to very steep. Most of the plateau region is mountainous with fertile valleys. Mt. Semeru or Mahameru (3,676 m) is the highest mountain on the island of Java. Mount Batok is very active and erupts almost every year.

Mt. Bromo and Mt. Semeru are still active. The region that includes and surrounds the park also has 50 rivers and four lakes; Ranu Darungan Lake (0.5 ha), Ranu Regulo Lake (0.75 ha), Ranu Pani Lake (1 ha) and Ranu Kumbolo Lake (14 ha).

History

In 1919 the Laut Pasir Tengger area (5,287.33 ha) was declared a protected area. It covers the caldera base which has a diameter stretching to 10 km (96 ha) and is called the Laut Pasir Tengger.

In 1921 the Ranu Kumbulo Nature Reserve, with an area of 1,340 ha, was declared.

On the 6th of March 1980, the Minister of Agriculture declared the Bromo Tengger Semeru National Park of 57,606 ha.

In 1981 Laut Pasir Tengger, Ranu Pane/Ranu Regulo and the Ranu Darungan were established as a Tourism Park.

On the 14th of October 1982, Decree No. 736/Mentan/X/1982, was released by the Ministry of Agriculture to declare the Bromo Tengger Semeru region as a National Park.

On the 23rd of May 1997, Decree No. 278/Kpts-VI/97, was released by the Ministry of Forestry to reinforce the declaration of the Bromo Tengger Semeru region's status as a National Park with an area of 50,276.20 ha.

Biodiversity and Ecosystems

The ecosystems in Bromo Tengger Semeru National Park consist of mountain rainforests, and associations of pines and grassland, that can be grouped based on altitude, as follows:

- 2,200-2,400 m asl. The characteristic tree species here are mountain pine, *Casuarina junghuhniana*, introduced black wattle, *Acacia decurrens*, sengon or Mollucan albizia, *Paraserianthes falcataria*, suren or Indonesian mahogany, *Toona sureni* and pasang, *Quercus lineata*.
- Above 2,400 m asl. The most common species found here are: kemlandingan or cape wattle, *Albizia lophanta* and bayberry, *Myrica javanica*. On the steep slopes there are pioneer species such as the senduro or Javanese edelweiss, *Anaphalis longifolia*. At Pananjakan there is a population of the heath, *Styphelia punguens*, which cannot be found elsewhere on the Island of Java.

This park is also the habitat of a diverse orchid flora. In the forest south of Semeru alone, there are 157 species of orchids, including *Malaxis purpureonervosa, Malaxis tenggerensis, Maleola wiiteana, Dendrobium jacobsonii* and *Liparis rhodochila*. *Malaxis tenggerensis* and *Dendrobium jacobsonii* are orchids endemic to Java, and *M. tenggerensis* can only be found in the mountain area of Semeru and Tengger.

Wildlife in this park include mammals such as the Javan leopard, *Panthera pardus melas*, leopard cat, *Prionailurus bengalensis*, deer, *Cervus timorensis*, barking deer, *Muntiacus muntjak*, ebony leaf monkey, *Trachypithecus auratus* and long-tailed macaque, *Macaca fascicularis*.

This area is the wintering ground of migrating birds of prey such as the elang-alap Cina or Chinese sparrowhark, *Accipiter soloensis* and elang-alap Nipon or Japanese sparrowhak, *Accipiter gularis*. In addition to being an important destination of migratory birds, Bromo Tengger is also an important habitat for birds endemic to Java such as Javan eagle, *Spizaetus bartelsi*, Puyuh Gonggong Jawa or Javan hill partridge, *Arborophila javanica*, Opior Jawa or Javan grey-throated white-eye, *Lophozosterops javanicus*, white-flanked sunbird, *Aethopyga eximia*, flame-fronted barbet, *Megalaima armillaris*, and the Tepus Pipi Perak or crescent-chested babbler, *Stachyris melanothorax*.

Local Communities and Culture

The communities surrounding the National Park are of the Tengger tribe, an ethnic sub-group of Javanese who practice Hinduism. Their population is approximately 50,000. There are Hindu temples at Wonokiri, Pasuruan and in Senduro, Lumajang. The temple in Wonokiri stages Balinese dancing for visitors. The temple in Senduro is visited by Hindus from various regions, especially Bali.

The Tengger tribe still maintains a traditional way of life. Their indigenous beliefs influenced many legends of Mount Bromo and Mount Semeru, which is a holy and sacred place for them. The Tengger tribe perform a ceremony called Yadnya Kasada, an annual event on the 14th day of the Kasada month in the traditional Hindu lunar calendar, which usually falls in the month of December. This takes place at exactly 00:00 local time during the full moon. The ceremony is usually held to invoke an abundant harvest and good health.

The event is very well-known and a major attraction for visitors. As part of the cereminy, offerings are thrown into the crater of Mount Bromo and participants have to scramble to retrieve them in order to be granted their request. During the ceremony, the number of visitors and participants can reach hundreds to thousands of people. The presence of the Kasada ceremony is a unique feature of Bromo Tengger Semeru National Park.

Tourism

- Cemorolawang is a location often visited from the Probolinggo Regency. Activities here include camping and viewing the natural panoramas of the coast and the Mt. Bromo complex.
- Laut Pasir Tengger and Mount Bromo are within the Tengger caldera. Other mountains within the caldera are Mount Batok (2,470 m), Mount Kursi (2,581 m), Mount Watangan (2,601 m) and Mount Widodaren (2,650 m). Activities here include climbing Mount Bromo using the 249 steps to see an active crater, which produces smoke and the smell of sulfur.
- The caves at Mt. Widodaren have water springs that never run dry and are often used for meditation.
- Wonokitri village, approximately 2 km from Tosari city, is located in the highlands with beautiful views of the natural surroundings.
- The watchpost at Mt. Pananjakan (2,774 m) has a caldera, a sea of sand (laut pasir) and beautiful views, which can include mountains encircled by fog, and dramatic sunrises.
- Ranu Pani, near the peak of Mount Semeru, is an isolated hamlet 2,200 m above sea level that is cold and always foggy, usually used as a stopover by hikers and nature lovers who are on the way to the summit. Facilities available here include cabins and an information center. There are two adjacent lakes (ranu), Ranu Pani and Ranu Regulo.
- Ranu Darungan is located south of Mount Semeru and is surrounded by rain forest rich in species. Activities here include camping, fishing and enjoying the beautiful natural scenery with cool air and comfortable conditions.
- Visitors to this park are advised to bring along winter clothing as temperatures can get as low as $3°C$.
- The best time to enjoy this park is during the dry season.
- Visitors who want to witness the Kasada ceremony can obtain information from the provincial government or the local Tourism Department.
- Visitors who want to climb to the top and camp will need appropriate warm clothing and their own camping gear, as well as a climbing permit from the Mount Bromo Tengger Semeru National Park Office in Malang.

- Permits to enter the area and to buy entry tickets can be obtained at the four park entrances, Cemorolawang-Probolinggo, Wonokitri-Pasuruan, Ngadas-Malang or in Burno-Lumajang.

Access and Trasnportation

There are flights from Jakarta to Malang city. Then from Malang there are busses or rental cars to many differect cities around the park such as Lumajang, Probolinggo and Pasuruan.
- From Malang to the Klakah Cabin (28 km) using public transport, then to Ngadas (10 km) using a four-wheel drive then to Ranu Pani, a small beautiful place at the edge of the national park.
- From Lumajang city towards Senduro (20 km) using public transport, then to Ranu Pani using a four wheel-drive vehicle, horseback riding or walking.
- From Lumajang to Pronojiwo (37 km) by public transportation, then to Ranu Danungan on foot. Ranu Darungan is located on the South side of Mt. Semeru.
- From Probolinggo to Ngadisari (47 km) using public transport, then to Cemoro Lawang and Mount Bromo on foot or on horse back. Cemoro Lawang is ideally located on the edge of the caldera (2.5 km). Visitors here can watch the sunrise from Mount Bromo and stay the night at Ngadisari and Cemorolawang. This route is the most impressive, winding through mountain passes and canyons with cool air and green vegetable gardens.
- From Pasuruan to Wonokitri (47 km) by public transportation, then to Mt. Pananjakan (11 km) by four wheel drive, on foot or on horseback. Visitors from this direction can witness the sunset from Mt. Pananjakan and stay overnight at Wonokitri.

Mt. Semeru peak (Mahameru) can be reached by the following routes:
- Probolinggo - Ngadisari - Burno - Ngadas - Ranu Pani - Ranu Kumbolo to Semeru Peak
- Malang - Tumpang - Gubuk Klakah - Ngadas - Ranu Pani - Ranu Kumbolo – to Semeru Peak
- Lumajang - Burno - Glagaharum - Ranu Pani - Ranu Kumbolo - to Semeru Peak

Facilities

Travel agencies located in East Java, especially Surabaya generally organize trips to the region with several tour packages.

- At Probolinggo minibus tours to Bromo Tengger Semeru National Park can be purchased.
- At Pasuruan and Lumajang there are several hotels, inns and tour packages available.
- Facilities within this park include: Camping grounds (Cemoro Lawang, Ranu Pani, Ranu Regulo, Ranu Kumbolo and Nangkojajar), work cabins, guesthouses, observation towers and watch posts and a trail 1-1.5 meters wide and 28 km long from Ranu Pani to Mahamer.

Park Office

Balai Taman Nasional Bromo Tengger Semeru
Jl. Raden Intan No. 6 PO. BOX 54 Malang- 65101
Jawa Timur
Telp. (+62-341) 491828, FAX: +62-341-490885
Call Center: 0851 5889 1828; 0813 3144 0630
Instagram: @ bbtn_bromo_ts
Email: tn.bromotenggersemeru@gmail.com

23. ALAS PURWO NATIONAL PARK

Geographical location and Size

Alas Purwo National Park is located between 8°26' and 8°47' South Latitude and 114°20' and 114°36' East Longitude. It occupies parts of Muncar Regency, Tegaldelimo Regency and Purwoharjo Sub-District, Banyuwangi Regency, East Java Province. The park borders Production Forest in the west, the Strait of Bali in the east and north, and the Indian Ocean in the south.

The park area is 44,337.3 ha, covering the farthest corner of the southeastern tip of Java Island.

Climate and Topography

Average rainfall ranges between 1,000 and 1,500 mm / year, which is relatively low. Average temperature ranges between 22° and 31°C, and in the dry season temperatures can reach 37°C. There is only one permanent river.

Altitude is between 0 and 332 m asl. In the western and southern parts, the topography is flat to gently sloping. In the northeast, around Cape Tanjung Sembulungan it becomes undulating to hilly. In the south around Sadengan and toward the center of the park, it is almost entirely hilly to undulating with the peak of Mount Linggamanis reaching 322 m asl.

The alluvial plains west of the park are planted with rice, watermelons and other crops. A wide buffer zone separates this area from the National Park, and the tea plantations which connect Alas Purwo National Park with Betiri Meru National Park in the west and Mount Raung in the north.

History

- On the 1st of September 1939, based on the Assessment Letter GB Stbl. No. 456 General Governor, Government of the Netherlands, the Alas Purwo region was designated to be the Banyuwangi WildLife Reserve with an area of 62,000 ha.
- On the 26th of February 1993, based on the Minister of Forestry Decree No. 190/Kpts-II/93, the status and name of the region was changed to Alas Purwo National Park, with an area of 43,420 ha.
- This National Park is a Sister Park, part of the Indonesia-Malaysia collaboration on National Parks.

Biodiversity and Ecosystems

The ecosystems of Alas Purwo National Park are mostly lowland tropical forest and dense coastal forest. These can be divided into
- Coastal forest, which occurs in the eastern part of the park. Prominent pland\t species include Ketapang, or sea almond, *Terminalia* sp., nyamplung, *Calophyllum inophyllum*, waru laut, *Hibiscus tiliaceus* and keben, *Barringtonia asiatica*.

- Mangrove Forest, which is found in a relatively good condition in the area of Slenggrong and Segoro Anak. The plant species include the mangroves *Rhizophora* sp. and *Avicennia* sp.
- Bamboo forest with 13 species covers 40% of the park.
- Lowland forest is a wet deciduous forest in which more than half of the tree species lose their leaves including rare species such as mahang, *Macaranga* sp. and *Trema* sp. Other trees such as figs, *Ficus sp.*, kepuh *Sterculia foetida*, kayu tahun, or guest tree, *Kleinhovia hospita*, bungur, a type of crape myrtle, *Lagerstroemia flos-reginae*, binong, *Tetrameles nudiflora* and jambu-jambuan, *Eugenia* spp. They are all below 30 m in height.
- Savanna grasslands occur on the hills near Trianggulasi, Kali Pancur, with native grass species such as balung, *Arundinella Setosa*, *Dischantium caricosum*, Lamu, *Polytrias Blanco* and meraken, *Heteropogon contortus*, as well as elephant grass, *Pennisetum purpureum*.
- The park also contains the largest natural forest of sawo kecik tree, *Manilkara kauki,* in Indonesia.

Alas Purwo National Park has approximately 50 species of mammals, including buffalo, *Bos javanicus*, Javan rusa deer, *Cervus timorensis*, kijang, or barking deer, *Muntiacus muntjak*, ajag, a wild dog, *Cuon alpinus*, and Javan leopard, *Panthera pardus*. Primates include ebony leaf monkey, *Trachypithecus auratus* and long-tailed macaque, *Macaca fascicularis*.

Bird diversity is high, with 302 species having been recorded in the park. The Green Peacock, *Pavo muticus*, is the primadonna species and can be easily found. Peacock activity is very interesting to watch, especially when they are displaying during the courting process. The best time to see them is from August to October. Other birds include green jungle fowl, *Gallus varius*, cekakak, or black-capped kingfisher, *Halcyon pileata*, sea eagle, *Haliaeetus leucogaster*, Javanese eagle, *Spizaetus bartelsi* and the tongtong, or lesser adjutant stork, *Leptoptilos javanicus*.

Alas Purwo is one of the most important bird areas in Java because it is on the migration path from the Asian mainland to Australia and back. Between November and January about 20 species of migrant birds from Australia as well as three species of hornbills can be found in the Anak Segoro area.

The aquatic fauna includes green turtle, *Chelonia mydas*, leatherback turtle, *Dermochelys coriacea*, hawksbill turtle, *Eretmochelys imbricata* and Ridley turtle, *Lepidochelys olivacea*, as well as sharks, dolphins and dugongs.

Local Communities and Culture

Residents located in the area surrounding the park are mostly migrants from Central Java and local East Javanese. In the west there the Segoro Anak lagoon and the mangrove forests are traditional fishing grounds for oysters, crabs, milkfish, clams, shrimp and others. Traditional art that can be found in this area includes the Jejer gandrung dance, Jaranan dance, the Hardah Kuntul dance, and sculpture art of shadow puppets and other wooden crafts.

Alas Purwo National Park has a reputation for mystery and spirituality. This is demonstrated by the many visitors who come in the month of *Suro* according to the Javanese calendar or Muharram according to the Islamic calendar, both from around Banyuwangi as well as from other provinces, to meditate. Meditation is generally carried out in places that are considered to have magical or supernatural power places, such as Kucur, Pancur, Goa Istana and Goa Padepokan.

Tourism

- Sadengan pasture is the feeding ground of banteng, deer, wild hog and the green peacock in large numbers. They can be seen from the watch towers located at the edge of the pasture, between 05:00 and 08:00 in the morning and 15:30 and 18:00 in the afternoon. This area also has a camping site.
- Plengkung is popular for surfing, especially between March and October. Camping sites are available. It is said to be the most admired left coral break in the world, and the waves can reach a height of 7 meters.
- Pancur or Parang Ireng is famous for making black gotri, which is made from sand or rocks. This place is often used for meditation, especially in the month of *Suro*.
- There is a turtle hatchery and nursery facility on the Ngagelan coast, which is within 6 km of the Post Rawabendo, and can be reached in 25 minutes.
- From Cape Selakah to Cape Pasir there are views of dolphins and dugongs.
- Trianggulasi beach has beautiful white sand and incredible sunsets.

It is recommended to visit the Wisma Cinta Alam and/ or the National Park Information Center to obtain general information on the National Park in the form of pictures, writings, maps, photographs and also on the regulations and facilities present.

Visitors who need a guide can contact the National Park Management Office.

Alas Purwo National Park can be visited all year round. But those who come to enjoy the atmosphere and the natural panoramas while walking, should plan to visit during the dry season because during the wet season the paths are difficult.

Access and Transportation

Banyuwangi is on the east coast of Java across the strait from Gilimanuk, Bali. The ferry arrives at Ketapang, a short distance from Banyuwangi. Coming from the west, it is 360 km east of Surabaya, Java.
- From Banyuwangi to Dambuntung take public transportation (bus) or rental car
- From Dambuntung to Pasranyar take public transportation (minibus) or rental car.
- From Pasaranyar to Trianggulasi using public transportation or rental car will take 15 minutes, or to Grajagan using public transportation or rental car will take 1 hour.
- From Triangulasi to Pantai Plengkung on foot will take 4 hours, or from Grajagan to Pantai Plengkung using a motorboat will take 1-3 hours.

Facilities

The facilities in the Alas Purwo National Park include work cabins, camping sites, watch posts and an information center. There are many hotels in Banyuwangi, the closest city to the park (1-2 hrs journey). Banyuwangi is the major city on the east coast of Java where there is a car ferry to Bali (30 minutes).

Park Office

Balai Taman Nasional Alas Purwo
Jl. Brawijaya No. 20 Banyuwangi 68416
Tlp. (0+62-333) 410857
Fax. (+62-333) 428675
Call Center: 081336893993
Instagram: @ btn_alaspurwo
Email: btnap@tnalaspurwo.org

24. MERU BETIRI NATIONAL PARK

Geographical Location and Size

Meru Betiri National Park is located between $8°22'$ and $8°32'$ south latitude and $113°38'$ and $113°57'$ East Longitude. Administratively it is part of the Jember and Banyuwangi Regencies, East Java Province. It lies on the south coast (Pantai Selatan) of East Java, approx. 250 km southeast of Surabaya and 60 km southwest of Banyuwangi. The park covers 58,000 ha with 37,626 ha in Jember Regency and 20,374 ha in Banyuwangi Regency, of which total, 57,139 ha are land, and 861 ha are water. Before World War II, the eastern part was connected by forests with Alas Purwo National Park.

Climate and Topography

The climate is distincly seasonsal. The wet season occurs from November to March when the wind blows from the northeast, and the dry season from April to October. The rainfall is generally heavier than in other areas of East Java, but there is a large variation. In the west it is generally drier, with average rainfall in Bandealit about 2,500 mm/year and in the east wetter, with about 4,000 mm/year in Sukamade. Humidity ranges between 65% and 80%.

The Meru Betiri region is derived from tertiary sediments that create very steep but low mountains, mostly less than 500 m asl. The topography is generally undulating, hilly and mountainous, with an altitudinal range from 0 to 1,223 m asl at the peak of Mount Betiri. In the south it is hilly. The mountains in the west include Mt. Permisan (587 m), Mt. Meru (343 m) and Mt. Betiri (1,223 m). In the south are Mt. Sumbadadung (520 m), Mt. Sukamade (363 m), Mt. Rajegwesi (181 m) and Mt. Benteng (222 m). In the east are Mt. Gendeng (893 m) and Mt. Lumberpacet (760 m).

The main rivers, which flow throughout the year, are the Sukamade and Meru. They rise on Mt. Betiri and Mt. Meru respectively and merge into one to form the Sukamade River. Along the coast there are many large and small bays.

History

In 1929 the Dutch government issued a policy to protect the Meru Betiri region.

In 1938 the Dutch government, through the Director of Economic Affairs, issued Decree No. 571 to establish the Meru Betiri Forest region as a Forest Reserve.

From 1961 – 1972, the Meru Betiri forest region was managed by the Perhutan, Forest Estate Company.

On the 6th of June 1972, based on the Minister of Forestry Decree No. 276/Kpts/Um/6/1972, the status was changed to Wild Life Reserve, with an area of 50,000 ha, to protect the Javanese Tiger and its habitat.

On the 21st of July 1982, based on the Ministry of Forest Decree No. 529/Kpts/Um/7/1982, the area was increased to 58,000 ha, consisting of 57,139 ha of land and 861 ha of sea.

On the 14th of October 1982, through the declaration of the Minister of Agriculture No. 736/Mentan/X/1981, at the 3rd World National Park Congress in Bali, Meru Beriti was declared a National Park. On the 30th of

April 1994, Directorate General Forest Protection and Nature Conservation Decree No. 68/Kpts/Dj-IV/94, detailed the zoning system of the Meru Betiri National Park.

On the 23rd of May 1997, a decree from the Forest Minister, No. 277/Kpts - VI/97, was issued to reinforce the status of Meru Betiri National Park with an area of 58,000 ha.

Biodiversity and Ecosystems

The ecosystems that can be found in Meru Betiri National Park include
- Mangrove forests, found in the Rajegwesi bay area and in the estuaries of the Sukamade and Permisan rivers.
- Coast or beach forest along the coastline in narrow segments
- Swamp forest, which is not large in extent, found at the back of the mangrove forest at Sukamade, Permisan and Magelang.
- Lowland Rainforest and the lower mountain forests located between 1,000-1,500 m asl.
- Non-indigenous vegetation types, such as the Jati plantation of 1,100 ha, high grasslands, banana plantation in the region of Meru Bay and grass meadows managed by the park.

In this park 364 species of plants have been recorded, including: mangroves such as, *Rhizophora* sp., tancang, *Bruguiera* sp., api-api, *Avicennia marina* and *Sonneratia ovata*. Other plants include nipah palm, *Nypa fruticans*, rengas, *Gluta renghas*, bungur, or crepe-myrtle, *Lagerstroemia speciosa*, pulai, *Alstonia angustifolia*, bendo, *Artocarpus elasticus*, *Barringtonia speciosa*, *Derris* sp., beach morning glory, *Ipomoea pes-caprae*, *Spinifex squarratus*, pandan, *Pandanus tectorius*, kapasan, *Hernandia* sp., waru laut, or sea hibiscus, *Hibiscus tiliaceus*, cemara-cemara, *Cycas rumphii*, kepuh, or Java olive, *Sterculia foetida*, *Cerbera manghas*, gelagah, *Saccharum malacensis*, *Saccharum spontaneum*,, nyamplung, *Calophyllum inophyllum* and ketapang, *Terminalia catappa*. Also found here are medicinal plants such as lada panjang or Javan cayenne, *Piper retrofractum* and kemukus, *Piper cubeba*.

Large trees of the canopy include gintung, *Bischofia javanica*, Putat, *Planchonia valida*, kayu tahun, *Kleinhovia hospita*, a type of bungur, *Lagerstroemia flos-reginae*, bayur, *Pterospermum javanicum*, kedondong, *Spondias pinnata*, kepuh, *Sterculia sp.* and banyan, *Ficus sp*. On the ground there is rattan, ginger and dense growths of bamboo. *Rafflesia zollingeriana* is also found here, which is parasitic on *Tetrastigma sp.*, and *Balanophora fungosa* a flower that is also a parasite living on the roots of *Ficus sp.* trees.

There are 29 species of mammals and 180 species of birds. Mammals include banteng, *Bos javanicus*, longtailed macaque, *Macaca fascicularis*, Javan leopard, *Panthera pardus melas*, wild dog, *Cuon alpinus javanicus*, Javan Leopard cat, *Prionailurus javanensis*, deer, *Cervus timorensis* and the flying redtailed squirrel, *Iomys horsfieldii*. The national park is known worldwide because it was the last know location of the Javan Tiger, *Panthera tigris sondaica*, an endeimic, rare and protected species. Unfortunately, it has not been seen since the end of the 1970s.

The most commonly seen birds in this park are peacock, *Pavo muticus*, Bubut Jawa, or Sunda coucal, *Centropus nigrorufus*, wreathed hornbill, *Aceros undulates*, snake eagle, *Spilornis cheela*, and sea eagle, *Haliaeetus leucogaster*.

Meru Betiri is also famous as a breeding ground of five of the seven species of sea turtle in the world. They are leatherback turtle, *Dermochelys coriacea*, hawksbill, *Eretmochelys imbricata*, green turtle, *Chelonia mydas*, Ridley turtle, *Lepidochelys olivacea* and loggerhead turtle, *Caretta caretta*.

Local Communities and Culture

The people living in the areas surrounding the park generally come from the Solo, Yogyakarta, Madura and Ponorogo areas.

There are two rubber plantation enclaves within the National Park along the fertile alluvial valleys, the Bandealit plantation in the west and the Sukamade plantation in the east, both plantation around 2,500 ha.

Tourism

- Sukamade Beach is a turtle nesting beach where people can also swim.
- On Bandealit, Rajegwesi, Plengkung, Meneng beaches and from Merah Island, there are beautiful views of the beaches and good swimming.
- On grazing meadows at Nganggelan, 50 ha, Pringtali, 35 ha, Rajegwesi, 31 ha and Sumbersari 192 ha, see banteng cattle and deer
- At Hijau Bay there is green colored sea, untouched forests and beautiful views.
- There is a 25 km path from Sukamade to Bandealit beach, lined with beach forest.

- Damai (Peace) Bay on the road between Rajegwesi at the eastern entrance and Sukamade beach is called Damai Bay because the waters are very calm with no strong or high waves.
- Visitors who need a guide can contact the National Park Management Office.
- Visitors who want to see turtles laying eggs at the Sukamade beach are advised to plan their visit during the months of December to March (wet season) and to hire a four-wheel drive vehicle or truck belonging to the Sukamade plantation. The track to Sukamade beach is sometimes muddy with standing water. Visitors can stay at the plantation's guest house owned by PT. Sukamade Baru.

Access and Transportation

Meru Betiri National Park can be reached from Surabaya via two routes.
- First Route
  - Surabaya - Probolinggo - Jember (200 km) by train or bus
  - Jember - Ambulu - Curahnongko - Bandealit (30 km) by public transportation
- Second Route
  - Surabaya - Probolinggo - Jember (200 km) by train or bus
  - Jember - Glenmore - Jajag (84 km) by bus
  - Jajag - Pesanggaran - Sarongan (44 km) by public transportation
  - Sarongan - Sukamade (18 km) by taxi, minibus or truck, which can travel further on to Rajegwesi.

Facilities

- In Sukamade there are camping sites, a laboratory and turtle aquarium, and accommodation called the "Wisma Sukamada".
- At Bandealit there is a Camping Site
- Between Sukamade and Bandealit there is a 25 km long hiking path.
- At Rajegwesi there is accommodation called "Wisma Cinta Alam" and an information center.

Other facilities in Meru Betiri National Park include: the office, work cabins, home stay (Pondok Wisata), canopy bridge, Shelter, Bus, Campgrounds, Observation Tower, and guard-post.

Outside the park there are many hotels and motels, mainly in Jember and Banyuwangi, which are medium sized cities. The National Park itself also provides translators and even food for visitors and researchers.

Park Office

Balai Taman Nasional Meru Betiri
Jl. Sriwijaya 53 Kode Pos 269 Jember 68101
Telp. (+62-331) 335535
Fax. (+62-331) 335384
Call Center: 085749912052
Instagram: @ btn_merubetiri
email   : meru@telkom.net

25. BALURAN NATIONAL PARK

Geographical Location and Size

Baluran Naitonal Park is located between $7^0 29'$ and $7^0 56'$ south latitude and $114^0 29'$ and $114^0 39'$ east longitude. It is on the most eastern end of the island of Java, in the northeastern part of East Java Province, bordering the Madura Strait in the north and the Strait of Bali in the east. Administratively, the park falls under the jurisdiction of Situbondo and Banyuwangi Regencies, East Java Province.

Baluran National Park covers 25,000 ha, of which 85% is land and 15% is water. In the northern and eastern parts there is a 42 km coastline consisting of irregular shaped capes and bays.

Climate and Topography

The climate in Baluran National Park is a savanna type with a long dry period and a short wet season. The area is influenced by a strong southeasterly airflow from April to October/November. The dry period ranges between 4 and 9 months/year. The wet season averages 3 to 4 months/year. Rainfall is 900 to 1600 mm/year.

Average temperatures range between 27 and 30°C. The Southern slope is the wettest and most mountainious landscape in the park. The long dry season and low rainfall in the northeast produce a savanna landscape.

During the dry season, the fresh water at the surface is limited to springs. In April water can sometimes only be found in the high mountainous parts of Widuri, Kacip and Talpat.

Baluran National Park has variable topography, from flat plains to mountains, the highest of which is Baluran at 1,247 m asl. Predominantly, the landscape is low and flat to slightly undulating consisting of rocks and gravel, which are remainders of Baluran eruptions in the past. In the west, east and south of Mt. Baluran the topography is very steep. There is a cladera at Mt. Baluran, which is similar to the famous caldera of Mt Ijen, 35km to the south. However, it is not an active volcano. The walls of the crater are between 900 and 1,247 m asl.

Some small islands are found along the east coast between Cape Karang and Candibang Bay. There are reefs in places along the coast.

History

In 1928, A.H. Loebour, a Dutch hunter, submitted a formal request to establish the Baluran area as a WildLife Reserve.

In 1930, K.W. Dammerman, Director of the Botanical Garden Bogor, proposed that the Baluran area be designated a Protected Forest.

In 1937, the Governor General of the Dutch East Indies declared it a Wildlife Reserve, as a result of the Dutch Government Decree No. 9 Year 1937 (Stbld. No. 544 year 1937).

In 1962, the Indonesian Minister of Agriculture released a Decree reitearating the status of Baluran Wildlife Reserve.

On the 6th of March 1980, coinciding with World Conservation Strategy Day, the Minister of Agriculture announced a change of status from Baluran Wildlife Reserve to Baluran National Park

On the 14th of October 1982, the Minister of Agriculture statement No. 736/ Mentan/X/1982, reiterated the status of Baluran National Park.

On the 23rd of May 1997, a Minister of Forestry statement, No. 279/Kpts-VI/1997, announced that the area of Baluran National Park was 25,000 ha.

Biodiversity and Ecosystems

The Ecosystems can be classified into seven types, as follows.

- Coastal forest, consisting of plants such as *Ardisia humilis, Glochidion rubrum,* clammy cherry, *Cordia oblique, Pemphis acidula,* Pandan, *Pandanus tectorius* and buta-buta, *Exoecaria agallocha.*
- Mangrove forest, consisting of species such as mangroves, *Rhizophora apiculata, Bruguiera gymnorrhiza,* kelor wono/dadap biru, *Erythrina eudophylla,* manting, or Asian bay, *Syzygium polyanthum,* poh-pohan, *Buchanania pubescens,* api-api, also a mangrove, *Avicennia* sp. and *Excoecaria agallocha.*
- Savannah occupies approx. 40% of the park and is located mostly along the coastline and low hills up to 50m asl. Baluran National Park is the only part of Java Island which has natural savannah plains. Plant species include palm, *Borassus* sp., alang-alang, a grass, *Imperata cylindrica,* rumput lamuran, *Dichantium caricoum, Heteropogon contorcus, and Surghum nitidus.* Meadows in the Bekol area are dominated by prickly acacia, *Acacia nilotica.* In other meadows there is pilang, *Acacia leucophloea,* kemiri, *Aleurites moluccana,* lantana, *Lantana camara,* and a kind of jukut, *Vernonia cinerea.*
- Decidious forest with species such as asam Jawa, or tamarind, *Tamarindus indica,* Indian screw tree, *Helicteres isora,* strychnine tree, *Strychnos lucida,* gebang, *Schoutenia ovata* and *Corypha utan,* kapas hutan, or common mallow, *Thespesia lampas,* and kepuh, or Java olive, *Sterculia foetida.*

- Mountain dry evergreen forest is dominated by species such as the walikukun, or gebang, *Schoutenia ovata*, pancal kijang, or golden shower, *Cassia fistula* and gliseng, or gia, *Homalium foetidum*.
- Forest that has evolved along the intermittent stony streambeds, called "curah", includes climbing plants such as gadung, a yam, *Dioscorea hispida*. Teak forest can be found at Bitakol, on the western part of Baluran.
- Crater forest dominated by introduced species such as prickly acacia, *Acacia nilotica*, open plains, jati forest and secondary seasonal forest, which can also be found adjacent to the savannah forest on the slopes southwest of Mount Baluran.

In the northern part of the park along the coastline, there are mangrove forests, coastal forests and coral reefs that include some different plants such as mimbo, or neem, *Azadirachta indica*, lontar, *Borassus* sp., little gooseberry, *Buchanania arborescens*, fishtail palm, *Caryota mitis*, spurred mangrove, *Ceriops tagal*, dadap biru, or coral tree, *Erythrina eudophylla*, milky mangrove, *Excoecaria agallocha*, mentigi, *Pemphis acidula*, mangrove, *Rhizophora ayinge*, kesambi, *Schleicera oleosa*, Indonesian bay leaf, *Syzygium polyanthum*, asam, or tamarind, *Tamarindus indica* and widoro bekol, or jujube, *Ziziphus rotundifolia*.

There are 27 species of mammals, 14 of which are rare and protected. These include banteng, *Bos javanicus*, wild dog, *Cuon alpinus javanicus*, barking deer, *Muntiacus aying*, Javan leopard, *Panthera pardus melas*, kanchil, or Javan mouse-deer, *Tragulus javanicus pelandok*, fishing cat, *Prionailurus viverrinus* and long-tailed macaque, *Macaca fascicularis*.

There are at least 155 species of birds, including aying-layang api, or barn swallow, *Hirundo rustica*, tuwuk, or Asian koel, *Eudynamys scolopacea*, peacock, *Pavo muticus*, red jungle fowl, *Gallus gallus*, kangkareng, or oriental pied hornbill, *Anthracoceros convexus*, rhinoceros hornbill, *Buceros rhinoceros* and bangau tong-tong, or lesser adjutant stork, *Leptoptilos javanicus*.

Local Communities and Culture

The Baluran National Park is located between settlements on the borders of Banyuwangi and Situbondo Regencies. The communities of this area are mostly from Madura Island, especially in Banyuwangi Regency where they speak a language called Oasing.

The communities in the north fish for Bandeng, or milkfish, which they catch at Kalitopo beach north of Bama during the months of October to January. These communities also rely on Baluran National Park for firewood and illegal cattle grazing.

Tourism

- Bama beach is gently sloping and is covered with astonishingly white sand. The beach vegetation is dense and is inhabited by the long-tailed macaque, *Macaca fascicularis*. These monkeys use their tails to fish for crabs. There is an extraordinary sea garden full of coral reefs and various ornamental fish species making it very interesting for divers.
- During the rainy season the large flying fox, *Pteropus vampyrus*, can be seen in huge numbers in the surroundings of Bama.
- The coral reefs in the Gatal and Bama-Kelor area are suitable for marine tourism, such as diving or snorkeling.
- Along the road between Batangan-Bekol-Bama and other open spaces, it is possible to see peacock dances during the mating season in the months of October and November.
- In the Talpat, Bekol, and Semiang area one can see banteng, deer, wild kerbau, or water buffalo, *Bubalus bubalis* and long-tailed monkeys from the Observation Tower, or along an adventure pathway.
- Observe male deer fighting, and the deer flocking in large numbers during the mating season from July to August.
- Climb Mt. Baluran (1,247 m asl.) by following the hiking path.
- Take a Safari Tour using the vehicles provided in the National Park to see wildlife.
- Here one can also visit the perennial waterspring in Manting, views over the mangrove forest at the pier, plus the beautiful landscape presented when standing at the peak of Palpat, and along the hiking trail at Bekol.
- It is recommended to visit the Wisma Cinta Alam and/or the National Park Information Center to obtain knowledge related to the National Park including photographs, books and reports, maps, rules and regulations and facilities provided.

- Visitors who need a guide can contact the National Park Management office.

Access and Transportation

Baluran National Park can be reached from two major airports, one at Surabaya and the other at Denpasar as follows:
- From Surabaya
  - Surabaya to Banyuwangi (250 km) by bus, rental car or train takes 7 hrs.
  - Banyuwangi to Bekol by public transportation or rental car (32 km) takes 1 hr.
  - From Bekol to Batangan, a village north of Wonorejo, along a porly maintained road of 12 km takes 45 minutes driving through savannah plains and grass meadows on the eastern side of the road. From Bekol to Bama, a distance of 3 km along the beach, takes 15 minutes.
- From Denpasar
  - Denpasar - Gilimanuk by public transport or rental car takes 4 hrs.
  - Gilimanuk - Banyuwangi By ferry crossing takes 30 minutes.
  - Banyuwangi-Bekol by public transportation or rental car takes 1hr.
  - From Bekol by public transportation or rental car entering via the Batangan or Bama entrance.

Facilities

The facilities include the park office, work cabins, guard post, Wisma Cinta Alam Guesthouses at Batangan, Bekol, and Bama, Scuba diving gear and snorkels for hire, camp site at Batangan, observation tower and hiking path at Bekol and the park Information Center complete with multi image presentations to give a short overview of Baluran National Park.

Hotels are available in Banyuwangi approximately 30 km from the park. The best hotels can be found along the coastal area between Banyuwangi and the park, mostly resorts. Cars can be rented from hotels. Peak seasons are during school holidays especially before and after christmas day.

Park Office

Balai Taman Nasional Baluran
Jl. K.H.Agus Salim No. 132 Mojopanggung- Banyuwangi 68425
Telp. (+62-333) 461650
Fax. (+62-333) 463864
Call Center: 085319389646
Instagram: @ btn_baluran
Email: balurannationalpark@gmail.com

26. BALI BARAT NATIONAL PARK

Geographic Location and Size

Bali Barat National Park is located between 8°05' and 8°13' South Latitude and 114°26' and 114°35' East Longitude. It is in the Gerokgak Sub-District of Buleleng Regency and the Melaya Sub-District of Jembrana Regency, Bali Province, about 60km Northwest of Denpasar.

Bali Barat National Park covers 19,002.89 ha. Two villages and a coconut plantation with an area of 618 ha are enclaves within the park. To the south it is bordered by farmland. In the north and north-west it is bordered by beaches with reefs, including Menjangan Island.

Climate and Topography

Rainfall varies between 972 and 1,560 mm/ year and averages 1,480 mm/ year. The wet season is from January to March, and the dry season is from April to September. Humidity averages 85% and the average temperature from November to April is 28° to 29°C.

The topography is hilly to mountaneous, with only a small part flat to undulating. Elevation ranges from around 200 m to 1,144 m asl. at the highest peak, Mt. Sangiang. In the North is the low peak of Mt. Prapat Agung at 310 m and in the South are the other low peaks of Mt. Panginuman, 816 m, Mt. Bakungan, 803 m, Mt. Ulu Teluk Terima, 603 m, Mt. Nyangkrut, 347 m and Mt. Malaye, 332 m.

History

On the 13th of August 1947, based on the Bali Board of Kings No. E/1/4/5/47, the region was declared as the Bali Protected Nature Park, covering the Banyuwedang complex, an area of 19,365.8 ha.

In 1970, based on the decree of the Minister of Agriculture No. 40/Kpts/Um/8/1970, the Jalak Bali, or Bali Starling, *Leucopsar rothschildi*, was declared to be a rare and protected bird species.

On the 14th of October 1982, the Bali Barat National Park was declared during the World National Parks Congress in Denpasar, Bali, and through the decree of the Minister of Agriculture No. 736/Mentan/X/1982, given an area of 77,727 ha consisting of the Bali Barat WildLife Reserve, Nature Preserve (2,250 ha), Protected Forest (55,312.5 ha) and coastal waters (6,280 ha).

On the 15th of September 1995, a decree was released by the Minister of Forestry No. 493/Kpts-II/95, re-defining Bali Barat National Park with an area of 19,003 ha. The park has also been declared a World Heritage Site.

Biodiversity and Ecosystems

Bali Barat National Park was one of the locations of the Bali Tiger, *Panthera tigris balica*, that unfortunately has been extinct since the 1950s.

The present vegetation is strongly influenced by the monsoon, especially in the lowlands. There are disctinct wet and dry seasons.

- Tropical rainforest occurs in some lowlands, with the following species. Duren-duren, *Aglaia argentea*, kesambi, or Macassar oil tree, *Schleichera oleosa*, anjring, *Drypetes* sp., bayur, *Pterospermum diversifolium*, ketangi, a crepe-myrtle, *Lagerstroemia speciosa*, laban, *Vitex pubescens* and klampok, or Java apple, *Eugenia javanica*.
- Savannah at the Cekik region is dominated by the tall grass, alang-alang, *Imperata cylindrica*, in the understory with tegakan lontar, a palm, *Borassus flabellifer*, another palm, gebang, *Corypha* sp. and sawo kecik, *Manilkara kauki*. At Semenanjung Prapat Agung savannah is dominated by gebang and tegakan lontar, other species include bidara, or jujbe, *Zizyphus jujuba*, kesambi, the Macassar oil tree, *Schleichera oleosa*, pilang, *Acacia leucophloea, Acacia lebbekioides*, the grass, *Desmostachys bipinnata,* and mimba, or neem, *Azadirachta indica*.
- Mangrove forest is spread across Teluk Gilimanuk, Batugondang beach, Teluk Terima beach and Banyuwedang. The species include tancang, *Bruguiera cylindrica, Bruguiera gymnorrhiza*, spurred mangrove, *Ceriops tagal, Excoecaria agallocha*, red manrove, *Rhizophora stylosa, Rhizophora apiculata*, mangrove apple, *Sonneratia alba*, nipah palm, *Nypa fruticans* and api-api, or grey mangrove, *Avicennia marina*. There is also a member of the myrtaceae family generally found in Eastern Indonesia and Australia, the mangrove myrtle, *Osbornia octodonta*.
- Seasonal Forest is located in the western part of the park and includes a kind of talok, *Grewia koordersiana*, asam, or tamarind, *Tamarindus indica*, pilang, *Acacia leucophloea*, berasan, a laurel, *Cryptocarya* sp., timoko, or guest tree, *Kleinhovia hospita*, bidara, or jujube, *Zizyphus jujuba*, gelagah, or wild suger cane, *Saccharum spontaneum*, teki, or purple nut-sedge, *Cyperus rotundus*, tumpeng, or corn spurry, *Spergula arvensis* and merakan, or tussock grass, *Andropogon contortus*.
- Swamp Forest is found at Tegal Bunder where the vegetation transitions between mangrove and swamp. The milky mangrove, *Excoecaria agallocha*, found here is a favoured food plant of the critically endangered Jalak Bali, or Bali starling.
- There are some areas of plantation or conversion forest that include jati trees, or teak, *Tectona grandis*, paperbark, *Melaleuca leucodendron*, cendana, or sandalwood, *Santalum album*, sono keling, or rosewood, *Dalbergia latifolia*, sono siso, another rosewood, *Dalbergia sissoo*,

earpod wattle, *Acacia auriculiformis*, ebony, *Diospyros celebica*, sawo kecik, *Manilkara kauki*, kemlandingan, *Leucaena leucocephala*, burahol, *Steleochocarpus burahol* and murbei, a mulberry, *Morus* sp.
- Wet mountain forest covers the mountains and in general is undisturbed. The southern mountain slopes include species such as putat, *Planchonia valida*, nyatoh, *Palaquium javense*, takir, *Duabanga moluccana*, bayur, *Pterospermum javanicum*, *P. diversifolium* and a kind of kitiwu, *Meliosma ferruginosa*. To the east, the dominant trees are laban, *Vitex pubescens*, suren, *Toona sureni* and bayur, *Pterospermum javanicum*, *P. diversifolium*.

The Bali Starling, locally called Jalak Bali, *Leucopsar rothchildi*, needs special management and very careful monitoring because its population is extremely low and it is easily captured. It is listed by the IUCN as critically endangered meaning that the population in the wild is so low that it is nearing extinction. Bali Barat National Park is essential for the survival of this species.

Other than the Bali Starling, the birds here include, Jalak Putih, or black-winged starling, *Sturnus melanopterus*, jalak suren, or pied myna, *Sturnus contra*, the black headed white ibis, *Threskiornis melanocephalus*, cerek Jawa, or Javan plover, *Charadrius javanicus*, wili-wili besar, or beach stone-curlew, *Esacus magnirostris* and the Tongtong, or lessar adjutant stork, *Leptoptilos javanicus*.

Other wildlife include banteng, *Bos javanicus*, barking deer, *Muntiacus muntjak*, marbled cat, *Pardofelis marmorata*, pangolin, *Manis javanica*, Sunda porcupine, *Hystrix javanicus*, kanchil, or Java mouse-deer, *Tragulus javanicus*, fishing cat, *Prionailurus viverrinus*, and Asian water monitor, *Varanus salvator*.

Wildlife that can be encountered in the surrounding waters include dugong, *Dugong dugon*, great white shark, *Carcharodon carcharias*, longfin batfish, *Platax pinnatus*, hawksbill turtle, *Eretmochelys imbricata*, green turtle, *Chelonia mydas* and giant clam, *Tridacna gigas*.

Tourism

Bali Barat National Park is a rewarding place to visit. The most frequently visited sites are as follows.
Not within the park, but just to the east of it about 6 km from Cekik village there is the large Pulaki temple, built on the edge of a rock cliff overlooking the

The National Parks of Indonesia

sea with a white quartz sand beach. Built in the 16th century, in connection with the life of the priest Sang Hyang Nirartha, whose eldest son was given a palace in the Pulaki temple called "Bhatari Malanting". This temple is located on the highway between the Gilimanuk city and Singaraja. It is much visited by local and foreign tourists. In the vicinity are many long-tailed macaques, *maccaca fascicularis*.

In the area of the Banyuwedang mangrove forests there are hot springs, which are frequently visited by people to cure skin diseases.

- In the coastal waters adjacent to the park, especially around Terima Bay there are outstanding coral reefs used for recreational diving. On the shore there are tame long-tailed macaques. The beach opposite Menjangan Island is a turtle nesting ground.
- Also at Terima Bay you will find the Taman Sari shrine and the tomb of Jayaprana and Layur Sari, which is frequently visited by Hindu people on pilgrimage, especially on sacred days. There is also a number of small temples along the road here.
- On the southwest side of the park there are small temples with several palmleaf decorations and Balinese offerings, which are still in use.
- At Cekik is the sacred Bakungan temple.
- At Tegal Bunder is the Jalak Bali Breeding Project, managed and run by the Forestry Departemen (Dit.Gen. PKA).
- At Klatakan, 400 m asl, is a panoramic view and the possibility to see deer and hornbills.
- At Menjangan Island in the north there are coral reefs rich in species, and black volcanic soil, with the opportunity for recreational activities such as swimming, diving and boating. The beach is free from dangerous currents. Tabuan Island north of Menjangan can also be visited. There is a dock in Menjangan Bay.
- Bahari Island is a marine tourism site that can be reached by a boat with a glass bottom for marine views.
- There is a 25 km path that runs along the Prapat Agung coast between Sumber Klampok and Tegal Bunder.
- Visitors or tourists who want to snorkel or dive can rent equipment at the management office in Labuan Lalang.

- Visitors can use facilities such as the Pondok Wisata at Labuan Lalang, the Observation Towers in Banyuwedang and Sumberrejo, motorboat, shelters and toilets in Terima Bay and Labuan Lalang, and there is overnight accommodation at Labuan Lalang.

Access and Transportation

- Go to Bali by air or road and ferry. Bali Barat National Park can be reached by the following routes.
- From the northeast along the highway from Singaraja to Gilimanuk along the north coast of Bali Island.
- From the South, along the highway between Denpasar and Gilimanuk
- Using the higway Gilimanuk – Cekik – Negara (43.3 km) or the Cekik-Seririt-Singaraja Highway (85 km).
- From Kuta, Denpasar to the Ubung bus terminal by taxi, continue to Gilimanuk – duration approx. 3 hrs; then on to the National Park office.
- From Gilimanuk to Labuan Lalang, rent a motorboat to Menjangan Island – duration 30-40 minutes.

Facilities

The facilities available in and adjacent to Bali Barat National Park include , work cabins, Watch Posts, Guesthouses, field stations, car rentals, motorbikes, speedboats, and motorboats. Other facilities Information Center, Pondok Wisata, Restaurants, Observation Tower, Motorboat, Shelter, Toilets, hicking tracks/ path, Diving Equipment, Camping sites and Bali Starling Breeding Station.

Other facilities include two 3 to 4 hour walking tour packages, during which it is possible to encounter the Bali Starlings, deers, long tailed macaques and jungle fowls, and to dive in the Menjangan Bay.

Accommodation is available in the Gilimanuk area and the Labuan Lalang area, including several resorts.

Park Office

Balai Taman Nasional Bali Barat
Jl. Raya Cekik Gilimanuk-Jembrana - Bali 82253
Telp. (+62-365) 61060
Fax. (+62-365) 61479
Call Center: 082247475988
Instagram: @ btn_balibarat
email   : tnbb09@telkom.net

CHAPTER 3
KALIMANTAN

Kalimantan is the Indonesian part of the island of Borneo, the third largest island in the world (534,890 km^2). This island also includes the Malaysian states of Sabah and Sarawak in the north and northwest and Brunei Darussalam, a tiny country surrounded by Sarawak and the coastline, also in the north. Approximately 30 years ago, 75% of the island was covered in tropical forest. Now there is only 25-30% left. Kalimantan does not have volcanoes. Its floral and faunal diversity is very high, with approximately 221 mammals and 450 species of birds. Of these, 29 mammals and 20 birds are endemic to Kalimantan.

Kalimantan is divided into 5 provinces:
- West Kalimantan. Pontianak is the capital city
- East Kalimantan. Samarinda is the capital city
- Central Kalimantan. Palangka Raya is the capital city
- South Kalimantan. Banjarmasin is the capital city
- North Kalimantan, a new province. Tanjung Selor is the capital city

Kalimantan has 8 National Parks:
- Gunung Palung National Park. 108,043 ha (West Kalimantan)
- Bukit Baka Bukit Raya National Park. 181,090 ha (West Kalimantan and Central Kalimantan)
- Danau Sentarum National Park. 129,700 ha (West Kalimantan)
- Betung Kerihun National Park. 800,000 ha (West Kalimantan)
- Tanjung Puting National Park. 415,040 ha (Central Kalimantan)
- Kutai National Park. 198,629 ha (East Kalimantan)
- Kayan Mentarang National Park. 1,360,500 ha (East and North Kalimantan)
- Sebangau National Park. 568,700 ha (Central Kalimantan)

The National Parks of Indonesia

A Gunung Palung National Park
B Bukit Baka Bukit Raya National Park
C Danau Sentarum National Park
D Betung Kerihun National Park
E Tanjung Puting National Park
F Sebangau National Park
G Kutai National Park
H Kayan Mentarang National Park

Forest & Landscape ecosystem at Gunung Palung & River at Bukit Baka Bukit Raya National Park National Park (Photo by Sandy Leo)

River Utik at buffer area near Betung Kerihun National Park; River at Sebangau National Park & Boat at Kayan Mentarang National Park (Photo by Sandy Leo & KLHK)

Rhyticeros undulates at river Utik - buffer area near Betung Kerihun National Park & *Buceros rhinoceros* at Gunung Palung National Park (Photo by Sandy Leo)

Nycticebus borneanus at Kalimantan & *Pongo pygmaeus* at Gunung Palung National Park (Photo by Randi Syafutra & Tri Wahyu Susanto)

The National Parks of Indonesia

Nasalis larvatus (Sofyan Iskandar), *Hylobates funereus* & *Presbitys hosei* at Kayan Mentarang National Park (Photo by Seksi Pengelolaan TN Wilayah II Long Alango & Yohanis Oreng); *Presbitys chrysomelas* at Danau Sentarum National Park (Photo by Asri Ali)

27. GUNUNG PALUNG NATIONAL PARK

Geographic Location and Size

Gunung Palung National Park is located between 01°03'-01°22' South Latitude and 109°54'-110°28' East Longitude. The area falls under the Sukadana Sub-District and Simpang Hilir Sub-District, Ketapang Regency, West Kalimantan Province. It covers an area of 108,043 ha.

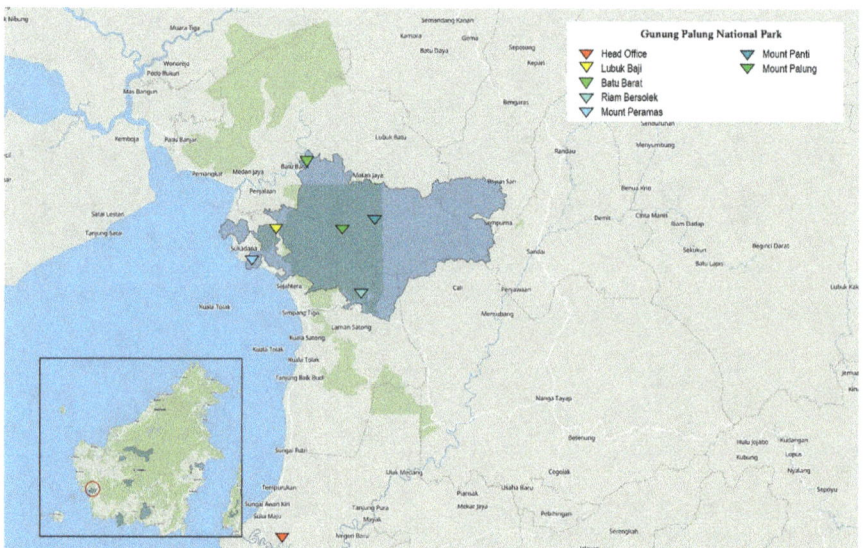

Climate and Topography

Annual rainfall is approximately 3,000 mm, with rain falling between 180 and 190 days/year. The temperature is between 25^0 and $35^0 C$.

The park climbs from 0 to 1,116m asl. The majority, 80%, is hilly and mountainous and 20% is swampy lowland. The two highest mountains are Mt. Palung (1,116 m asl) and Mt. Panti (1,030 m asl). The geology consists largely of acid intrusive rocks. The lowland surrounding this is composed of alluvium.

History

In 1937 the Dutch government declared Mt. Palung Nature Monument or Nature Preserve, with an area of 30,000 ha.

In December 1981, based on the Agriculture Minister Decree No. 10014/Kpts/Um/12/1981, the Mt. Palung area was declared to be a Wildlife Reserve, with an area of 60,000 ha.

On the 3rd of June 1990, based on the Minister of Forestry decree No. 448/MenHut/VI/1990, Mt. Palung Wildlife Reserve was changed to Gunung Palung National Park, with an area of 90,000 ha.

On June 10th 2014, this park was extended to 108,043 ha based on Ministerial decree 4191/Menhut-VII/KUH/2014.

Biodiversity and Ecosystems

The park contains the following ecosystems all of which have a high level of biodiversity.
- Mangrove forest, with a variety of mangroves such as *Bruguiera sp.*, *Lumnitzera sp.*, *Rhizophora sp.*, *Avicennia sp.* and *Sonneratia alba* as well as the nipah palm, *Nypa fruticans*.
- Coastal forest is found along the coastline and adjacent steep rocky areas. Plant species include mastwood, *Calophyllum inophyllum*, *Terminalia* sp., *Morinda* sp. and waru laut or beach hibiscus, *Hibiscus tiliaceus*.
- Swamp forest and freshwater swamp forest, with the characteristic plant species pulai, or blackboard tree, *Alstonia scholaris*, jelutung, *Dyera lowii*, ramin, *Gonystylus bancanus*, rengas, *Melanorrhoea wallichii*, bintangur, *Calophyllum* sp., pisang-pisang, *Mezzettia parvifolia*, *Litsea amara*, *Parastemon* sp., *Hopea sangal*, listed as critically endangered by IUCN, red meranti, *Shorea leprosula* and nyatoh, *Ganua motleyana*
- Lowland tropical forest includes a variety of species of the Dipterocarpaceae family such as red meranti, *Shorea leprosula*, kapur, or Borneo camphor,

Dryobalanops aromatic, also listed as critically endangered, keruing, *Dipterocarpus* sp., *Vatica* sp., and *Hopea ferruginea*. Other species include *Calophyllum grandiflorum, Palaquium gutta, Agathis borneensis*, ulin, Bornean ironwood, *Eusideroxylon zwageri* and kayu gaharu, or agar wood, *Aquilaria malaccensis*.

- Mountain forests are characterized by species of the conifer family Podocarpaceae such as *Dacrydium* sp. and many others. In the middle and lower canopy there are tea tree, *Leptospermum* spp. and the carnivorous pitcher plant *Nepenthes* spp., as well as orchids, ferns, small flowering herbs and mosses in the ground layer. The mountain forests are always covered by fog.

Mammals of Gunung Palung National Park include the primates, bekantan, or proboscis monkey, *Nasalis larvatus*, western tarsier, *Tarsius bancanus*, silvered langur, *Trachypithecus cristatus*, maroon langur, *Presbytis rubicunda*, pig-tailed macaque, *Macaca nemestrina*, banded langur, *Presbytis femoralis* long-tailed macaque, *Macaca fascicularis*, Bornean gibbon, *Hylobates albibabris* and the Bornean orangutan, *Pongo pygmaeus*. Other mammals include sambar deer, *Cervus unicolor*, Bornean yellow muntjac, *Muntiacus atherodes*, lesser mouse deer, *Tragulus javanicus*, greater mouse deer or chevrotain, *Tragulus napu*, Malayan sunbear, *Helarctos malayanus*, binturong, or bearcat, *Arctictis binturong*, Bornean bay cat, *Catopuma badia*, clouded leopard, *Neofelis nebulosa*, marbled cat, *Pardofelis marmorata*, leopard cat, *Prionailurus bengalis*, flat-headed cat, *Prionailurus planiceps*, small-toothed palm civet, *Arctogalidia trivirgata*, otter civet, *Cynogale bennettii*, banded palm civet, *Hemigalus derbyanus*, masked palm civet, *Paguma larvata*, common palm civet, *Paradoxurus hermaphroditus*, banded linsang, *Prionodon linsang*, Malay civet, *Viverra tangalunga*, bearded pig, *Sus barbatus*, Sunda pangolin, *Manis javanica*, large flying fox, or kalong, *Pteropus vampyrus*, common short-tailed porcupine, *Hystrix brachyura*, thick-spined porcupine, *Hystrix crassispinis* and long-tailed porcupine, *Trichys fasciculata*.

There are also many species of birds and reptiles. The birds include white-crowned hornbill, *Aceros comatus*, black hornbill, *Anthracoceros malayanus* and Storm's stork, *Ciconia stormi*. Several species that are endemic to Kalimantan are found in this park including flying squirrel, *Aeromys thomasi*, tree squirrel, *Tupaia splendidula*, Bornean blue flycatcher, *Cyornis superbus*, and blue-banded pitta, *Erythropitta arquata*.

Local Communities and Culture

Around this park had been inhabited by various ethnic groups with the majority of Kayong Malay ethnic (64.84%) (Sudrajat et al., 2018). Most of the people close to the park are settled in villages near the coast. There are no known settlements within the national park. Population density varies from 4.2 persons/km^2 in the Simpang Hilir region to 31.9 persons/km^2 in the Matan Hilir region. A mixed farming was the most common livelihood around the park. Villagers mainly cultivated paddy rice, rubber and durian (as cash crop), other fruit trees, and other seasonal crops. They also reared small-scale livestock, such as goat and chicken, and had small-scale fisheries. Besides, some villagers living adjacent to the park had a traditional garden in the park. It was a mixed garden within the park zone consisting of various species of fruit plants, such as durian, banana, sugar palm, etc. Durian was a main seasonal fruit planted in the garden within the current park boundary. It was common that indigenous communities (i.e., the Dayak people) practice shifting cultivation, collect non-timber forest products, and hunt wildlife (Dove, 1988).

Tourism

- The main attraction in the park is the Bornean orangutan. Other attractions include other primates and wildlife and landscape panoramas from paths and tracks.
- In the western part of the park there are Pagar Mentimun Beach, Pulau Datok Beach and Bukit Lubang Tedong located beside Karimata Strait
- At the villages along the Matan river up to the junction with the Simpang river, there are remainders of the historical heritage of the local communities, such as Matan Palace, a wooden palace painted yellow located on the edge of Matan river.
- After climbing to the top of Mt. Palung or Mt. Panti, there are views over towns at the foot of the mountains such as the Teluk Melano and below the long shoreline leading into the Karimata Strait. There are several waterfalls on both mountains

Access and Transportation

Gunung Palung National Park can be reached as follows.

- From the airport in Pontianak to Ketapang by airplane, duration 1 hr and 15 min., then continue to the park by land and follow up by river transportation, duration 6-10 hrs.
- From Pontianak by an express motorboat to Ketapang, duration 6-7 hrs., then continue to Sukadana or Teluk Melano, duration 5 hrs. Once there, the rainforest is on your doorstep. Collect a park entry permit from the park service office here. There are also guides for hire.
- From Pontianak to Teluk Batang by motor boat, duration 4 hrs. then continue to Teluk Melano by motorbike, duration 1 hour.
- From Pontianak to Rasau by public transportation, duration 1.5hr. then continue to Teluk Melano by small boat approximately 5hrs.
- From Pontianak to Teluk Melano along the Kapuas River by motor boat, duration 14 hrs. From Teluk Melano rent a traditional boat to Simanjak. From Simanjak to the park by canoe or klotok (traditional boat with outboard motor)
- From Pontianak directly to Teluk Melano by rental boat, duration 9-10 hrs. or by traditional boat, duration 10-20 hrs. Continuing to Simpang then to the Matan river, duration 4 hrs. From here, the National Park can be reached on foot, duration 5 hrs.

Facilities

At the Cabang Panti area, on the western side of Mt. Palung, there is a research station established by Harvard University, with a study area of approximately 8,000 ha, including production forests. This station allows tourist visits.

Other supporting facilities can generally be found in the villages along the coastal area west of the park such as lodgings and motor boat rentals or tour guides. There is also accommodation at Teluk Melano.

Park Office

Balai Taman Nasional Gunung Palung
Jl. Gajahmada, Ds Kalinilam, Delta Pawan
Ketapang - Kalimantan Barat
Telp. (+62-534) 32720, 7707345
Fax. (+62-534) 32720
Call Center: 082253034343
Instagram: @ btn_gn_palung
Email: bTaman Nasional_gunungpalung@yahoo.co.id

28. BUKIT BAKA BUKIT RAYA NATIONAL PARK

Geographic Location and Size

Bukit Baka Bukit Raya National Park is located in the Regency of East Kotawaringin in Central Kalimantan province and the Regency of Sintang in West Kalimantan province. Geographically, this park is in 0° 47' 60"-0°78' 33" S, 112° 37' 0"-112° 61'67" E.

Bukit Baka Bukit Raya National Park covers 181,090 ha with 110,590 ha in Central Kalimantan and 71,500 ha in West Kalimantan provinces.

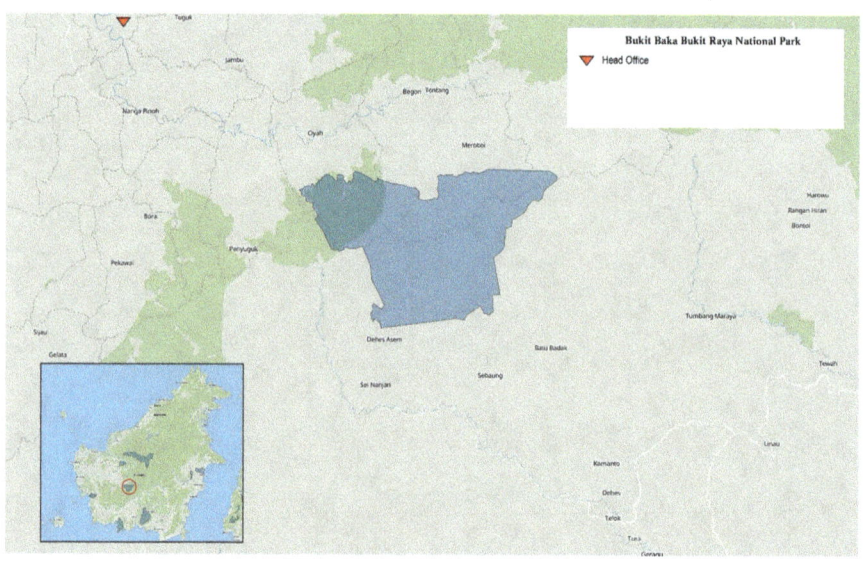

Climate and Topography

Rainfall ranges between 2,700 and 4,300 mm/year. The average temperature ranges between 20 and 32 °C, with a relative humidity of 86%. In the northern and western parts, the park is hilly with slopes of more than 30%, while in the south the topography is relatively flat with a height of between 150 to 300 m asl. There are several mountain ranges, namely Bukit Raya (2,278 m, the highest peak in Kalimantan), Bukit Asiang (1,750 m), Bukit Panjing (1,620 m), Bukit Baka (1,617 m), Bukit Lusung (1,600 m) and Bukit Panjake (1,450 m). There are two major rivers, Melawi River in West Kalimantan and Katingan River in Central Kalimantan.

History

On the 26th of June 1978, based on the Minister of Agriculture decree no. 409/Kpts/Um/6/1978, a region with an area of 110,950 ha located in Central Kalimantan was declared as the Bukit Raya Nature Preserve.

The Minister of Agriculture Decree No. 781/Kpts/Um/12/1979, was released reinforcing the status of the Bukit Raya Nature Preserve.

On the 9th of June 1989, based on the Minister of Forestry Decree No. 192/Kpts-II/1989, the forest area at Saruei Manukung, West Kalimantan was declared as Bukit Baka Nature Preserve, with an area of 70,500 ha.

On the 26th of February 1992, based on the Minister of Forestry Decree No. 281/Kpts-II/92, the Bukit Raya Nature Preserve and the Bukit Baka Nature Preserve were merged to become the Bukit Baka Bukit Raya National Park, with an area of 181,090 ha.

Biodiversity and Ecosystems

The ecosystems in the Bukit Baka Bukit Raya National Park include Lowland Tropical Rainforest, which can be found from the foothills to a height of 400m asl. They exhibit typical characteristics of lowland tropical rainforest, containing approximately 30% of the species in the Dipterocarpaceae family such as bangkirai, or red balau, *Shorea balangeran*, which is critically endangered according to IUCN, meranti, *Shorea virescens* and keruing, *Dipterocarpus oblongifolius*.

Tropical mountain forests, which extend from 400m to the top of the ranges. They are characterized by rhododendrons, *Rhododendron fortunans*,

Rhododendron exuberans and *Rhododendron verticillata* as well as *Vaccinium laurifolium*, *Diplycosia punctulata* and *Myrica* sp., and various species of orchids, rattan, and palm trees.

Other plant species present in forests, riverine vegetation, and the mountain vegetation include ulin, or Bornean ironwood, *Eusideroxylon zwageri*, kempas, *Koompassia malaccensis*, nyatoh, *Palaquium* sp., pasak bumi, *Eurycoma longifolia*, durian, *Durio kutenjensis*, rambutan hutan, *Nephelium mutabile*, Padma raksasa, *Rafflesia* sp. and Tabat barito, *Ficus deltoidea*, which is endemic to Kalimantan.

The wildlife present in this national park include honey bear, *Helarctos malayanus*, clouded leopard, *Neofelis nebulosa*, western tarsier, *Tarsius bancanus*, Sunda slow loris, *Nycticebus coucang*, beruk, or pig-tailed macaque, *Macaca nemestrina*, long-tailed macaque, *Macaca fascicularis*, maroon langur, *Presbytis rubicunda*, silvered langur, *Trachypithecus cristatus*, white-fronted surili, *Presbytis frontata*, Bornean gibbon, *Hylobates albibabris* and Bornean orangutan, *Pongo pygmaeus*.

Bird species include white-crowned hornbill, *Aceros comatus*, wreathed hornbill, *Aceros undulatus*, helmeted hornbill, *Buceros vigil*, green junglefowl, *Gallus varius*, kuau Melayu, or crested peacock-pheasant, *Polyplectron malacense* and paradise flycatcher, *Terpsiphone paradisi*.

Local Communities and Culture

The people of the indigenous settlements around the park are descendants of a group of Dayak Limbai, Ransa, Kenyilu, Danum, Malahui, Kahoi and Kahayan. Generally, they live in small groups scattered along the small rivers that flow into the Melawi River (West Kalimantan) and the Katingan River (Central Kalimantan). They generally follow a nomadic lifestyle, with shifting cultivation and livestock. Traditionally the Dayak people live in long houses called Betang. The village is governed by traditional institutions managed by the Head of community or Adat. Customary law regulates the social relationships among villagers such as marriage, divorce, land use arrangements, procedures for clearing fields and forests, tree felling and the utilization of forest products.

There are many customs, which can be observed like the typical marriage ceremony, tiwah. There are also dances and rebab music or fiddle. Cultural arts include the ancestral statues made of ulin wood, handicrafts of rattan, bamboo and *pandan* leaves. The local languages are Dohoi and Osa. However, Bahasa Indonesia is reasonably well understood and is used as the language of instruction in education and at public meetings.

Tourism

The main activities available are nature tourism and adventure tourism such as hiking, climbing, rafting, fishing, and trekking to ancient historical sites.

- At Nanga Juoi, which is in the vicinity of the *pondok wisata* (tourism lodge), there are scenic natural views such as waterfalls, and the clear water and strong currents of beautiful rivers. At the Bukit Rimban River located above Tumbang Tabulus village, there is a hot water spring.
- In the vicinity of Tumbang Gagu, Katingan Hulu Regency, there are Betang long houses.

Access and Transportation

Bukit Baka Bukit Raya National Park can be reached via two routes, as follows

To reach the Bukit Raya area travel from Palangkaraya, the capital of Central Kalimantan Province, to Kasongan by car, continue to Tumbang Samba by speedboat along the Katingan River – duration 3 hrs.

- From Tumbang Samba the park can be reached by three alternative routes, as follows:
- To Tumbang Habangoi by klotok (long boat) or Jukung (rowboat) – duration 2 days, then continue to the Bukit Raya region – duration 1 day.
- To Tumbang Hiran by motorboat – duration 6 hrs, continue to the village of Sebaung – duration 6 hrs., then by a little klotok to the Bukit Raya region – duration 1 day.
- To Tumbang Senamang via Tumbang Hiran, use a speedboat, duration 2.5 hrs, then continue to Kuluk Sepangei village on the border of the park by traditional long boat - duration 7 hrs.

To reach the Bukit Baka area in West Kalimantan Province is difficult. From the bus station at Bayu Layang in Pontianak to Sintang, duration 7-12 hrs. From there continue to Nanga Pinoh by bus then continue to Popai by motorboat. Here there is the PT Sari Bumi Kusuma Logging Company office. From Popai it is approximately 35km to the park entrance. It is possible to hitch a ride with the company's truck.

Facilities

The facilities present in the Bukit Baka – Bukit Raya National Park include:
- The Visitor Center located on Wahidin Sudirohusodo street, Sintang, West Kalimantan. This is the information center for visitors who want to enter from the Sintang Regency.
- There is a Watch Post in the vicinity of Beaban Ella village, which is manned with several forest guards.
- There is a small Lodge, which is available for visitors who come to see the waterfall and the diverse variety of plants and animals.

Park Office

Balai Taman Nasional Bukit Baka - Bukit Raya
Jl. Dr.W. Sudiro Husodo No.75 Sintang 78611, Kalimantan Barat
Telp./ Fax. (0565) 23521
Call Center: 082158564609
Instagram: @ btn_bukitbakabukitraya
e-mail : bukitbakabukitraya@gmail.com

29. DANAU SENTARUM NATIONAL PARK

Geographical Location and Size

Danau Sentarum National Park is located between $0°45'$-$1°02'$ North Latitude and $111°57'$-$112°20'$ East Longitude. The park includes parts of six sub-Districts; Badau, Batang Lupar, Embau, Empanang, Selimbau and Semitau, all in Kapuas Hulu District, West Kalimantan Province. The park covers 132,000 ha.

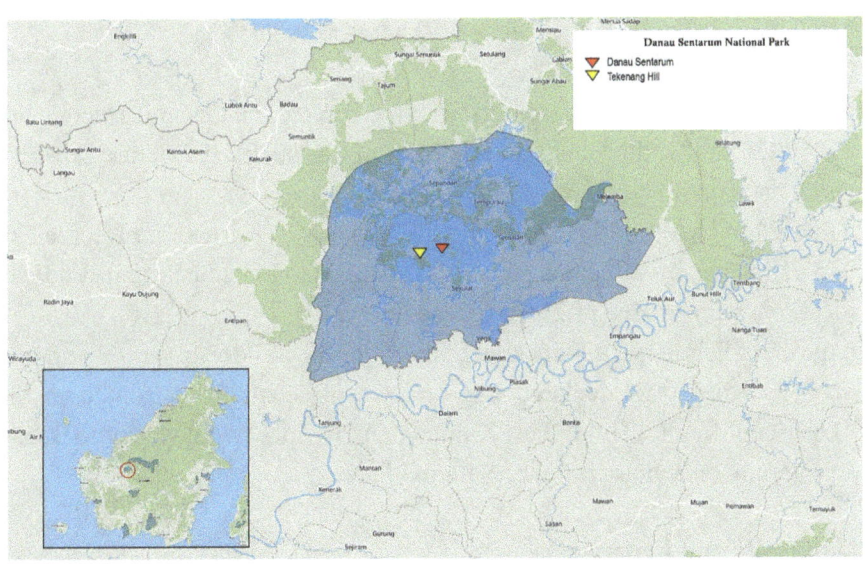

Climate and Topography

Average temperature varies between 23º and 31ºC. Rainfall varies between 4,000 and 4,730 mm/year, with rain falling on 180 to 184 days/year. For approximately nine months of the year the lakes in the park are full of water. Between the months of June and August, the water level decreases.

Danau Sentarum National Park consists mainly of lakes and swamps at a height of approximately 50 to 100m asl. The lakes are located above the upper basin of the Kapuas River approximately 700 km from where it flows into the South China Sea. The basin is surrounded by low mountains. Sentarum lake itself is a seasonal lake that during the wet season can reach depths of 6 to 8 meters, causing the entire region including the surrounding forest to flood. In the dry season, the water level in the Kapuas River gradually drops and the lake becomes smaller.

The presence of peat soil, which contains tannins, turns the lake water blackish in color with an acidity of between 4.0 to 5.5 pH. The lake is lined with dwarf plants and shrubs that gradually merge into a forest standing around 10 to 12 meters high, which in turn gradually becomes forest of between 25 to 30 meters high.

History

On the 15th of June 1981, based on Forestry Directorate General Decree No. 2240/DJ/I/1981, the Sentarum Lake region was declared a Nature Preserve with an aera of 80,000 ha.

On the 12th of October 1982, based on the Ministry of Agriculture Decree No. 757/Kpts/Um/10/1982, the status was changed to Lake Sentarum WildLife Reserve with an area of 73,906.25 ha.

On the 4th of February 1999, decree of the Ministry of Forestry and Plantation No. 34/Kpts-II/99, declared that an area of 132,000 ha located in the District of Kapuas Hulu, West Kalimantan Province, would become Danau Sentarum, or Sentarum Lake, National Park.

Biodiversity and Ecosystems

The dominant ecosystem is the wetlands, consisting of water and swamps. Three main swamp forest types were recognized by Giesen (2000) on the basis of structure: dwarf swamp forest, stunted swamp forest and tall swamp forest. These structural types are closely related to the depth and duration of flooding. Aquatic vegetation is virtually absent due to a combination of severe fluctuations in water levels and low nutrient levels in the water. The park also includes areas of lowland hill forests and riparian vegetation, which is found on the banks of the larger rivers. Plant species found here include rengas, *Gluta renghas* and *Dichilanthe borneensis*, an endemic, which represents a link between the coffee family (Rubiaceae) and the figwort family (Scrophulariaceae), incorporating characteristics from both. Another endemic is the palm, *Eugeissona ambigua*. There are only six species in the *Eugeissona* genus the only genus of its family, Eugeissonaea.

Of the 143 mammal species found in the park 23 are endemic to Borneo including the proboscis monkey, *Nasalis larvatus*. Other primates are the Bornean orangutan, *Pongo pygmaeus pygmaeus*, Abbott's gibbon, *Hylobates abbotti*, maroon langur, *Presbytis rubicundus*, Hose's langur, *Presbytys hosei*, Silvered langur, *Trachipithecus cristatus*, Pig-tailed macaque, *Macaca nemestrina*, Banded langur, *Presbytis chrysomelas*, Long-tailed macaque, *Macaca fascicularis*, Bornean slow loris, *Nycticebus menagensis* and Bornean tarsier, *Cephalopacus bancanus borneanus*. There is a relatively large population of the endangered orangutans present in the park. Other mammals include Honey bear, *Helarctos malayanus*, Wild boar, *Sus barbatus*, Clouded leopard, *Neofelis nebulosa*, Marbled cat, *Pardofelis marmorata*, Leopard cat, *Prionailurus bengalesis*, Flat-headed cat, *Prionailurus planiceps* and Squirrel, *Callosciurus notatus*.

There are at least 237 bird species, including Storm's stork, *Ciconia stormi*, kuau raja, or great argus, *Argusianus argus*, rhino hornbill, *Buceros rhinoceros*, sandang-lawe, or wooly-necked stork, *Ciconia episcopus*, tongtong stork, *Leptotilos javanicus* and layang-layang, or brown-backed needletail, *Hirundapus giganteus*. The 26 reptile species found include estuarine crocodile, *Crocodylus porosus*, and false gharial, *Tomistoma schlegelii*.

This park has a rich fish fauna with around 240 species having been recorded, including the Asian arowana, *Scleropages formosus*, linut, *Sundasalanx cf. Microps*, seluang, *Rasbora* sp., belida, *Notopterus borneensis*, baung, *Mystus nemuzus* and tebirin, *Belodontichthys dinema*.

Local Communities and Culture

The hills surrounding the park are settled by the Melayu, Dayak Iban, Embaloh and Kantu tribes, whose livelihoods are mostly derived from farming, fishing, hunting and, collecting forest products. Traditionally, the lake is fished in the dry season when the water is low during July, August and September with people returning to their villages once the lake floods again. They follow a traditional fishing system, which allows every village to have their own fishing ground. There are also small lakes that they will not fish in order to maintain stocks. This traditional law helps maintain the ecosystem balance and sustainability (Sofian and Hartinie, 2011). There are more than 50,000 people in the region of the park.

Tourism

Tourism activities that can be enjoyed in the park include the beautiful natural views across the wide lake, rowing in boats and swimming, as well as the culture and history of the region. The border of Malaysia is around an hour away by car on a good paved road, so if an immigration office was established there, international tourists could use this road to access the park.

The Iban people build unique traditional longhouses and one is located in the Pelaik river, an hour from Leboyan village. Some of the longhouses have ten to twenty doors in the one long building. The women weave textiles such as Pua and tanun, which are used to make traditional clothing worn by Iban women for traditional dancing. The men make fish traps, nets, and fishing spears and also a unique basket to carry the fish.

Local delicacies are also worth trying, especially the best fish found in the lakes, Sultan Fish or Jelawat, *Leptobarbus hoevenii* that can be as large as 15-20 kg. There is also salted and smoked fish

Access and Transportation

Danau Sentarum National Park can be reached from Pontianak to Sintang by bus, duration 8 hrs, or plane, duration 1hr, then continue to Semitau by minibus, duration 3 hrs, speedboat, duration 4hrs or longboat, duration 7hrs. From Semitau to Sentarum Lake by speedboat takes approximately 1.5 hrs.

From Pontianak to Putussibau by plane takes approximately 2 hrs, then continue to Nanga Suhaid by public transportation or rental car to the park, duration 4 hrs. This journey is through beautiful scenery with few people but there are some Iban tribal longhouses on the way. The road to the Sarawak gate, close to the park, has been paved recently and there is a small hotel close to Bada city near the border.

From Sarawak, Malaysia crossing the border to Bada city. Then from the border with a rental car to the park takes approximately an hour and half.

Facilities

There are hotels in the city of Bada only an hour by car from the lake. At the lake there are small bed and breakfast facilities.

Park Office

Balai Taman Nasional Danau Sentarum
Jl. Hj. Fatimah Rt11 RW 02, Ds Sui, Ana
Sintang , 78611, Kalimantan Barat
Tel : +62-565-2020009 Fax: +62565-2020010
Call Center: 082158794140
Instagram: @ bbtn_bentarum
Email:balai_tnds@yahoo.com, balai_tnds@gmail.com

30. BETUNG KERIHUN NATIONAL PARK

Geographical Location and Size

Betung Kerihun National Park is located between 0°40'-1°35' North Latitude and 112°15'-114°10' East Longitude. The park is in the headwaters of the Kapuas River. This park borders Lanjak Entimau Nature Preserve (200,000 ha) of Sarawak State, Malaysia. This park and Entimau Nature Preserve, together form the first Transfrontier Park in South East Asia.

The size of the Betung Kerihun National Park is around 800,000 ha or approximately 5.5% of the total area of West Kalimantan Province.

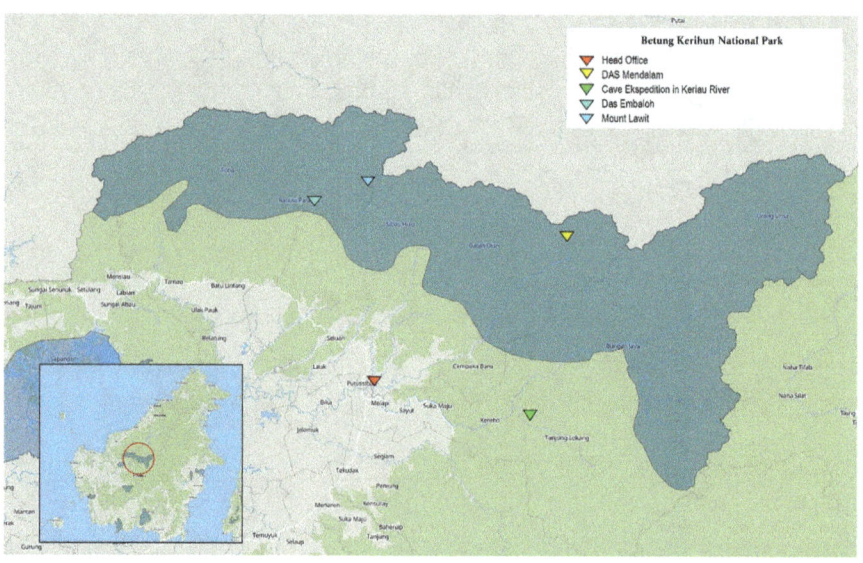

Climate and Topography

Rainfall in this region is high, between 2,860 and 5,520 mm/year with between 120 and 309 rain days/year.

The topography is hilly to mountainous, from 150 to 1,790 m asl. The park connects Mount Betung with Mount Kerihun, and forms part of the border between Indonesia and Sarawak, Malaysia. It is part of the Muller cordillera that stretches from Kapuas Hulu in the north near Sarawak south to near the border of East Kalimantan. This park has approximately 179 mountain peaks. The highest is Mt. Kerihun, 1,790 m, closely followed by Mt. Betung, 1,770 m, then Mt. Dayang, 1,645 m, then Mt. Jemuki, 1,357 m and Mt. Cemaru, 1,180 m. There are also has hundreds of large and small rivers, which are tributaries of the Kapuas River. The rivers are the means of access to the park.

History

On the 12th of October 1982, the Minister of Agriculture declared the Benuang Karimun Nature Preserve, with an area of 600,000 ha.

On the 11th of February 1992, based on the Minister of Forestry Decree No. 118/Kpts-II/1992, the area of Bentuang Karimun Nature Preserve was increased to 800,000 ha.

On the 5th of September 1995, the decree of the Minister of Forestry No. 467/Kpts-II/1995, changed Bentuang Karimun Nature Preserve into the Bentuang Karimun National Park.

On the 30th of June 1999, based on the Minister of Forestry & Plantation Decree No. 510/Kpts-II/1999, the name Bentuang Karimun National Park was changed to Betung Kerihun National Park. The name originates from Mt. Betung located in the west and Mt. Kerihun in the east, due to it being popular and well-known, to the West Kalimantan population, specifically the community in Kapuas Hulu Regency.

Biodiversity and Ecosystems

The ecosystems in this park can be divided into Dipterocarpaceae lowland forest, old secondary forest, Dipterocarpaceae hill forest, limestone forest, swamp forest, moss forest, sub-montane forest and montane forest. Approximately 1,216 species of plants have been recorded, of which 75 are endemic to Kalimantan. Characteristic plant species include jelutong, *Dyera costulata*, pulai, or devil tree, *Alstonia scholaris*, ulin/belian, or Bornean ironwood, , *Eusideroxylon zwageri*, pelawan, *Tristania* sp., cemara gunung, *Casuarina* sp., pandan, *Pandanus* sp. and gaharu, *Aquilara malaccensis*, which is now endangered in the wild.

There are approximately 48 species of mammals including honey bear, *Helarctos malayanus*, rusa sambar, *Cervus unicolor*, 18 species of bats, *Chiroptera* and 17 species of pengerat, *Rodentia*. At least 8 species of primates are present. They are Bornean orangutan, *Pongo pygmaeus pygmaeus*, Abbott's gibbon, *Hylobates abbotti*, white fronted langur, *Presbytis frontata*, maroon langur, *Presbytis rubicunda*, pig-tailed macaque, *Macaca nemestrina*, long-tailed macaque, *Macaca fascicularis*, Bornean tarsier, *Cephalopacus bancanus borneanus* and Bornean slow loris, *Nycticebus menagensis*. There are approximately 301 species of birds from 151 genera and 36 families, 63 of them are protected and 24 of them are endemic. A recently recorded species in the park is the brown-chested jungle flycatcher, *Rhinomyias brunneata*, a rare winter visitor.

There are also many species of reptiles and amphibians. So far 103 species have been identified; 51 species of amphibians, 26 lizards, 2 crocodiles, 3 tortoises and 21 snakes.

Local Communities and Culture

The indigenous people present in the region are Dayak Iban, Tamambaloh, Taman Sibau, Kantu Kayan Mendalam, Bukat Mendalam, Bukat Metelunai and Punan Hovongan. In general, they gain their livelihoods from hunting, collecting forest products and shifting cultivation farming.

The Dayak Iban perform a special dance of Adat Pantak Pemarang to welcome guests. The ritual commences with a dance by both women and men to music from their traditional instruments and is led by the tribal leader (Tetua Adat). This is followed by the Sialo dance, in which guests are invited to participate. Prior to that, guests perform the Umpang ritual, which is the slicing of the *tebu* (sugarcane stalk). There is also a ritual offering in the form of a rooster and yellow rice.

Tourism

There are three main tourism sites in the park, Embaloh, Sibau and Bungan. Tourism activities include mountain climbing, enjoying the beautiful and unique scenery, rafting, fishing, water springs and caves.

Outside the park are various tribes of the Kapuas Hulu Regency who are ethnic Dayak. The Dayak's traditional Longhouses are called Betang in West Kalimantan and Uma in East Kalimantan. Tourist attractions include handicrafts, local food and drink, especially the famous jelawat, a fish of the *Tor* genus, dances, adat rituals, local wisdom on nature, farming and traditional medicine. There are beautiful views of nature at Sentarum Lake, Bukit Lanjak, Ukit-Ukit village, the protected forest at Sungai Ulu Palin village and the winding road in the north. Most of the road from Penggal Nanga Badau bridge to Mataso at Banua Martinus is sealed.

White water canoeing is a potential tourist activity. The Dayak tribes not only have longhouses but also long boats that can navigate into the upper parts of rivers including the largest and longest of all Kalimantan rivers, Kapuas.

Access and Transportation

Access to Betung Kerihun National Park is difficult but it can be reached in the following ways
- From Pontianak by land (bus), water and air transportation to Sintang. The bus leaves from Batu Layang bus terminal and will take 7-8 hours.
- Hire a boat from Sintang to Putussibau. From Putussibau, the National Park can be reached by air, flying with a MAF flight to Tanjung Lokang takes 25 minutes or by boat along the Kapuas river to Tanjung Lokang. The trip will pass through the Nanga Lapung area, with its white-water rapids.
- Garuda airlines and Lion air fly from Pontianak to Putussibau.

Facilities

Facilities that can be found in the Betung Kerihun National Park include the Watch Post and Klotok (boat with outboard motor). There is accommodation in Putussibau including good hotels.

Park Office

Balai Besar Taman Nasional Betung Kerihun
Jl. Banin no.6, Kelurahan Kedamin Hilir, Putussubau Selatan, Kalimantan Barat
Tel /Fax: +62-567-21935
Call Center: 082158794140
Instagram: @ bbtn_bentarum
Email: tn_betungkerihun@yahoo.com

31. TANJUNG PUTING NATIONAL PARK

Geographical Location and Size

Tanjung Puting National Park is located between $2°35'$-$3°24'$ South Latitude and $111°31'$-$112°15'$ East Longitude. It is on the south coast of Kalimantan, in the south-west of Central Kalimantan Province and includes parts of Kumai districts, in West Kotawaringan Regency and Pambuang Hulu districts, Sebuluh and Seruyan Hilir districts in East Kotawaringin Regency. The park is bordered by several logging companies, by the headwaters of the Kumai River in the north, the Seruyan River in the east and the coast of the Java Sea in the South and West. Tanjung Puting National Park covers 415,040 ha, most of which consists of the peninsula between Kumay bay and Seruyan Bay.

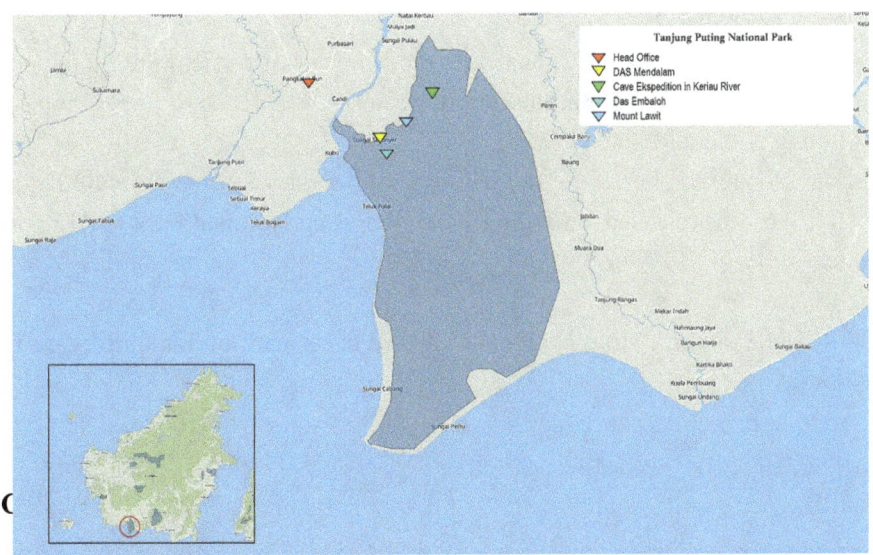

The rainfall in West Kotawaringin averages 2,417 mm/year with an average of 115.3 rain days/year. The relative humidity in the morning is between 70%-98% and during the afternoon between 55%-70%. The temperature in the morning is 22^0-28^0C and during the rest of the day 30^0-33^0C. In the East Kotawaringin Regency, the average rainfall is around 2,400 mm/ year, with an average of 97.7 rain days/year. From July to September the dry season is accompanied by strong winds, creating high waves around Teluk Sampit and disrupting sea transportation.

The topography varies from lowlands to hilly and gently undulating, with an altitude of 0-100 m asl. It consists of a peninsula that lies between low marshy lands that are inundated at certain times, fed by three major rivers; Sekonyer River, Buluh Kecil River and Buluh Besar River. All of the water is acidic and dark, containing dissolved humus and peat. Geologically, the area is composed of tertiary and pre-tertiary alluvium.

History

In 1939 the Dutch government issued Decree No. 39 regarding the establishment of the Kotawaringin-Sampit Wildlife Reserve.

In 1977 an area of 205,000 ha in this region was declared a Biosphere Reserve by UNESCO.

In November 1978, based on the Minister of Agriculture Decree No. 698/Kpts/Um/11/1978, Tanjung Puting Wildlife Reserve was declared with an area of 300,040 ha.

On the 14th of October 1982, based on the Minister of Agriculture Statement No. 736/Mentan/X/1982, the wildlife reserve was declared to be Tanjung Puting National Park.

On the 25th of October 1996, the Minister of Forestry Decree No. 687/Kpts-II/1996, was released, enlarging Tanjung Puting National Park to 415,040 ha

The Tanjung Puting National Park is a sister park of Taman Negara in Malaysia.

Biodiversity and Ecosystems

The ecosystems in the park are tropical lowland forest, dry land heath forest with coarse grasses, freshwater swamp forest, peat swamp forest, mangrove forests, coastal forest and secondary forest.

The main vegetation in the north is dry land heath forest, found in areas of white quartz sand that is characterized by small trees with a dense lower layer of tree saplings including palms and pines. Other characteristics are epiphytic vegetation, a lack of large woody vines, an abundance of microphyll phenol compounds for protection and insectivorous plants such as *Nepenthes* spp.

The true peat swamp forests are found in the central part of the park and on the banks of some rivers. They are marked by a wide-open canopy and are adapted to water logging, with aerial roots called pneumatophores. The banks of all rivers in the park are bordered by freshwater alluvial swamp forests. This forest has a complex species composition consisting of large and small woody vines, epiphytes and ferns in large quantities.

In the south there are plains covered largely with scrub vegetation, which is the result of deforestation and burning. These are commonly found in pockets along the Sekonyer River and its tributaries. The vegetation in the headwaters of the major rivers consists of grassy marsh areas, dominated by *Pandanus* spp. and stretches of floating aquatic plants. The coastal vegetation includes mangroves, which grow near the Kumai River estuary. Further inland in brackish estuaries along the major rivers, there are stands of the nipah palm, *Nypa fruticans*, which grows widely as far inland as the Arutebal River, and marks the intrusion of brackish water. Coastal vegetation on the sandy beaches includes species from the genera of *Casuarina, Pandanus, Podocarpus, Scaevola* and *Barringtonia*

Other plant species in the park include meranti, *Shorea* sp., ramin, *Gonystylus bancanus*, jelutung, *Dyera costulata* and *Dyera lowii*, gaharu, kayu lanan, keruing, *Dipterocarpus* spp., the rare ulin, or ironwood, *Eusideroxylon zwageri*, durian, *Durio* sp., jambu, *Eugenia* sp., *Imperata cylindrica, Crinum* sp., *Sonneratia* sp. and *Rhizophora* sp.

The wildlife found in Tanjung Puting National Park include Bornean Orangutan, *Pongo pygmaeus*, Proboscis monkey, *Nasalis larvatus*, Bornean white bearded gibbon, *Hylobates albibabris*, Maroon langur, *Presbytis rubicunda*, pig-tailed macaque, *Macaca nemestrina*, long-tailed macaque, *Macaca fascicularis*, Bornean tarsier, *Tarsius bancanus* and honey bear,

Helarctos malayanus. Bird species include Bubut besar, or greater coucal, *Centropus sinensis*, crimson-headed partridge, *Haematortyx sanguiniceps*, Dusky Munia, *Lonchura fuscans*, which is endemic to Borneo, Bulwer's Pheasant, *Lophura bulweri*, blue-headed Pitta, *Pitta baudii*, bristlehead, *Pityriasis gymnocephala* and flowerpecker, *Prinochilus xanthopygius*.

Local Communities and Culture

The people in the surroundings of the park are mostly of the Kumai (Malay) and Dayak tribes. Others include people from, Banjar, Madura, Bugis and Java. In general, they are Moslem, and their livelihoods are farming, fishing, handicrafts, and collecting forest products. However, the Dayak tribes have a different culture than the others except Kumai. The Dayak are mostly Christian and they live in several villages around the park with mostly agriculture and hunter gatherer activities.

Tourism

Several popular tourist locations in Tanjung Puting National Park include
- Camp Leakey. This is an Orangutan research and rehabilitation Center that began operating in 1971 initiated by Prof. Birute Galdikas. It is located in the center of the park. It is the second largest orangutan rehabilitation center after the Mt. Leuser Rehabilitation Center in Sumatra. At the Center, visitors can see orangutans in rehabilitation and various other wildlife such as gibbons, various species of birds, lowland forest vegetation, and river vegetation of pandan and swamp grass.
- Kumai Tanjung Harapan. This is a nature tourism site where visitors can see numerous species of wildlife such as orangutan, proboscis monkey, various other primates and birds, mangrove, nipah palm forest, secondary forests and dolphins who show themselves on the surface of the Kumai River. There is also the orangutan rehabilitation site, complete with a wildlife clinic.
- Natai Lengkuas. This is also a nature tourism site located on the west side of the Sekonyer River in the north of the park. This is the location of a research station to study proboscis monkey. Here, visitors can also see orangutans, plus deer and other primates including long-tailed macaques,

The National Parks of Indonesia

and can follow a hiking trail. Along the Sekonyer river various birds can be seen such as hornbills, Asian paradise flytchatcher and flocks of loud parrots. The Sekonyer river can also be used for river cruising.
- Tanguii cottages. This is a special utilization zone that was opened for orangutan rehabilitation activities. It includes deer and wild boar habitat and there are various species of birds. Tanggui huts are connected by walking paths leading to Tanjung Harapan (25 km), so this area is also suitable for hiking.
- Buluh Besar River. A special permit is needed to visit this place. This is the migratory grounds of various water birds specifically during the months April-June.
- Danau Burung. Here there is a great variety of birds, as well as crocodiles and the arowana fish.
- Aspai. This is a gold mining site, which can be reached by a klotok boat from the Camp Leakey pier, duration 2 hrs. It is located on the border of the National Park.
- Kubu Beach is decorated with long lines of beautiful coconut trees. There is fishing, and the Tanjung Keluang Nature Recreation Park with its white sand and clear waters, suitable for swimming.

Access and Transportation

There are flights from Jakarta, Bandung or Semarang in Java, and from Palangkaraya, Banjarmasin, Sampit, Pontianak and Ketapang in Kalimantan to Pangkalan Bun, the capital of West Kotawaringin Regency. From there, continue to Kumai by car, which will take 1 hour or by motorbike, which will take 1.5 hours.
- Camp Leakey, within the park, can be reached from Kumai by speedboat, duration 1.5 hrs, or by klotok (a traditional boat with outboard motor), duration 4-5 hrs. It can also be reached from Tanjung Harapan by klotok, duration 2 hrs.
- Tanjung Harapan, can be reached from Kumai by klotok, duration 1.5-2 hrs., or by speedboat, duration 0.5-1 hr.
- Natai Lengkuas, can be reached from Kumai by speedboat, duration 1.5- 2 hrs.

- A special permit is needed to be able to enter the Buluh Besar River and Burung Lake. It is difficult to reach and is limited to researchers, and adventurer tourists because it is a long journey by boat.
- At Natai Lengkuas there is a Research Station, walking trail and pier.

Another route is from Sampit to Kuala Pembuang Sub-Regency and from there take a motorboat along the Seruyan river for 8 hours, and then along the Baung River to the National Park, another 6 hours in a jukung, a traditional outrigger fishing boat.

Facilities

- In Kumai there are several good hotels.
- In Camp Leakey there are watch posts, student accommodation (research stations), a guesthouse which can accommodate 40 people, the Orangutan Rehabilitation Center covering 35,000 ha ($2°45'-2°48'$ South Latitude and $111°57'-112°01'$ East Longitude), and a walking trail in the form of a 200m long bridge crossing swamps of the Sekonyer Kanan river.
- At Tanjung Harapan there is a work cabin, watch post, an orangutan rehabilitation site, shelters, walking trails, observation tower, pier, information center, and the Rimba Hotel on the bank of the Sekonyer River.
- On the southeast bank of the Sekonyer River is the bekantan, ecology, vegetation and forest recovery research site.

Park Office

Balai Taman Nasional Tanjung Puting
Jl. HM Raf'i Km 2 Pangkalan Bun - Kalimantan Tengah
Telp./ Fax. (062-532) 23832
Call Center: 08115211040
Instagram: @ btn_tanjungputing
e-mail : btntp_arsiparis3@gmail.com or sakpabtntp@gmail.com

32. SEBANGAU NATIONAL PARK

Geographical Location and Size

Sebangau National Park is geographically in $2^0\,28'\,36.84"$ -$2^0\,28'\,61.41"$ S and $113^0\,41'\,35.30$-$113^0\,41.5'$ E. This park is located in Central Kalimantan Province covers 568,700 hectares. It is located between the Sebangau and Katingan Rivers. The park takes in parts of Katingan Regency, Pulang Pisau Regency, and Palangka Raya Municipality. It is the last huge remaining peat swamp forest in Central Kalimantan.

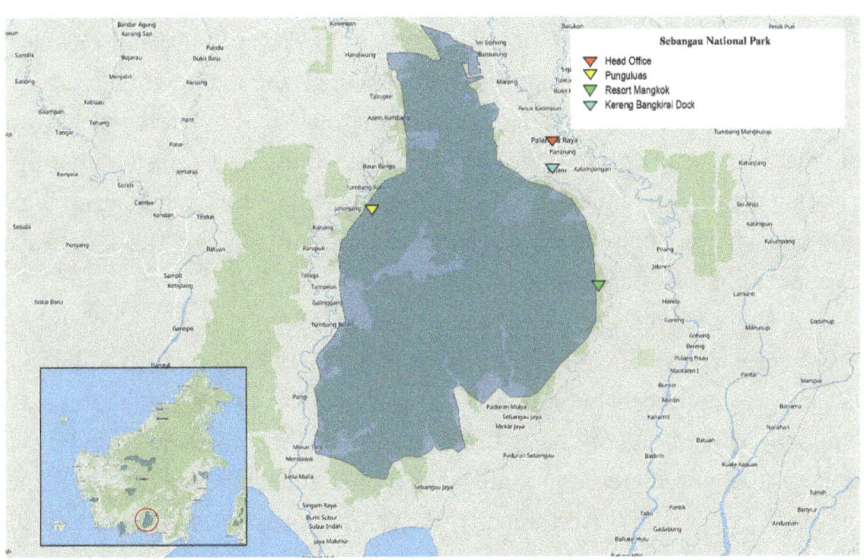

Climate and Topography

The climate in in the park is included in type A with a humid and hot tropical climate. The maximum air temperature reaches 35.1 degrees Celsius in the average air temperature in is 27.5 degrees Celsius. It is high humidity and the rainfall is between 2400 to 3200 mm.

The national park is relatively flat and centered on Sebangau River, a black water river. It flows through the Kelompok Hutan Kahayan or Sebangau peat swamp forest (5,300 km^2), between the Katingan and Kahayan rivers. The peat swamp forest is a dual ecosystem, with diverse tropical trees standing on a 10 m-12 m layer of peat—partly decayed and waterlogged plant material—which in turn covers relatively infertile soil.

History

Over the past decade, during the Mega Rice Project (MRP), the government of Indonesia drained over 1 million hectares of peat swamp forest for conversion to agricultural land. Between 1996 and 1998, more than 4,000 kilometers of drainage and irrigation channels were dug, and deforestation accelerated in part through legal and illegal logging and in part through burning. The water channels, and the roads and railways built for legal forestry, opened up the region to illegal forestry. In the MRP area, forest cover dropped from 64.8% in 1991 to 45.7% in 2000, and clearance has continued since then. It appears that almost all the marketable trees have now been removed from the areas covered by the MRP. What happened was not what had been expected: the channels drained the peat forests rather than irrigating them. Where the forests had often flooded up to 2 meters deep in the rainy season, now their surface is dry at all times of the year. The Indonesian government has now abandoned the MRP.

A study for the European Space Agency found that up to 2.57 billion tons of carbon were released to the atmosphere in 1997 as a result of burning peat and vegetation in Indonesia. This is equivalent to 40% of the average annual global carbon emissions from fossil fuels, and contributed greatly to the largest annual increase in atmospheric CO_2 concentration detected since records began in 1957. Additionally, the 2002-3 fires released between 200 million to 1 billion tons of carbon into the atmosphere. After the MRP was abandoned, this area plus others that had not been cleared for the rice project were proposed as a national park The Minister of Forestry declared Sebangau to be Indonesia's 50th national park on October 19, 2004, by issuing ministerial decree SK.423/Menhut-II/2004.

Biodiversity and Ecosystems

The peat forests here are unusual ecosystems, with trees up to 70 m high—vastly different from the peat lands of the north temperate and boreal zones, which are dominated by Sphagnum mosses, grasses, sedges and shrubs. The spongy, unstable, waterlogged, anaerobic beds of peat can be up to 20 m deep with low pH (pH 2.9-4) and low nutrients, and the forest floor is seasonally flooded. The water is stained dark brown by the tannin that leaches from the fallen leaves and peat—hence the common name blackwater swamps. During the dry season, the peat remains waterlogged and pools remain among the trees.

There are at least 26 species of mammals, 115 species of birds, 38 species of fishes and many other species are found here. Some of the mammals include Sambar deer, *Cervus unicolor*, Bornean yellow muntjac, *Muntiacus atherodes*, Lesser mouse deer, *Tragulus javanicus*, Greater mouse deer or chevrotain, *Tragulus napu*, Malayan sunbear, *Helarctos malayanus*, Binturong, *Arctitis binturong*, Bornean bay cat, *Catopuma badia*, clouded leopard, *Neofelis nebulosa*, marbled cat, *Pardofelis marmorata*, leopard cat, *Prionailurus bengalis*, flat-headed cat, *Prionailurus planiceps*, small-toothed palm civet, *Arctogalidia trivirgata*, otter civet, *Cynogale bennettii*, Malay civet, *Viverra tangalunga*, bearded pig, *Sus barbatus*, Sunda pangolin, *Manis javanica*, common short-tailed porcupine, *Hystrix brachyura* and thick-spined porcupine, *Hystrix crassispinis*.

The forest here is home to the world's largest orangutan population, estimated at 6,910 individuals in 2003, and other rare or unique species. The total agile gibbon, *Hylobates agilis,* population in the Sebangau catchment is estimated to be in the tens of thousands but is declining rapidly. Vulnerable bird species include the large green pigeon, *Treron capellei* and possibly Storm's stork, *Ciconia stormi* and lesser adjutant stork, *Leptoptilus javanicus*.

Local Communities and Culture

Sebangau National Park is close to the big city of Palangkaraya, the capital of Central Kalimantan province. There are at least 3 dominant tribes in the area close to the park; Dayak, Banjar and Malay. Some have transmigrated from Java and Madura Island. Some Dayaks are located inside the park. Dayak Ngaju is the largest dayak tribe in Central Kalimantan and villagers are found along the Katingan river. Banjar villages are also found along the Katingan river. Other Dayaks found in the Central Kalimantan include Bakumpai, Manyaa, Danum, Siang Murung, Tabayon, Lawangan and Dusun.

Tourism

This park, together with Tanjung Puting National Park, is the largest tropical peat swamp forest ecosystem in the world. Tourists come to see the black river with its waters stained by peat. People from local communities will escort them along several small rivers in several Regencies. During the boat ride, many charismatic primates can be seen in the riverine forests such as orangutans and the Bornean endemic proboscis monkey, *Nasalis larvatus*. In early morning, the resounding calls of gibbons and maroon langurs can be heard. Many birds are found here, sometime more than 20 individual hornbills along the river. Many good hotels are available in Palangkaraya.

Returning from the park to Palangkaraya, tourists may also visit the orangutan rehabilitation center in Nyaru Menteng. This center is run by the Borneo Orangutan Society Foundation (BOSF) to care for and rehabilitate orangutans orphaned as a result of clearing forest, mostly for oil palm plantations. Almost 1000 orphans have been rehabilitated and some have been released back into forests where orangutan have become scarce.

Access and Transportation

From Jakarta or Surabaya, fly to Palangkaraya, Central Kalimantan. At least four airlines fly every day to Tjilik Riwut airport in Palangkaraya.

From Palangkaraya, a hire a car with a driver to Kareng Bengkirai, or the bus from Palangkaraya bus station, takes approximately 20 minutes. Another entrance to the park is a 90 minute ride along the Katingan river.

Within the park, travel is mostly by boat. Taxi boats are long (12 m) wooden boats managed and operated by local communities. The cost is not fixed and can be bargained if you want to charter. But tourists can also go on public taxi boats along with local people and their many goods and chattels. From Kareng Bangkirai cruising into the Sebangau river downstream to Pegatan Hilir port, takes approximately 8 hours. From Pegatan Hilir on to the Katingan river and landing at Mendawai port takes approximately one hour. Since that whole trip will take 9-10 hours, it is recommended to travel in the early morning, from around 6 am.

Facilities

The park is very close to Palangkaraya, the capital city of Central Kalimantan. There are many hotels available from lodges and bed and Breakfasts, to 5 star hotels. Small hotels close to the park are available in Katingan Regency.

Office Park

Balai Taman Nasional Sebangau
Jl. Mahir Mahar KM.1,2 Kotak Pos 65
Palangka Raya - Kalimantan Tengah 73113
Telp. (+62-536) 3327093, 3359595, Fax: +62-536-324877
Call Center: 0811945545
Instagram: @ btn_sebangau
e-mail : btnsebangau@yahoo.com , sebangaukalteng@gmail.com

33. KUTAI NATIONAL PARK

Geographical Location and Size

Kutai National Park is located between $0°7'54"$-$0°33'53"$ North Latitude and $116°58'48"$-$117°35'29"$ East Longitude. The park is in the Kutai Regency of East Kalimantan Province, include North Bontang, South Bontang, Marang Kayu and Muara Badak districts. The park is approximately 90 km north of Samarinda, the Capital of East Kalimantan Province. It is bordered by coastal beaches in the east, the Sanggatta River in the north, to the headwaters of Sedulang River in the west, and in the south by the beaches of Bontang. Kutai National Park covers 198,629 ha.

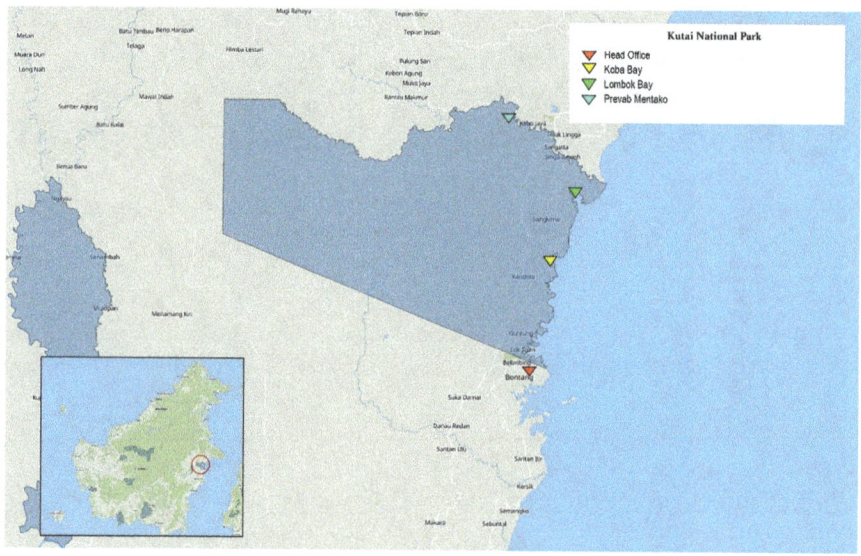

Climate and Topography

Average annual rainfall is 2,200 mm/year. Daily temperatures range between 24^0-26^0C minimum and between 30^0-34^0C maximum, with a humidity level between 67% -98%. Wind speeds normally range between 2-4 knots/hour.

The park goes from 0-300m asl. Topography varies from flat to hilly. The big rivers are the Sangkimah River and the Sanggata River. There are approximately 60 hills, the highest of which is Bukit T42. The park also includes Lake Maau, Lake Santan, Lake Besar and Lake Sirapan.

History

Kutai National Park was originally Kuta Game Reserve, which used to be the largest and most diverse example of protected coastal tropical heath and peat swamp forest covering approximayely 2 million ha of eastern Borneo until mid-1970. Then it was reduced several times until the size was less than 200,000 ha. At that stage the Ministry of Forestry declared it to be Kutai National Park (KNP). on the 29th of June 1995, according to Decision Letter of the Ministry of Forestry No. 325/Kpts-II/1995, with a total area of 198,629 ha. While the Park has a chequered history of weak protection, nonetheless, it remains substantially wild and natural. It is a complex mosaic of diverse lowland habitats described in more detail below. The history can be summarized as follows:

In 1932, based on the recommendation of Ir. H. Witcamp, a mining engineer, the Kutai Wild Reserve (Wilderservaat East Kutai), was established covering approximately 2,000,000ha.

On the 7th of May 1934, based on the decree of the Dutch East Indies government (GB) No. 3843/AZ, the area was named the Kutai WildLife Reserve.

On the 10th of July 1936, based on the decree of the Kutai Kingdom Government (ZB) No. 80/22-ZB/1936, the Kutai WildLife Reserve with an area of 306,000 ha, was declared to protect the Sumatran Rhinoceros (by then already extinct), banteng and orangutan.

On the 20th of June 1971, the area was reduced to 200,000 ha.

On the 14th of October 1982, based on the Minister of Agriculture's statement No. 736/Mentan/X/1982, the area was declared as the Kutai National Park covering 310,000ha.
On the 29th of June 1995, a Decree was issued by the Minister of Forestry No. 325/Kpts-II/95, to reinforce the Kutai National Park declaration.

The Friends of Kutai organisation, popularly known as Mitra Kutai, was created in 1994 by an agreement between the Directorate General of Conservation of Natural Resources and Ecosystem, Kutai National Park and eight companies operating in and around the park. The eight companies included two coal mining companies (PT Kaltim Prima Coal and PT Indominco), three timber concessions (PT Porodisa, PT Kiani Lestari and PT Surya Hutani Jaya), and three state oil, gas and fertilizer companies (PT Pertamina, PT Badak NGL and PT Pupuk Kalimantan Timur). In 1999 the two timber concessions went out of business and were no longer members of the Friends of Kutai, and in 2010 two new companies joined the Friends: PT Pama (a mining operator) and PT Kaltim Parna Industry (a bio-energy company). The Friends of Kutai was established with the aim of assisting the development of Kutai National Park both through fundraising and infrastructure development. The Friends of Kutai provides funding for biodiversity studies, community and buffer zone development, boundary demarcation, park facilities and infrastructure, ecotourism development, outreach programmes and the park secretariat.

The area of Kutai National Park has now been reduced to 198,629 ha

Biodiversity and Ecosystems

Kutai National Park contains a variety of ecosystems including coastal forest, mangrove forest, freshwater swamp forest, seasonal swamp forest, heath forest and lowland mixed forest. In the lowland mixed forest there is a variation of the Ironwood, dipterocarp and mixed dipterocarp forest (there are about 263 species in the Dipterocarpaceae family). The Kutai region has the largest intact forest dominated by ulin, or Bornean ironwood, *Eusideroxylon zwageri*, remaining in Indonesia. Other plant species include mangroves, *Rhizophora apiculata*, *Rhizophora mucronata*, tancang, another mangrove, *Bruguiera sexangula*, cannon ball mangrove, *Xylocarpus granatum*, cemara laut, *Casuarina equisetifolia*, nipah, *Nypa fruticans*, sempur, *Dillenia* sp., meranti merah, *Shorea parvifolia*, *Shorea smithiana*, *Shorea polyandra*, benuang,

Octomeles sumatrana, kapur, or camphor *Dryobalanops aromatica*, puspa, *Schima wallichii*, keruing, *Dipterocarpus cornutus*, menggeris, *Koompassia excelsa*, kayu arang, *Diospyros borneensis*, *Lithocarpus celebicus*, *Parinari oblongifolia*, *Duabanga moluccana*, *Antiaris toxicaria*, *Dracontomelon dao*, mangga hutan, *Mangifera caesia*, durian, *Durio acutifolia* and *Durio dulcis*, *Barringtonia asiatica*, *Ficus* spp., manggis hutan, *Garcinia mangostana*, various species of rattan and the black orchid, *Coelogna pandurate*, as well as various other species of wild orchids.

The Dipterocarpaceae sandy lowlands forests include many tropical timber trees of the genus *Shorea*, such as *Shorea leptoclados*, *Shorea pauciflora*, *Shorea ovalis*, *Shorea argentifolia*, *Shorea xanthophylla*, *Shorea acuminatissima*, *Shorea faguetiana*, *Shorea* cf. *superba*. Other plants include *Dipterocarpus caudiferus*, *Dipterocarpus warbughii*, *Dryobalanops lanceolata*, and *Irvingia malayana*. There are palms of the genera *Licuala*, *Pinanga* and *Korthalsia* and orchids such as the tiger orchid, *Gramatophyllum speciosum* and the Bornean endemic, *Phalaeonopsis gigantea*. Large woody vines include members of the families *Annonaceae*, *Leguminosae* and *Loganiaceae*.

Plants of the mangrove and coastal forests include the mangroves, *Rhizophora apiculata*, *Rhizophora mucronate* and *Bruguiera sexangula*, nipah, *Nypa fruticans*, cemara laut, *Casuarina equisetifolia*, waru, sea hibiscus, *Hibiscus tiliaceus* and pandan, *Pandanus odoratissimus*. The riverbanks are lined with *Octomeles sumatrana*, *Pterospermum javanicum*, *Patserifolium* sp. and *Barringtonia* sp. The limestone forest vegetation is dominated by the plants *Magnolia sp.*, *Diospyros sp.*, *Ixonanthes sp.*, *Dipterocarpus humeratus*, *Hopea* sp., *Chisocheton sp.*, *Tristania sp.* and *Dillenia sp.*

Wildlife include primates such as the Bornean orangutan, *Pongo pygmaeus*, Müller's gibbon, *Hylobates muelleri*, pig-tailed macaque, *Macaca nemestrina*, long-tailed macaque, *Macaca fascicularis*, proboscis monkey, *Nasalis larvatus*, Hose's langur, *Presbytis hosei*, Miller's grizzled langur, *Presbytis canicrus*, silvered langur, *Trachypithecus cristatus*, Bornean tarsier, *Tarsius bancanus borneanus* and Bornean slow loris, *Nycticenus menagensis*. Other mammals include clouded leopard, *Neofelis nebulosa*, honey bear, *Helarctos malayanus* and wild boar, *Sus borneoensis*.

The reptiles include estuarine crocodile, *Crocodylus porosus*, false gharial, *Tomistoma schlegelii*, monitor lizard, *Varanus salvator* and green crested lizard, *Bronchocela cristatella*. Over 300 species of birds have been recorded in Kutai National Park, which is 80% of all bird species found on the island of Borneo.

Local Communities and Culture

In and around Kutai National Park, there are many tribes of Malays, Dayaks and migrants. In Tenggarong, the Regency capital, located 40 km north-west of Samarinda and close to the park, there are Malay tribes, with ancient sites and palaces that show the grandeur of the kingdom of Kutai during its heyday.

There is a lot of controversy as to when the influx of migrants began – especially the Buginese communities from south Sulawesi– and what drove them. The story of a senior Buginese who settled in the Teluk Pandan area suggests that they first arrived in the early 1920s and settled in the Sangkimah area in the north-eastern part of the park, long before the area was established as a reserve by the Dutch Governor General. People claim that they came in search of a better life in East Kalimantan, with plenty of land available and employment opportunities. In the early 1960s other groups of Buginese followed, often due to political pressure in their homelands resulting from the controversial Kahar Muzakkar's rebellion against the central Indonesian government. A third influx of Buginese occurred in the mid-1970s when the government developed liquid natural gas and fertilizer industries in Bontang, and they became the workforce of these industries. This group of people associated themselves with other labourers from Java and other regions and occupied the Kandolo and Selimpus areas in the southeastern part of the park.

Tourism

Visitors come to Kutai National Park primarily to see wildlife, in particular orangutans and forest birds. In the past, orangutans and other wildlife could easily be observed along the Bontang-Sangata road corridor, but when the corridor became the main road with many illegal settlements springing up along it, wildlife moved deeper into the forest and became more difficult to see.

Bukit Enggang Lok Tuan is 6 km from Bontang City, the capital of SubRegency Bontang. It is a relatively high point where there are beautiful views of the coastline and Bontang City. A cross country walk of 6 km goes between Bukit Enggang Lok Tuan, Sidrap and Guntung areas. Here the wildlife that can be seen includes gibbon, pig-tailed macaque, sambar deer, orangutan, silvered leaf monkey and various bird species.

At Teluk Kaba and muara Sungai Sangkimah there are mangrove forests with beautiful coastal scenery and various species of coral and fish. Boats can be hired to look for orangutan, proboscis monkeys, sambar dear, kijang, kancil, grey monkeys and various species of birds. Caves with stalactites and stalagmites can be explored along the path. Banteng, *Bos javanicus lowi*, can be seen from the trail about 9 km from the Working cottage Teluk Kaba and proboscis monkeys can be seen in the Sangkimah River estuary.

The Gulf of Lombok and the Sangatta River estuary is a place for beach tourism. It is covered with white sand. Mangrove forests with the proboscis monkey wildlife rehabilitation Center can be seen from a boat.

At Mentoko there are natural forests with various species of wildlife such as orangutan, gibbons, silvered leaf monkey, otter civet and various species of birds. Rafting is available on the Sangatta River. Up until 2005 Dr. Suzuki from Kyoto University in Japan employed local people to monitor and record the movements of every single orangutan in the Mentoko area daily. In 1990, with permission from the park's management, he constructed a wooden two-storey house equipped with a simple library, herbarium and storage space as a base for his work. Mentoko and its limited facilities gradually attracted domestic and international attention, particularly from Japan and embassy staff in Jakarta who wanted to see orangutans. The park, East Kalimantan provincial government and Kutai Timur regency local government have consequently promoted the Mentoko area in the north of Kutai National Park as an ecotourism destination.

Access and Transportation

Fly to Balikpapan from many parts of Indonesia. It is 2 hrs from Jakarta. the trip on to Kutai National Park then continues as follows:
- From Balikpapan to Samarinda (125km) by land transportation – 2 hrs; or by plane – 15 minutes.
- From Samarinda to Bontang by land transportation – 4 hrs.

- From Bontang to Sanggatta (50km) a chartered flight or a motor boat from Lok Tuan harbor, following the east coast and then through Tanjung Panda, Teluk Kaba, Teluk Lombok to Sanggatta.
- From Sanggata, Mentoko can be reached by *Ketinting*, a traditional canoe.
- From Balikpapan directly to Bontang by a chartered flight from PT Badak LNG and PT Pupuk Kaltim – 40 minutes.

Several tourism sites are present in this location:
- Bukit Enggang Lok Tuan, can be reached:
- From Bontang to Bukit Enggang Lok Tuan by vehicle.
- From Bontang to the PC VI Perumahan PT Pupuk Kaltim entrance by taxi or public transportation.
- From PC VI Perumahan PT Pupuk Kaltim to Lok Tuan by taxi owned by the PT Pupuk Kaltim Coop., stop at the Km 4 Jalan Koridor PT Kayu Mas, and then continue to Bukit Enggang.

Teluk Kaba, can be reached:
- From the port of Tanjung Limau at Bontang or Loktuan to the Gulf of Kaba using a motor boat rental for 2 hours.

Teluk Lombok, can be reached:
- From the Tanjung Limau at Bontang harbor or Lok Tuan to Teluk Lombok by motor boat – 3 hrs. Muara Sungai Sanggatta can be reached from Teluk Lombok by motorboat – 2 hrs.

Facilities

The facilities in Kutai National Park include office, watch posts and work cabins (at Bontang, Sanggatta, Telok Lombok, Teluk Kaba and Mentoko), Research Stations (Teluk Kaba and Mentoko), Shelter (Lok Tuan) and hiking trails, which altogether are 150km long at Lok Tuan, Teluk Kaba, Teluk Lombok, Sanggatta and Mentoko.

There are several accommodation options here:
- Hotels in Bontang
- Hotels in Sangatta
- At Prevab-Mentoko (West Sanggatta) there is an orangutan research base camp, with research assistants from Japan. It can be reached by using a

ketinting (small canoe with outboard motor) duration 15-30 minutes from the camping site on the bank of Sanggata River.
- Other supporting facilities in Kutai National Park include a main road between Bontang and Sangatta that is partly paved. This road begins at Bontang's Protected Forest, which is also under the management of Kutai National Park.

Park Office

Balai Taman Nasional Kutai
Jl. Awang Long Tromol POS I Bontang, Kalimantan Timur 75311
Telp. (+62-548) 27218
Fax. (+62-548) 22946
Call Center: 082151192021
Instagram: @ btn_kutai
e-mail : tn_kutai@yahoo.com, kutaitourism@gmail.com

34. KAYAN MENTARANG NATIONAL PARK

Geographical Location and Size

Kayan Mentarang National Park is located between 1°59′ -4°24′ North Latitude and 114°49′-116°16′ East Longitude. It is deep in the interior of North Kalimantan Province and stretches between the Kayan River and the Mentarang River, bordering Sarawak and Sabah, East Malaysia. This park is mostly located in the Regency of Malinau (Districts of Bahau hulu, Pujungan, Kayan Hilir, Mentarang Hulu, Mentarang and Malinau Selatan) with a small portion in the Regency of Nunukan (Districts of Lumbis, Krayan, and Karayan Selatan).

The park covers 1,360,500 ha, making it the largest conservation area in Kalimantan. Almost half of it consists of Dipterocarpaceae forest.

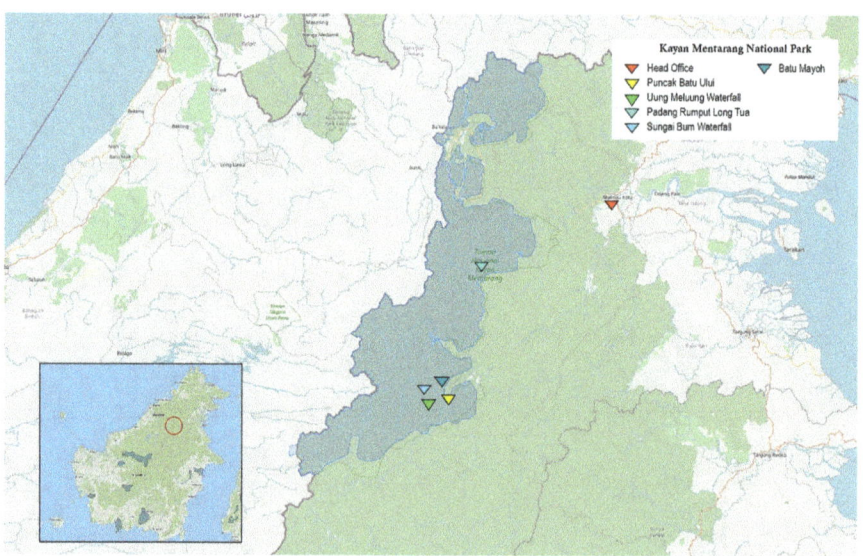

Climate and Topography

The climate is tropical with an average rainfall of 3,000mm/year and a temperature range of approximately 20.5°-33°C.

Topography varies from flat to hilly to mountainous, between 200 – 2,558m asl. The highest peak is Mt. Longnawan. Most of the park is higher than 1,000m asl.

History

On the 25th of November 1980, based on the Minister of Agriculture's decree No. 847/Kpts/Um/1980, the Kayan Mentarang Nature Preserve was established in the Berau Regency of East Kalimantan Province, which now is part of North Kalimantan Province with an area of 1,600,00 ha.

On the 7th of October 1996, based on the Decree of the Minister of Forestry No. 631/Kpts-II/1996, the status was changed to the Kayan Mentarang National Park with an area of 1,360,500 ha.

The Kayan Mentaran National Park has been declared as a "Wetland of International Significance" under the RAMSAR Convention. The park also shares a border with Sarawak (Malaysia) and has been declared to be a Transfrontier Park along with Pulong Tau National Park in Malaysia.

Biodiversity and Ecosystems

There has been a thorough ecological survey of this remote park by a team from Worldwide Fund for Nature (WWF) (Wulfraat et al., 2006). The park is in a region of very high biodiversity and has a variety of vegetation types ranging from lowland forests to mountain forests which are always covered in thick fog. The ecosystems include lowland forest, dipterocarpaceae mixed forest, low mountain forest, high mountain forest that lies at an altitude of 2,500 m above sea level, bush, grassland, *Fagaceae* forests, heath forest, *Tristania* forest and moss forest in the high mountain areas. Fruit plants that are found here include salak, *Salacca borneensis*, durian, *Durio ssp.*, mangosteen, or manngis, *Garcinia mangostana*, and rambai, *Baccaurea motleyana*. At Mt. Lunjut a previously unknown vegetation type dominated by *Anisophyllea ismailii* and *Sonerila verticillata* has been found. Other plants in the park include pulai, *Alstonia scholaris*, jelutong, *Dyera lowii*, ramin, *Gonystylus bancanus*, damar, *Agathis borneensis*, rengas, *Melanoorhoea wallichii*, meranti, *Shorea* spp., keruing, *Dipterocarpus* sp., gaharu, or agarwood, *Aquilaria sp.* and ulin, *Eusideroxylon zwageri*. Endemic plants in the park include forest salak, *Salacca borneensis*, Damar, *Agathis borneensis*, *Anisophyllea ismailii* and *Sonerila verticillata*.

An important and interesting plant found in this park is *Rafflesia pricei*, a parasitic plant with a very large flower, which can be seen in Paraye village and Buduk Tumu in Krayan subdistrict. Other threatened plants are the many species of the genus *Amorphophallus* that can be found in many parts of the park.

Wildlife found here include Banteng, *Bos javanicus lowi*, Sambar deer, *Cervus unicolor*, Bornean yellow muntjac, *Muntiacus atherodes*, lesser mouse deer, *Tragulus javanicus*, greater mouse deer or chevrotain, *Tragulus napu*, Malayan sunbear, *Helarctos malayanus*, Binturong, or bearcat, *Arctictis binturong*, otter civet, *Cynogale bennettii*, banded palm civet, *Hemigalus derbyanus*, masked palm civet, *Paguma larvata*, common palm civet, *Paradoxurus hermaphroditus*, banded linsang, *Prionodon linsang*, Malay civet, *Viverra tangalunga*, bearded pig, *Sus barbatus*, Sunda pangolin, *Manis javanica*, large flying fox or kalong, *Pteropus vampyrus*, common short-tailed porcupine, *Hystrix brachyura*, thick-spined porcupine, *Hystrix crassispinis* and long-tailed porcupine, *Trichys fasciculate*. Wulffrat et al.(2006) also recorded many other cats such as small-toothed palm civet, *Arctogalidia trivirgata*, Bornean bay cat, *Catopuma badia*, clouded leopard, *Neofelis nebulosa*, marbled cat, *Pardofelis marmorata*, Leopard cat, *Prionailurus bengalensis* and flat-headed cat, *Prionailurus planiceps*. There were also some reports on the occurence of Sumatran rhino, *Dicerorhinus sumatraensis* in this park.

Primates found in this park are white-fronted langur, *Presbytis frontata frontata*, East Bornean gray gibbon, *Hylobates funerus*, Hose's langur, *Presbytis hosei*, maroon langur, *Presbytis rubicunda*, Silvered langur, *Trachypithecus cristatus*, and Bornean slow loris, *Cephalopachus bancanus borneanus*.

There is also a rich bird fauna. At least 240 species have been recorded at one research station. These include 6 species of hornbill, plus many other birds. Similarly, the herpetofauna of this park is so rich because of the diversity of landscapes and the large size of the park.

Local Communities and Culture

Approximately 27,000 people from 13 ethnic Dayak tribes are scattered along the rivers within the park. The Dayaks in the vicinity of the Bahau River utilize approximately 200 species of plants for their various daily needs. Most of the Kenyah population and a small part of the Punan population practice shifting cultivation, and the other communities are padi farmers mostly in Lun Dayeh and Lun Bawang in the northern part of the park.

In the Long Pujungan SubRegency, North Kalimantan, 10 languages are used by the Kenyah Dayaks. Several traditional dances are based on the way the Enggang (hornbill) bird moves.

Most people live in 2 zones within the park called "Utilization" or "traditional" zones where people can hunt, fish and collect non-timber forest products for their daily use. Some forest products such as Gaharu (Sandal wood) can also be collected in the Core zone.

Tourism

- Cruising along the Bahau River, there are several large and small rapids that are remarkable for their beautiful clear water. Rafting in this river is also possible.
- Graves similar to European megaliths and tools made of stone relics can be seen at Bungaran and Wa Yagung villages. These graves look like a stone water jars. Graveyards of the Dayak Ngorek at Long Berini village also contain stone megaliths. These graveyards may be hundreds or thousands of years old. In Apo Kayan, at Long Bawan village, Kayan Hulu of Malinau Regency, there is Raja Lencan Ingan graveyard. This is recent but has an historic statue commemorating a war against the colonial Dutch.
- Many Dayak handicrafts can be found in the villages along the many rivers including the Kayan and the Mentarang.
- In many villages there are also traditional dayak dances at night. In Long Berini village, Malinau Regency, there are special dances and Sampe music made with a traditional guitar-like instrument.
- At Long Tua grassland close to Apau Ping village, Banteng, wild cattle, can be seen. Many primates such as Bornean gray gibbon, Hose's langur and maroon langur can be seen and heard calling in the morning. The high biodiversity of this park is reflected in the many different kinds of singing frog that can be heard and there are many birds to be seen. The spectacular hornbill, *Buceros rhinoceros*, in flocks of hundreds, occurs in the forest.

- There are many waterfalls such as U'ung Melu'ung on the Pujungan River that has a 50 meter fall and Sungai Bum, with a 70 meter fall on a tributary of the Pulungan river.

Access and Transportation

From Balikpapan by plane to Tarakan takes 1 hr. Then continue the journey by a smaller plane (Susie or Kalstar airlines) to Karayan, Pujungan or Apokayan. From Tarakan it is also possible to go by motorboat, called klotok, on the Mentarang River. To go all the way to Pujungan in the center of the park will take 3 days.

Facilities

The facilities present in the Kayan Mentarang National Park include the Birai Sea Conservation Station, Information Center, rest point, watch post, walking trails, Klotok and a badminton field. There are no hotels close to the park but there are home stays in several villages.

Park Office

Balai Taman Nasional Kayan Mentarang
Jl Pusat Pemerintahan, Malinau 77554 Kalimantan Utara
Telp. (062-553) 2022 758
Fax. (062-553) 2022 757
Kantor Perwakilan (sementara) :
Jl. Flamboyan No 6 RT 27, Karang Anyar
Tarakan 77111, Kalimantan Utara
Call Center: 08115991991
Instagram: @ btn_kayanmentarang
Email: balai_Taman Nasionalkm@yahoo.com, bTamanNasionalkm@dephut.go.id

CHAPTER 4
NUSA TENGGARA

Nusa Tenggara, also known as the Lesser Sunda Islands, covers both land and sea, an area of approx. 89,770 km^2 stretching 1,300 km east of the island of Bali. It has five large islands. Lombok and Sumbawa Islands make up West Nusa Tenggara, and Sumba, Flores and part of Timor Island make up East Nusa Tenggara. The Nusa Tenggara region is separated from Asia and Australia by deep sea troughs. The flora and fauna of both continents is represented, but overall species diversity is low. There are 66 endemic bird species, 8 endemic mammals and 17 endemic reptiles. There is an informative book on the ecology and communities of Nusa Tenggara written by Monk et al. (2000).

The Nusa Tenggara islands are divided into two provinces:
- West Nusa Tenggara, Province. The capital is Mataram
- East Nusa Tenggara, Province. The capital is Kupang

Nusa Tenggara has 7 National Parks:
- Mt. Rinjani National Park with an area of 40,000 ha on Lombok Island, West Nusa Tenggara
- Komodo National Park, with an area of 173,300 ha on Flores and some smaller adjacent islands, East Nusa Tenggara
- Mt. Kelimutu National Park, with an area of 5,000 ha on Flores Island, East Nusa Tenggara
- Laiwangi – Wanggameti National Park, with an area of 47,014 ha on Sumba Island, East Nusa Tenggara
- Manupeu-Tanah Daru National Park, with an area of 87,984.09 ha on Sumba Island, East Nusa Tenggara

- Mt Tambora National Park, with an area of 71,644 hectares on Sumbawa Island, West Nusa Tenggara
- Moyo Satonda Island, West Nusa Tenggara with an area of 31,200 ha.

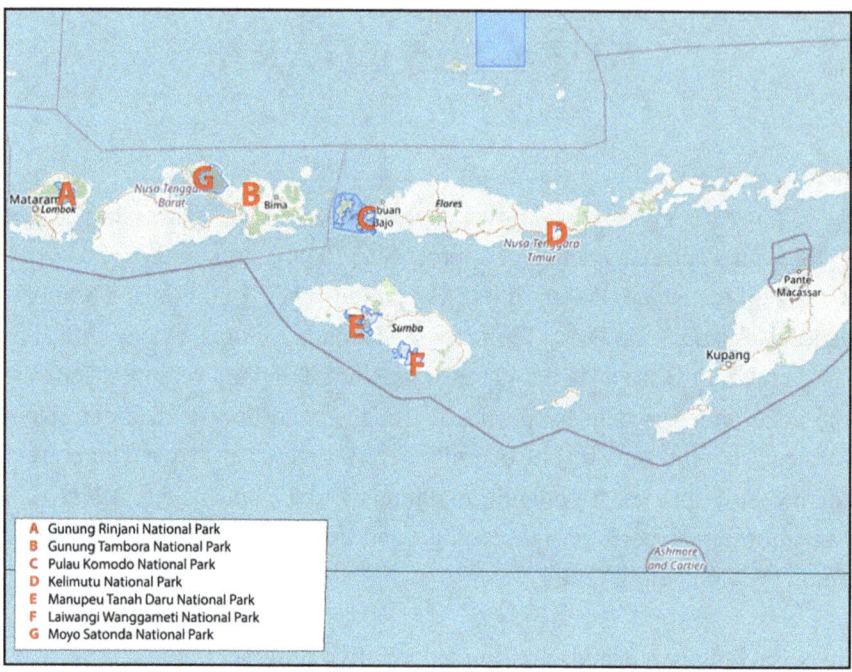

A Gunung Rinjani National Park
B Gunung Tambora National Park
C Pulau Komodo National Park
D Kelimutu National Park
E Manupeu Tanah Daru National Park
F Laiwangi Wanggameti National Park
G Moyo Satonda National Park

Laibola Hill & Wakainabu Waterfall at Laiwangi Wanggameti National Park (Photo by Memen Suparman)

Pindu Harani; Mondu Lambi Beach; & Lapopu Water Fall at Manupeu Tanah Daru National Park (Photo by Memen Suparman)

Matajitu Waterfall, river, & view landscape at Moyo Satonda National Park (Photo by Mirwan)

Three Color Lake at Kelimutu National Park (Photo by Arief Setiawan); Traditional House at Sumba (Photo by Sri Mariati) & Padar Island at Komodo National Park (Photo by Jatna Supriatna)

Landscape at Gunung Rinjani National Park & Landscape MATALAWA National Park (Photo by KLHK) & Landscape at Gunung Tambora National Park (Photo by Sri Mariati)

Komodo (*Varanus komodoensis*) at Komodo National Park (Photo by KLHK)

Megapodius reinwardti & Cacatua sulphurea at MATALAWA National Park (Photo by Memen Suparman)

Nemo fish & Coral Reef at Moyo Satonda National Park (Photo by BKSDA NTB)

35. GUNUNG RINJANI NATIONAL PARK

Geographic Location and Size

Gunung Rinjani National Park is located between $8^018'$-$8^032'$ South Latitude and $116^021'$-$116^034'$ East Longitude. The park falls within parts of North Lombok Regency, Central Lombok Regency and East Lombok Regency. This Park covers 40,000 ha.

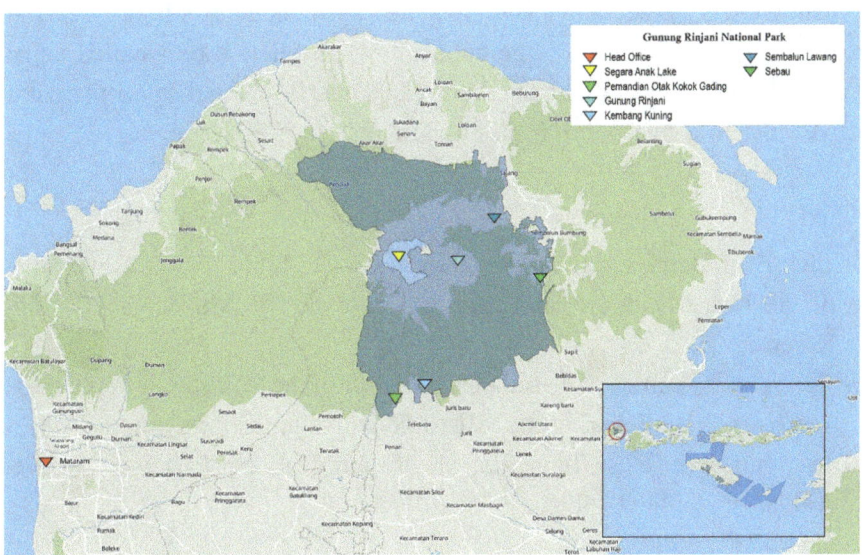

Climate and Topography

The climate is strongly seasonal, influenced by the monsoon. The dry season with predominantly eastern winds is from April to September and the wet season with westerly winds is from October to April. The rainfall is between 2,000-4,000 mm/year in the south and center, whilst in the eastern part it is drier with an average rainfall of approximately 1,200 mm/year. The number of days of rain is approx. 93 days/year.

The topography is mountainous varying in height from 300-3,726 m asl. at the peak of Mt. Rinjani, the third highest mountain in Indonesia and still an active volcano. The region's topography can be classified by slope, as follows: Medium (9%-15%), Steep (16%-45%), very steep (more than 45%).

Mount Rinjani is the second highest volcano in Indonesia (after Sumatra's Mount Kerinci) and easily one of the finest and most popular treks in Asia. The volcano complex, previously known as Samalas, is now thought to have been the location of an enormous eruption in 1257 which caused climatic change across the globe (the 'little ice age') and left the huge caldera which we see today.

There are several mountains in the park as well as Mt. Rinjani (3,726 m), Mt. Sangkaraeng (2,914 m), Mt. Buangmangge (2,895 m), Mt. Baru (2,376 m), and Mt. Manuk (m). The mountains are separated by large valleys and very steep cliffs, with steep rocky slopes. At the most western edge of Mt. Rinjani there is a caldera that is 6 km across surrounded by a high steep wall, and in the valley there is a large but shallow lake. The lake is called Segara Anak Lake and is located at a height of 2,008 m asl. Within the lake there is an active volcano named Mt. Barujari (2,376 m asl).

History

On the 17th of March 1941, based on Decree No. 15 Stbld. 77, the Government of the Netherlands declared the region to be the Mt. Rinjani Wildlife Reserve.

On the 24th of March 1990, based on the Decree of the Minister of Forestry No. 448/Kpts-II/1990, during the "III National Nature Conservation Week" at Mataram, West Nusa Tenggara, Mt. Rinjani National Park was declared, with an area of 40,000ha.

Biodiversity and Ecosystems

The ecosystems of the park include primary forest (40%), savannah plains (40%), barren plains (10%) and shrubs, plantation forests, and secondary forest (10%). The species of plants present can be grouped based on altitude, as follows:

- Lower than 1,000 m asl. The species include beringin, a strangler fig, *Ficus superba*, garu, *Dysoxylum* sp., jambu-jambuan, *Syzygium sp.*, deduren, *Aglaia argentea*, wild nutmeg, *Myristica fatna*, manyering, mimba, or neem, *Azadirachta indica*, bayur, *Pterospermum javanicum*, randu hutan, *Gossampinus heptaphylla*, lambudu, *Lasianthus* sp., harendong, *Melastoma* sp. and ratan, *Calamus* sp.
- Above 1,000 m asl. The species include meneni, malaka, *Phyllanthus emblica*, kayu jukut, *Eugenia polyantha*, garu, *Dysoxylum* sp., Sentul, *Aglaia* sp., beringin, *Ficus superba*, pandan, *Pandanus* sp., gelagah, or wild sugercane *Saccharum spontaneum*, alang-alang, *Imperata cylindrica*, edelweiss plant (*Anaphalis javanica*), orchids, e.g. *Vanda* sp., lumut jenggot, *Usnea* sp., mountain cemara, *Casuarina junghuhniana*, bangsal, *Engelhardtia spicata*, kayu raksasa, malela, *Podocarpus* sp. and jambu-jambuan (*Syzygium sp*).

On the mountain peaks the vegetation is sparse grass, due to the soil consisting mostly of sand and gravel. In the South reforestation has occurred with plants such as cinnamon, *Cinnamomum zeylanicum*, suren, *Toona sureni*, mahagony, *Swietenia macrophylla*, Jackfruit, *Artocarpus integra* and candle nut, *Aleurites moluccana*.

Wildlife in the park includes Timor deer, or Javan Rusa, *Cervus timorensis*, barking deer, *Muntiacus muntjak*, wild pig, *Sus vittatus*, Javan porcupine, *Hystrix javanicus*, long-tailed macaque, *Macaca fascicularis*, sebony leaf monkey, *Trachypithecus auratus* and fruit bat, *Macroglossus minimus*. An endemic subspecies is the Rinjani civet, *Paradoxurus hemaproditus rinjanicus*. There are also various bird species such as lesser yellow-crested cockatoo, *Cacatua sulphurea occidentalis*, red forest hen, *Gallus gallus*, green forest hen, *Gallus varius* and the endemics, scaly-crowned honeyeater, *Lichmera lombokia*, ruddy cuckoo-dove, *Macropygia emiliana*, black-backed fruit dove, *Ptilinopus cinctus* and chestnut thrush, *Zoothera dohertyi*.

Local Communities and Culture

Surrounding the park there are several villages. The people here include the mainly Hindu Balinese, and the mainly Muslim Sasak Tribe and

Sumbawan. They are influenced by animism-dynamism beliefs, and have routine rituals for the following reasons:
- The thanksgiving ritual at or just before the rice harvesting season. They release thin gold leaf fish into the Segara Anakan Lake.
- Bathing in the hotwater spring (sulphur) upstream of the Kaliputih River near the Segara Anakan Lake, to cure various diseases.
- The belief that Mt. Rinjani and its surroundings are inhabited by guard spirits.

The education level of the local communities is relatively low. The agriculture sector is the main income generator. However, the communities are generally poor and need welfare support. The interactions amongst communities are through firewood collecting, making charcoal, logging or felling trees, grazing livestock, hunting and farming.

Tourism

Mt. Rinjani is a very famous climb. It can be climbed by two routes, the north route and the south route. The north route is much easier but takes a bit longer. It is recommended to climb up following the north route and climb down using the south route. The duration of climb is approximately three days, as follows:

Day One: Batu Koq-Senaru - III Post

Day Two: III Post - Pelawangan I - Segara Anak Lake

Day Three: Segara Anak Lake - Pelawangan II – Padebelong – Sembalun.

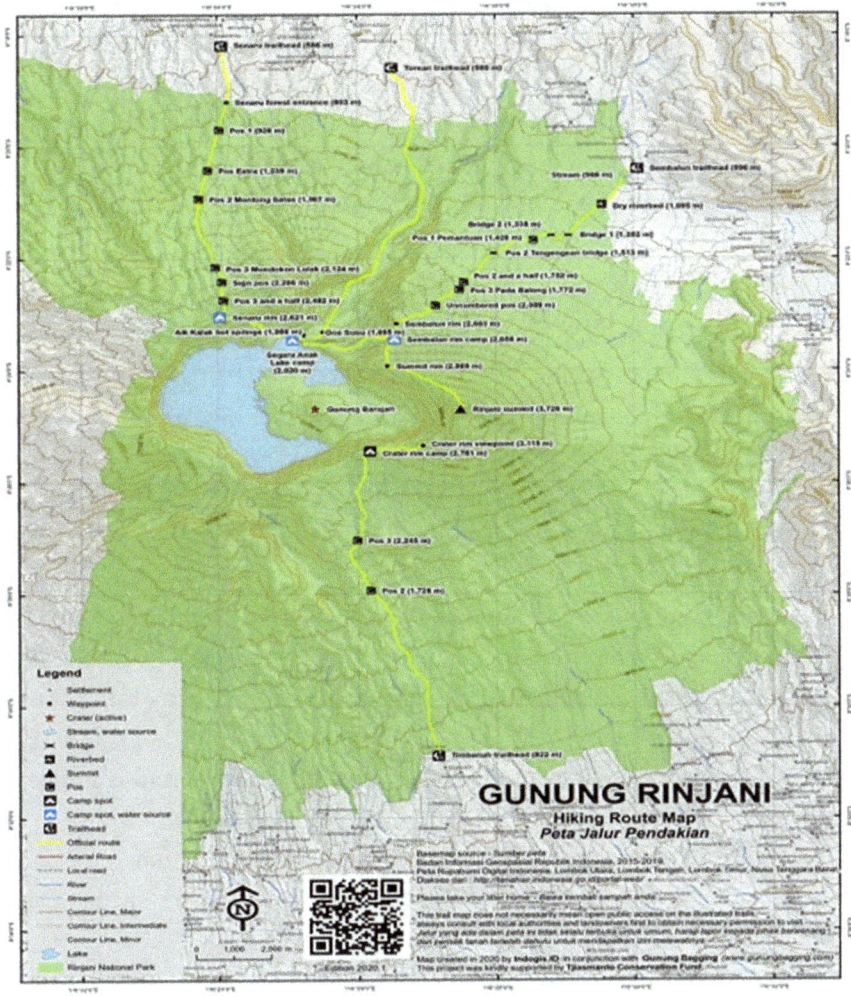

Segara Anakan Lake located on the west side of Mt. Rinjani at a height of 2,008 m asl, is surrounded by mountains ridges and colorful caldera walls. The water smells of sulphur, and the temperature varies from cold, to warm to hot depending on the presence of hot magna below the lake. In the center of the lake Mt. Barujadi has emerged, which is an active volcano within the Mt Rinjani caldera that is still growing and forms a magnificent and beautiful natural panorama. During the hike, there are waterfalls at Kembang Kuning, Diotak Kokok Gading and Sendang Gili with beautiful views of nature. At Sebak, located between the Pesugalan and Sembalun road, there are hot springs for bathing.

Visitors who need a guide can contact the National Park Management Office. People planning to climb the mountain have to report first to the officer in charge, because there are many dangerous locations, and it is recommended they be accompanied by a guide and undertake the climb through either Bayan or Sembalun. Visitor permits can be obtained from the National Park offices

After completing the climb visitors must report back to the National Park office.

Access and Transportation

Gunung Rinjani National Park can be reached by first going to Mataram, Lombok, by air or by land and sea. From Mataram continue to nearby villages by car along sealed roads. There are several routes, including:

- From the Bus Terminal at Sweta, Mataram to Aikmel by bus or rental car, continue to Sembalun Lawang or Sembalun Bumbung by minibus or on foot for 5 hrs.
- From Mataram to Selong, Sambelia and Sembalun Lawang (140 km) by public transportation or rental car for 4.5 hrs. Then continue on foot then on to Mt. Rinjani and Segara Anak Lake, a distance of 25km, following the path. Duration 9 hrs.
- From the bus terminal at Sweta to Bayan, stop at Anyar and continue by minibus to Batu Koq.
- Mataram - Bayan Senaru (82 km) by public transportation for 2.5 hrs. Continue on foot to Mt. Rinjani and Segara Anak Lake, a distance of 25km, duration 9 hrs.
- Mataram - Bayan - Torean (85 km) by public transportation or rental car for 2.5hrs. Continue on foot to Mt. Rinjani and Segara Anak Lake, distance 20 km., by following the path – duration 7.5 hrs.
- Mataram - Masbagik - Kotaraja - Tetebatu (60 km) – by public transportation or rental car 1.7 hrs. From Tetebatu to Otakkoko/ Kembang Kuning by public transportation and continue on foot to Mt. Rinjani and Segara Anakan Lake duration 30 hrs.

Facilities

Facilities provided in the park include the Resort Office, tourist guides (in the Batu Koq, Sembalun or Wisma Triguna at Ampenan), camping sites and hiking paths. There is accommodation at Batu Koq, Pondok Senaru, Homestay Guru Bakti, The Rinjani Sapit, and Hati Suci Homestay. But in Mataram City, the capital of West Nusa Tenggara, there are many quality hotels. Lombok island is almost the size of Bali Island and there are many other attractions including Kuta beach on the southern shore of the island, Gili-gili trawangan, just north of Mataram, and many others.

Park Office

Balai Taman Nasional Gunung Rinjani
Jl. Arya Banjar Getas LingkarSelatan - Mataram, NTB
Telp/Fax: (0370) 6608874
Call Center: 0811283939
Instagram: @ btn_gn_rinjani
email   : tngr@indo.net.id

36. KOMODO NATIONAL PARK

Geographic Location and Size

Komodo National Park is located between $8°17'$-$8°59'$ South Latitude and $119°30'08"$-$119°51'56"$ East Longitude, at the most western side of East Nusa Tenggara Province. Administratively it falls under the Komodo District (the capital is Labuan Bajo), Manggara Regency (the capital is Ruteng), East Nusa Tenggara Province. The region in which the park is located is bordered by Flores Island (East Nusa Tenggara) and Sumbawa Island (West Nusa Tenggara).

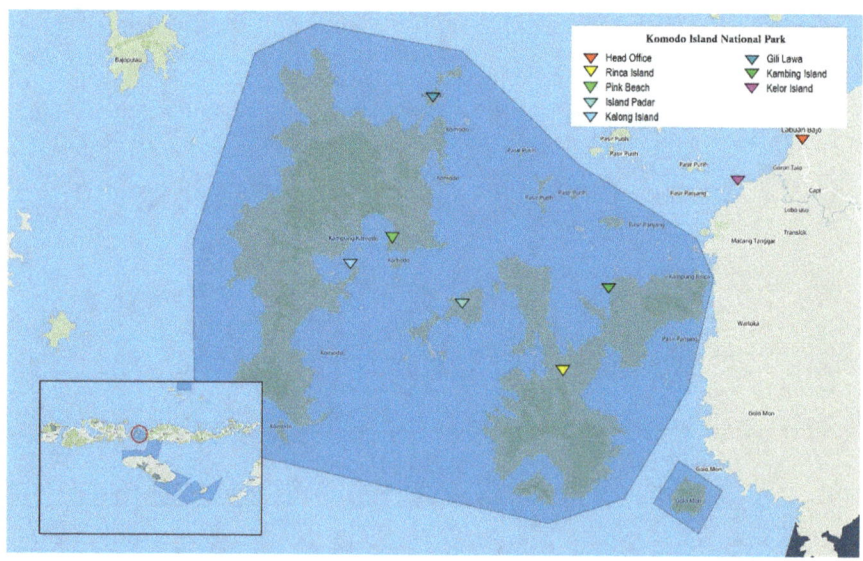

Komodo National Park covers 173,300 ha., made up of Komodo Island (33,937 ha), Padar Island (2,017 ha), Rinca Island (19,625 ha) and approximately 26 smaller islands in the surrounding waters (112,500 ha).

Climate and Topography

The climate is influenced by moist northwesterly winds from November to March, and strong dry southeasterly winds from April to October. The wet season is from January to April, and the average annual rainfall is 800 – 1,000mm/year. The temperature varies between 17^0-43^0C, with an average of approximately 30^0C.

In general, the islands are hilly, with flat areas mainly along the coast. Most of Komodo National Park consists of savannah plains dominated by the Lontar palm tree, *Borrasus flabellifer*. Komodo Island is approximately 35 km long and 15 km wide with an average altitude of 500-600 m above sea level. The highest peak is Mt. Satalibo (735 m asl) in the north, surrounded by broad plains covered by grasslands and the Lontar palm. Rinca Island is hilly, especially in the south. Its highest peak is Mt. Doro Ora (667 m asl), which is cone-shaped and covered by dense forest. It is generally lower in the north, but there are steep peaks such as Mt. Tumbah (187 m asl) and Mt. Doro Raja (351 m asl). Padar island is located between Komodo Island and Rinca Island. It is 9 km long and between 1-3 km wide. It is covered with grasslands and small trees. The topography is hilly, with an average height of 200-300 m asl.

History

In 1912 P.A Ouwens named the Komodo lizard, *Varanus komodoensis*

In 1912 Sultan Bima released a decree to protect the Komodo.

In 1938 the Flores Resident decreed the establishment of Rinca Island WildLife Reserve consisting of the southern and western parts of the Rinca island and Padar Island.

In 1965 the Flores Resident decreed the establishment of Komodo Island Wildlife Reserve, with an area of 31,000 ha.

In 1977 an area of 30,000 ha based on Komodo Island was declared a Biosphere Reserve by the UNESCO Man and the Biosphere (MAB) program.

On the 6th of March 1980, the Minister of Agriculture changed that status to of this park to Komodo National Park, including the surrounding small islands and waters.

In 1991 the International Union for Conservation of Nature and Natural Resources (IUCN) declared the region to be a World Heritage Site.

On the 26th of February 1992, the Minister of Forestry released Decree No. 326/Kpts-II/1992, reinforcing the status of Komodo National Park.

In 1992, based on the President's Decree No. 4 Year 1992, the Komodo dragon was declared to be a National Wildlife Emblem.

On the 26th of June 1995, the Minister of Forestry issued Decree No. 306/Kpts-II/95, to reinforce the status of Komodo National Park covering an area of 173,300 ha.

Biodiversity and Ecosystems

The ecosystems in Komodo National Park include:

- Tropical Rainforest, which is found in the mountains above 500m asl. Prominent plant species include rattan, *Calamus* spp., bamboo, *Bambusa* spp., *Terminalia zolingeri*, *Podocarpus neriifolius*, *Uvaria rufa*, a fig, *Ficus orupaceae*, *Calophyllum spectabile*, and orangeberry, *Glycosmis pentaphylla*.
- Tropical deciduous forest is deciduous during the dry season, and is dominated by tamarind, *Tamarindus indica* and kepuh, *Sterculia foetida*. Other plant species include *Jatropha curcas*, *Cladogynos orientalis*, *Hypoestes malacensis* and *Ziziphus horsfieldii*.
- Savannah forest is the most widespread forest in the park (70%) and is the habitat of the rusa deer, *Rusa timorensis*, water buffalo, *Bubalus bubalis* and wild pigs, and is also the main habitat of the Komodo dragon. Plant species include lontar palm, *Borassus flabellifer*, bidara, *Zizyphus jujuba* and tamarind, *Tamarindus indica*.
- Mangrove forest covers 5% of the park. It is found in the sheltered bays at Loh Sebita and Soro Lawi on Komodo Island, and at Loh Kima and Loh Buaya on Rinca Island. The species include Asiatic mangrove, *Rhizophora mucronata* and the white-flowered black mangrove, *Lumnitzera racemosa*.
- The Coral Reefs of the park are highly diverse and rich in species. There are more than 1,000 species of fish, at least 105 species of crustacean and 70 species of sponges, as well as 250 species of coral. The fishes include

mackerel, *Scomberomorus commerson*, tuna, *Thunnus* spp., bengkolo, *Caranx* sp., cakalang, Skipjack tuna, *Katsowanus pelamis* barracuda, *Sphyraena* sp. and tongkol, *Euthynnus* sp. Other marine species include whitetip reef shark, *Triaenodon obesus*, teripang, or sea cucumber, mata tujuh or abalone, pearl oyster, grouper, *Epinephelus* sp., ikan sunu, *Plectropomus* sp., napoleon fish, *Cheilinus undulatus* and humpback grouper, *Cromileptes altivelis*.

The floral diversity is relatively low, at approximately 102 species. As well as the dominant lontar palm trees, other prominent species include nyamplung, *Callophyllum inophyllum*, gebang, or cabbage palm, *Corypha utan*, kepuh, *Sterculia foetida*, tamarind, *Tamarindus indica*, and Ketapang, *Terminalia catappa*.

The Komodo dragon, *Varanus komodoensis*, is the dominant charismatic wildlife species of this national park. The Komodo is found on Komodo Island, Padar Island, Rinca Island, Gili Motang, Gili Dasami and Wae Wuul in east Flores. It is estimated that there are more than 3,000 of them left in the wild. The mature Komodo can reach a length of 3 meters and weigh 90 kg, and can run 30 km/hour. They prefer lowland shrubs bordering savannah plains but can also be found at heights of 400-600m asl. The Komodo is a carnivore, which also eats carrion, and sometimes can be a cannibal. It often preys on deer, wild hog, wild buffalo and horse. Sometimes prey is only injured and later dies of infection. Then the Komodo will eat it. The wild Komodo is difficult to encounter during the mating season from July to August because it moves deep into the center of the forest or digs a hole to go into. The female Komodo can lay up to 20 eggs/nest. The eggs hatch in the month of April. The Komodo hatchlings move into a tree and feed on small animals such as geckoes and other lizards, insects and birds.

There are 185 other wildlife species including water buffalo, *Bubalus bubalis*, rusa deer, *Cervus timorensis*, wild hors, *Equus caballus*, long-tailed macaque, *Macaca fascicularis*, rat dragon, *Komodomys rintjanus*, fruit bat, *Pteropus vampirus*, dugong, *Dugong dugon*, green turtle, *Chelonia mydas* and saltwater crocodile, *Crocodylus porosus*.

Komodo National Park has a diverse bird fauna. Among the 111 species, are burung gosong, or orange-footed scrubfowl, *Megapodius reinwardt*, yellow-crested cockatoo, *Cacatua sulphurea*, great-billed heron,

Ardea sumatrana, Cikalang, or great frigatebird, *Fregata minor*, elang-alap, or variable goshawk, *Accipiter hiogaster*, trinil, or wood sandpiper, *Tringa glareola*, green imperial pigeon, *Ducula aenea*, helmeted friarbird, *Philemon buceroides* and collared kingfisher, *Halcyon chloris*.

Local Communities and Culture

Kerora village, with a population of less than 600 in 2020, is situated within the park. The people are traditional fishermen, forest product collectors and cattle farmers. Most are Moslems but retain their own customs, which they still strongly uphold, such as the Dowry, land distribution, planting season and harvest season, and fishing season.

Another settlement located in the park is Komodo village, at Slawi Bay, approximately half an hour from Labuan Bajo following the coastline. Visitors can see and experience local tradition life style by taking a rental boat to the village. On Komodo island there are performances of traditional silat dance, and Komodo statues handcrafted from wood are available.

Tourism

Komodo National Park and the surrounding islands welcome travelers year-round with the best months being April to December, the dry season. The days have bright blue skies and a gentle breeze, the ideal weather for exploring the islands and marine seascape.

- Padar Island. The best trekking in Komodo National Park is on Padar Iskand. It will take 1-2 hours to the summit. There are beautiful landscapes and seascapes from the summit with blue water and small islands.
- Komodo Island. The main entrance for visitors is at Loh Liang, where there is also a cafeteria. Here visitors can see the culture and the daily life of the indigenous people, who at certain times present a silat dance, and sell wooden hand crafts. At Banu Nggulung it is possible to see up to 20 individual komodo dragons.

The hiking trails that can be used on Komodo Island include:
- Loh Liang - Komodo Village - Mt. Ara - Loh Serikaya - Mt. Ara - Komodo Village - Liang (30 km), duration 2 days. Overnight at Loh Serikaya

- Loh Liang - Komodo Village - Soro Masangga - Mt. Ara - Komodo Village - Loh Liang (26 km), duration 2 days with overnight at Soro Masangga
- Loh Liang - Mt. Poreng - Loh Sebita - Mt. Poreng - Loh Liang (24 km), duration 2 days with overnight at Loh Sebita
- Liang Lake - Komodo Village - Mt. Ara - Mt. Satalibo - Mt. Poreng to Loh Liang (27 km), duration 2 days with overnight at Mt. Satalibo
- Loh Liang - Waegulung - Loh Liang (7 km), duration 2 hrs.
- There are also excellent dive sites around Loh Liang.
- In August 2022, the government raised the entrance fee to Komodo island to Rp3.75 million rupiah/person or (almost of USD250). The idea is that the revenue raised will be used for conserving Komodo dragons. People can still see the Komodo in the wild on Rinca island for the regular price of USD20.
- Kalong and Loh Lasa islands. The attractions are swimming, camping and viewing corals and ornamental fish. There are cross country walks where fruits bats can be seen hanging in the mangrove forest.
- Rinca Island. There are Komodo dragons, wild horses and deer. In the Eastern part near Rinca village there are several caves which are connected by numerous tunnels. The roots of Euphorbias and figs hang down into the caves. They can also be used to climb down into the caves. Sometimes there may be encounters with komodos and bats. At Loh Kima beach in the northwest and Doro Tumbuh beach in the northeast, there is seasonal deciduous forest. Recently the government built a big platform for tourists to see komodos from the bridge, so that interaction between komodos and humans are minimized.
- At Mt. Ara (510 m asl) there is hiking and climbing while enjoying the savannah forest with its spectacular views
- Marine tourism: The best places to snorkel and dive are Merah beach, Loh Namo, Loh Bo and Lasa Island. There are dive shops and glass bottomed boats.

Visitors who need a guide should contact the National Park Management office. Visitors who want to see Komodos up close must be accompanied by an official from the National Park, and should always take note of what that official says.

Access and Transportation

There are daily direct flights connecting Labuan Bajo Airport (LBJ) on Flores Island to Denpasar, Bali, duration in 1.30 hrs and to Jakarta duration 2.30 hrs.

You can also go Labuhan Bajo by sea from Denpasar, Bali. The ferry from Labuan Bajo goes to Komodo Island on Saturdays and Sundays but a motorboat can be rented on other days.

Transportation to Rinca Island, Padar Island, Gilimatong and Lasa Island, is by rental speedboat

Facilities

The facilities provided in Komodo National Park include an office and information center in Labuan Bajo and Sape (West Nusa Tenggara), a Tourism Cabin (Pondok Wisata) at Loh Liang, camping sites at Mt Ara and Loh Kima, shelters at Loh Liang, Komodo, Loh Lasa, Loh Kubu, Loh Punya, Giliawa and Loh Buaya, and many guest houses. Labuan bajo has a harbor and good hotels and restaurants.

Park Office

Balai Taman Nasional Komodo
Jl. Kasimo Labuan Bajo Ruteng, NTT 86554
Telp: (062-385) 41004, 41005
Fax. (062-385) 41005
Call Center: 082145675612
Instagram: @ btn_komodo
email: info@komodo-park.com

37. KELIMUTU NATIONAL PARK

Geographic Location and Size

Kelimutu National Park is located on Flores Island between $8°43'21"$-$8°48'24"$ South Latitude and $121°44'21"$-$121°50'15"$ East Longitude. Administratively, it falls under Detsuko, Wolowaru and Ndona Districts, Ende Regency, East Nusa Tenggara Province. The size of Kelimutu National Park is 5,356.5 ha.

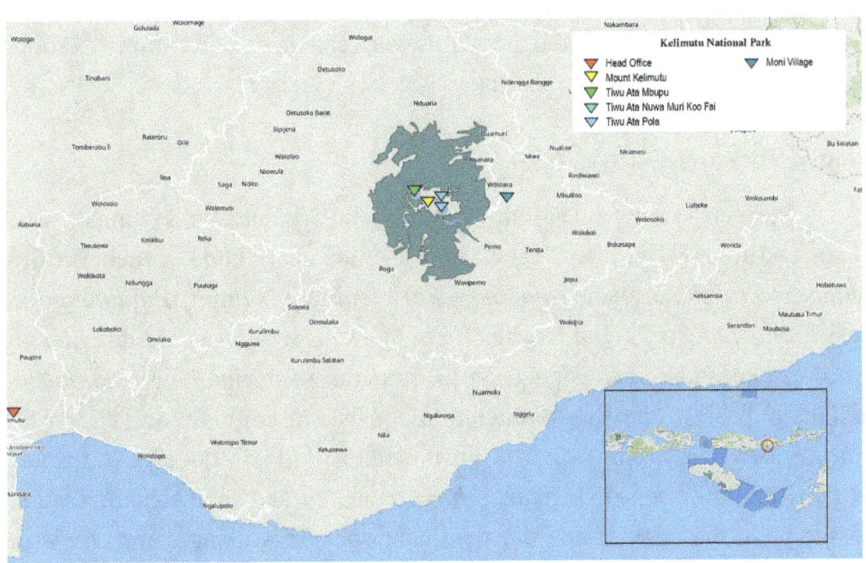

Climate and Topography

Average annual rainfall is 1,000 to 1,500 mm. The rainy season is from January to April. During July and August, the minimum temperature can reach 21.6°C. The average temperature is around 25.5°-31°C.

Topography varies ranging from undulating to rugged and steep slopes. There are several mountains, including Mt. Kelibara (1,731 m) and Mt. Kelimutu (1,690 m). At Mt. Kelimutu there are the three famous crater lakes known collectively as Lake Kelimutu. The waters of these three lakes change every year from red to dark green to dark red, dark green to light green and from blackish brown to sky blue, depending on the composition of minerals, chemicals and dissolved oxygen in each of the three lakes. Kelimutu Lake is one of the 9 Wonders of the World declared by UNESCO.

History

On the 4th of October 1984, based on the Decree of the Minister of Forestry No. 185/Kpts-II/1984, The Kelimutu Lake area was declared as the Kelimutu Lake Nature Reserve, with an area of 16ha and the Kelimutu Lake Recreation Park, with an area of 4,984 ha.

On the 26th of February 1992, based on the Decree of the Minister of Forestry No. 279/Kpts-II/92, the status was changed to Kelimutu National Park, with a total size of 5,356.5 ha.

Biodiversity and Ecosystems

The ecosystems in Kelimuti National Park include savannah, open plains and forest. The plants found here include ajang kode, a species of red cedar, *Toona* sp., kawah, *Neolamarckia cadamba*, kempo, *Palaquium* sp., kesi, *Canarium* sp., kodal, or black ebony, *Diospyros ferrea*, sita, *Alstonia scholaris*, kesambi, *Schleichera oleosa*, cemara, *Casuarina junghuhniana* and mountain daisies. The forest vegetation consists of empupu, or Timor white gum, *Eucalyptus urophylla*, with an understory of introduced exotics such as putri malu, or sensitive plant, *Mimosa pudica* and tembelekan, *Lantana camara*. On the open plains the dominant species is alang-alang, *Imperata cylindrica* along with other grasses.

The wildlife that can be encountered here include barking deer, *Muntiacus muntjak*, rusa or Timor deer, *Rusa timorensis*, long-tailed macaque,

Macaca fascicularis, kuskus, *Phalanger* sp., wild pig, *Sus scrofa*, and a variety of birds such as yellow-crested cockatoo, *Cacatua sulphurea*, green junglefowl, *Gallus varius*, black drongo, *Dicrurus macrocercus* and scaly-crowned honeyeater, *Lichmera lombokia*, which is endemic to Nusa Tenggara.

Local Communities and Culture

There are many tribes in the Ende regency where Kelimutu is located, each with their own culture and customs such as traditional wear, music and dance and high quality hand woven patterned cloth. Customary houses can still be seen in several villages.

The people in Kelimutu believe that the three crater lakes have different functions in their lives. According to the Lio tribe, the souls of dead people go to the craters, but stop over somewhere called Pere Konde. The Lio people will give an offering so that the soul will be ready to go to the craters. One crater is pictured as the place of hell, another is heaven and the third is the place for the cleansing of sin. Tiwu Ata Polo is the place for bad people, Tiwu Nuwa Muri Koofai is the place of young people who have behaved well, and Tiwu Ata Bupu is the place for old people who have behaved well.

A very interesting aspect of the local people is that they had already developed an agroforesty system called "Napu", even before the park was established. Muda (2005) studied this Napu system, in which coffee, oranges and salak are grown under a cover of Dadap trees, *Erythrina subumbrans*. The system, a mixture of forest and food trees, supports the renewal of soil by decomposing litter.

Tourism

- The three colored lakes, Tiwu Ata Polo, Tiwu Ata Bupu and Tiwu Ata Nuwa Muri Koofai in the Mt. Kelimutu crater (approx. 1.600 m asl) are the three colored lakes, included in UNESCO's "9 Wonders of the World"
- There is a hiking path to the summit of Mt. Kelimutu, which is 14 km long and takes about 4 hours to climb.
- Flores Island has beautiful beaches with an enormous diversity of corals.
- Mt. Kelimutu is in the Ende Regency. In Ende city there is the Bung Karno Museum, the house occupied by Sukarno, the nationalist leader exiled there by the Dutch in 1934, who later became the first president of Indonesia.

Visitors who need a guide can contact the National Park Management office. The best time to visit the area is during the dry season between July and September.

Access and Transportation

Kelimutu National Park can be reached in several ways
- By air from Denpasar to Kupang then either to Ende or Maumere airports. Then travel by public transportation or rental car to Mt. Kelimutu.
- By sea using the Kelimutu motorboat once every 2 weeks, from Semarang - Surabaya - Ujung Pandang - Padang Bay (Bali) - Lembar (Lombok) - Bima (Sumbawa) - Ende. Continue from Ende to Kelimutu by public transportation or rental car.

Facilities

Visitors can use the Flores highway (length 546 km) to the cities surrounding Mt. Kelimutu, including from Labuan Bajo to Ruteng distance 140 km, from Ruteng to Bajawa - distance 112 km, from Bajawa to Ende - distance 146 km From Ende to Maumere - distance 148 km

Hotels are available in Ende, approximately 3 hours from the park. Guest houses and bed and breakfasts close to the park are also available at Moni village. There are car and motorcycle rentals available at Moni Village (Wisma Kelimutu and Sao Risa Wisata and many small resorts) or in the surroundings of Koanara.

Park Office

Balai Taman Nasional Kelimutu
Jl. Eltari No. 16 Ende Flores, Nusa Tenggara Timur
Telp: +62-381-23405, Fax: +62-381-23892
Call Center: 08113829716
Instagram: @ btn_kelimutu
Email: btnkelimutu@gmail.com

38. LAIWANGI-WANGGAMETI NATIONAL PARK

Geographic Location and Size

Laiwangi-Wanggameti National Park is located between 120°00'-120°22' South Latitude and 9°58'-10°11' East Longitude. It is in the south-east of Sumba Island, East Sumba Regency, East Nusa Tenggara Province. The park covers 47,014 ha. This park includes all the forest area remaining in south-eastern Sumba Island.

Climate and Topography

The topography of the park is generally hilly to mountainous with steep tp very steep slopes, though less steep in the south and southeast. Height ranges from 0-1,255m asl at Mt. Wanggameti peak, which is the highest on Sumba. This park is the water source for much of Sumba and the deciduous forest is still in very good condition. The climate is tropical, hot all year round, with a rainy season from December to March and a dry season from June to October.

Temperatures are high and uniform throughout the year. They are slightly higher from October to May and a bit lower from June to August, when the southeast monsoon blows. At this time the minimum temperature can drop to 18/19 °C. In August 2012 it dropped to 15.5°C. The hottest days of the year, with highs about 35/36°C, are often recorded in October and November, before the rainy season. In the latter, high humidity makes the heat muggy.

History

On the 3rd of August 1998, based on the Decree of the Minister of Forestry No. 576/Kpts-II/1998 the Laiwangi-Wanggameti area in the East Sumba Regency, which is the largest forest area on Sumba Island still in relatively good condition, was declared the Laiwangi-Wanggameti National Park, with an area of 47,014ha.

Biodiversity and Ecosystems

This national park represents all types of forest present on Sumba Island, including the rare elfin, or dwarf, forest. Plant species found in this park include *Syzygium* spp, *Alstonia scholaris*, Figs, *Ficus* spp, *Canarium oelosum*, cinnamon, *Cinnamomum zeylanicum*, *Myristica littoralis*, *Toona sureni*. The blackboard, tree or devil's tree, *Alstonia scholaris*, is a dominant species.

The estimated number of bird species in the park is 215. Some of the birds are endemic, including Sumba hornbill, *Aceros everetti*, Sumba cicada bird, *Coracina dohertyi*, Sumba myzomela, *Myzomela dammermani*, Sumba flycatcher, *Ficedula harterti*, walik rawamanu, or red-naped fruit dove, *Ptilinopus dohertyi*, Sumba boobook, *Ninox rudolfi* and Sumba green pigeon, *Treron teysmannii*. Various other birds include gemak Sumba or Sumba buttonquail, *Turnix everetti*, cempaka, or citron crested, cockatoo,

Cacatua sulphurea citrinocristata, nuri, purple-naped lory, *Lorius domicella*, kepodang-sungu Sumba, or pale-shouldered cicadabird, *Coracina dohertyi*, and the apricot breasted sunbird, *Nectarinia buettikoferi*.

The park also has a variety of butterflies. At least 43 species have been recorded, of which five are endemic to East Nusa Tenggara. These are the halipron birdwing, *Troides haliphron naias, Elimnias amoena, Sumalia chilo, Ideopsis oberthurii*, and the *Athyma karita*

There are few large mammals. They include long-tailed macaque, *Macaca fascicularis*, and wild pig, *Sus sp*. Many small mammals are found here. Reptiles include Timor python, *Python timorensis*.

Local Communities and Culture

In this area there 15 villages, with an approximate population of 20,000 people. Most of them are farmers. To truly understand what makes Sumba special, plan a visit to a local village, like Lamboya or Wanokaka. Make sure to come with a local guide so that you can properly pay your respects to village chiefs and enter safely. If you have a strong stomach, some hotels can help you plan a visit to a village when a funeral or wedding is underway. Be aware that these ancient rituals often involve animal sacrifice, which provides food for the entire village and social standing for the hosting family. For a less confronting visit, come for the weekly markets. Farmers and craftsmen sell fresh fruit and vegetables, eggs, betel nuts, jewelry, woven baskets, traditional ikat (a style of Indonesian decorative fabric), and mamole—a traditional totem that honors fertility and feminine energy. A side note for textile lovers: if ikat is really what you are after, ask your hotel to arrange an expedition to East Sumba. There's very little tourism there, but it's where artists weave the world's most beautiful ikat.

Tourism

The most popular activity in this park is climbing the highest mountain on Sumba Island. Climbing Wanggameti Mountain from Wanggameti village takes approximately 4 hrs. Many endemic birds of Sumba Island can be seen along the track to the mountain. The other attraction is the waterfall more than 100 m high close to Laputi Village with very clean and blue water. There are many megalithic stones in villages around the park.

Access and Transportation

The National Park can be reached by using public transportation from the Waingapu Terminal, which leaves daily to Wanggameti. Car rentals are available in the city of Waingapu.

Facilities

At Larondja village there are hiking and climbing guides as well as local accommodation in the form of guest houses.

Wainggapu is equipped with several small hotels.

At Rua Beach there is the Nihiwatu Resort. Nihiwatu resort is one of the most expensive resorts in Indonesia. In 2017 it won the Indonesia Sustainable Tourism Award. It is not too far from the park with a white sandy beach that is good for surfing and other marine tourism activities.

Park Office

Balai Taman Nasional Laiwangi Wanggameti
Jl. Matawai Amahu, Kampung Baru
Kel. Hambala, Waingapu, Sumba Timur
Nusa Tenggara Timur - 87113
Tel and Fax: . +62-387- 61940
Call Center: 0811384408
Instagram: @ btn_matalawa
Email: laiwangi5@gmail.com

39. MANUPEU-TANAH DARU NATIONAL PARK

Geographic Location and Size

Manupeu-Tanah Daru National Park is located between 9°35' -9°53' south latitude and 119°29-199°53 east longitude. It is on Sumba Island, West Sumba Regency, East Nusa Tenggara Province. It is 87,984.09 ha in size.

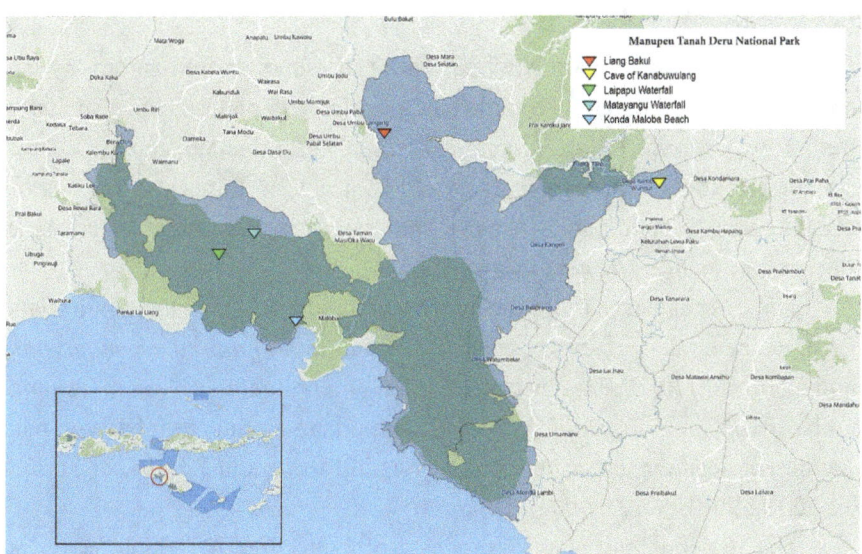

Climate and Topography

Most of the forest area in the national park is flat with steep cliffs, rising up from the sea surface to a height of 600m. Annual rainfall is between 500-2000 mm. The climate is tropical, hot all year round, with a rainy season from

December to March and a dry season from June to October. Temperatures are high and uniform throughout the year. The hottest days of the year, with highs about 36 °C, are often recorded in October and November, before the rainy season. They are slightly higher from October to May and a bit lower from June to August, when wind blows from south due to Australian winter. At this time the minimum temperature can drop to 18/19°C. In August 2012 it dropped to 15.5°C.

History

Originally, the area was described as Manupeu Protected Forest with an area of 12,000 ha.
The status was changed to Manupeu WildLife Reserve.
On the 3rd of August 1998, Decree No. 576/Kpts-II/1998, of the Forestry Minister declared the Manupeu-Tanah Daru National Park with an area of 87,984.09 ha.

Biodiversity and Ecosystems

Manupeu-Tanah Daru National Park represents the remaining lowland Forest on Sumba Island. The park has a variety of plant species including suren, *Toona sureni*, taduk, *Sterculia foetida*, kesambi, *Schleichera oleosa*, pulai, or blackboard tree, *Alstonia scholaris*, asam, or Tamarind, *Tamarindus indica*, kemiri, or candlenut, *Aleurites moluccana*, jambu hutan, *Syzygium* sp. and cemara gunung, *Casuarina* sp.

There are at least 87 bird species of which 7 are endemic to Sumba Island. These are cempaka, or citron cockatoo, *Cacatua sulphurea citrinocristata*, Sumba hornbill, *Rhyticeros everetti*, Sumba green pigeon, *Treron teysmannii*, Sumba flycatcher, *Ficedula harterti*, kepodang-sungu Sumba, or pale-shouldered cicadabird, *Coracina dohertyi*, and the madu Sumba, or apricot-breasted sunbird, *Nectarinia buettikoferi*.

The national park has 57 species of butterflies of which seven are endemic to Sumba Island. These are *Papilio neumoegenii, Ideopsis oberthurii, Delias fasciata, Junonia adulatrix, Athyma karita, Sumalia chilo*, and *Elimnia amoena*.

Local Communities and Culture

At the Manupeu-Tanah Daru National Park there are 23 villages. The people around the park practice a ritual ceremony called hamayang. Communities adopt different zones for different activities. One is for agriculture including the production of food, firewood and pasture. A second is for harvesting timber for housing and non-timber products such as medicinal plants. A third protects water sources for drinking and agriculture. The fourth is a sacred place for the Marapu communities to conduct rituals.

Tourism

- Matayangu and Lapopu. There is a waterfall near the villages of Waimanu and Katikutana.
- At Waikabubak, which is near Manupeu-Tanah Daru National Park, there is an ancient carved tomb. To the local communities, these ancient tombs are a symbol of social status and health.
- Pasola, is a very interesting and exciting ritual attraction. It consists of men riding colorfully decorated horses. The men try to knock their opponents off their horses using wooden sticks. This activity can be seen in the month of February at Lamboya and Kodi, and in the month of March at Gaura and Wanukaka.

Access and Transportation

There are flights from Denpasar, Bali, to Tambolaka airport on Sumba Island. It will then take approximately an hour by rental car to Waikabubak, a small city in the West Sumba regency. From Waikabubak to the park will take about an hour to travel the 30 kms. There are many rental cars available.

Facilities

The facilities available is homestay which is managed by the community around the national park. Most tourists who visit to Manupeu Tanah Daru National Park is generally associated with time remaining on the cultural tour on Sumba Island. In Waikabubak where the national park headquarters is located, here are many hotels.

Park Office

Balai Taman Nasional Manupeu Tanadaru
Jl. Adyaksa km 3, PO Box 108
Waikabubak, Sumba Barat, Nusa Tenggara Timur 87212
Tel +62-387-387) 2286, Fax: +62-387-22163
Call Center: 0811384408
Instagram: @ btn_matalawa
Email: manupeu.tanahdaru@gmail.com

40. GUNUNG TAMBORA NATIONAL PARK

Geographic Location and Size

Mount Tambora National Park is on a peninsula of Sumbawa Island, part of the Lesser Sunda Islands. It is a segment of the Sunda Archipelago, a string of volcanic islands that forms the southern chain of the Indonesian archipelago. To the north of the peninsula is the Flores Sea and to the south is Saleh Bay, 86 km long and 36 km wide. The park is located between 117° 47' -118°17' East Longitude and 8° 7' 0 0" South Latitude.

The park is dominated by the 2,851-meter-high Mount Tambora, a volcano, which is located in Doro Ncanga, Dompu Regency of Sumbawa Island. It covers a total area of 71,645.64 hectares, comprising three previous protected areas, a 23,840 hectare nature sanctuary, a 21,674 hectare wildlife reserve, and a 26,130-hectare hunting park.

Climate and Topography

According to a geological survey before the 1815 eruption, Tambora had the shape of a typical stratovolcano, with a high symmetrical volcanic cone soaring up to 4,300 m above sea level, and a single central vent. The diameter at the base was 60 km. The central vent emitted lava frequently, which cascaded down a steep slope.

Since the 1815 eruption, the lowermost portion contains deposits of interlayered sequences of lava and pyroclastic materials. Thick scoria beds were produced by the fragmentation of lava flows. Within the upper section, the lava is interbedded with scoria, tuffs, and pyroclastic flows and falls. At least 20 subsidiary or parasitic cones are known. Some of them have names: Tahe, 844 m; Molo, 602 m; and Kubah, 1,648 m. Most of these parasitic cones have produced basaltic lavas.

History

This is a relatively new park in Indonesia, declared in 2015. The decree was based on the Ministry of Environment and Forestry No: SK.111/MenLHK-II/2015 on the 7th of April 2015. This date coincided with the 200th anniversary of the eruption of Mount Tambora. President Joko Widodo inaugurated Mount Tambora National Park during a special tourism promotion event called "Tambora Greets the World", on April 11, 2015. The President emphasized the importance of Mount Tambora National Park, stating that it was an asset that must be preserved and developed to benefit the local people.

The local people are expected to conserve the region well and turn it into an ecotourism destination. The Tambora Festival is held annually, with funding from the central government, so that all know where Dompu, Bima, and West Nusa Tenggara are, as well as where Indonesia is.

Tambora mountain is one of the 129 active volcanos in Indonesia, some of which are classified as super volcanos. The Toba super volcano, in North Sumatra, which erupted in 74,000 BC, is now one of the largest craters in the world. Indonesia also has the most colossal eruptions that have been recorded, Tambora and Krakatoa. Tambora was a ten times greater eruption than the very famous Krakatoa.

According to many experts, the Tambora eruption was the most super colossal in the last 500 years. It ejected more than 100 km^2 of materials with a column of more than 20 km up into the atmosphere. The diameter of the caldera now is approximately 7 km with a depth of about 1.2 km. The Tambora eruption led directly to the death of 71,000 people. As a result of changes in the weather worldwide, it is predicted that indirectly more than 10 thousand others perished.

Biodiversity and Ecosystems

A scientific team led by a Swiss botanist, Heinrich Zollinger, arrived on Sumbawa in 1847. Zollinger's mission was to study the eruption scene and its effects on the local ecosystems. He was the first person to climb to the summit after the eruption. It was still covered by smoke. As Zollinger climbed, his feet sank several times through a thin surface crust into a warm layer of powder-like sulphur. Some vegetation had reestablished itself and he saw a few trees on the lower slopes. A *casuarina* forest was noted at 2,200-2,550 m. Several *imperata cylindrica* grasslands were also found.

Resettlement of the mountain began in 1907. A coffee plantation was started in the 1930s on the northwestern slope of the mountain, in the village of Pekat. A dense rainforest, dominated by the pioneering tree *Duabanga moluccana*, had grown at an altitude of 1,000-2,800 m covering an area of 80,000 ha. The rainforest was explored by a Dutch team, led by Koster and de Voogd, in 1933. From their accounts, they started their journey in "fairly barren, dry and hot country", and then at 1,100 m they entered "a mighty jungle" with "huge majestic forest giants". Above 1,800 m, they found shrubs of broadleaved hopbush, *Dodonaea viscosa,* with an overstory of *Casuarina sp.* trees. On the summit, they found sparse daisybush, *Anaphalis viscida* and bluebells, *Wahlenbergia sp.*

There are at least 90 bird species in the park. They include crested white-eye, *Heleia dohertyi,* lesser yellow-crested cockatoo, *Cacatua sulphurea,* hill mynah, *Gracula religiosa,* green junglefowl, *Gallus varius,* and rainbow lorikeet, *Trichoglossus moluccanus,* which is hunted for the cagebird trade by the local people. Orange-footed scrubfowl, *Megapodius reinwardt,* is hunted for food. This exploitation has resulted in a decline in the bird population. The yellow-crested cockatoo is nearing extirpation on Sumbawa Island.

Local Communities and Culture

Sumbawa has historically had two major linguistic groups who spoke languages that were unintelligible to each other. One group on the western side of the island speaks Basa Semawa (Indonesian: *Bahasa Sumbawa*) which is similar to the Sasak language from Lombok. The second group in the east speaks Nggahi Mbojo (*Bahasa Bima*). They were once separated by the Tambora people, who spoke a language related to neither. After the eruption of Mt. Tambora and the demise of the Tambora people, the kingdoms located in Sumbawa Besar and Bima were the two cultural focal points of Sumbawa. This division remains today; Sumbawa Besar and Bima are the two largest towns on the island

Tourism

Three ascent routes can be used to reach the Tambora caldera. The first route starts from Doro Mboha village south of the mountain and follows a paved road through a cashew plantation until it reaches 1,150 m asl. The end of this route is the southern part of the caldera at 1,950 m, reachable by a hiking track. This location is usually used as a base camp to monitor the activity of the volcano, because it takes only one hour to reach the caldera. From here. The second route is southwest of the mountain, starting from Doro Peti village; the Tambora volcanic monitoring station is in Doro Peti. The third route starts from Pancasila village northwest of the mountain and passes through a coffee plantation. Using the third route, the caldera is accessible only by foot. The highest point of Tambora is on a hill near the western rim of the caldera.

Two zones have been declared: the dangerous zone and the cautious zone. The dangerous zone is an area that would be directly affected by an eruption: pyroclastic flow, lava flow and other pyroclastic falls. This area, including the caldera and its surroundings, covers 58.7 km². Habitation of the dangerous zone is prohibited. The cautious zone includes areas that might be indirectly affected by an eruption. The size of the cautious zone is 185 km², and includes Pasanggrahan, Doro Peti, Rao, Labuan Kenanga, Gubu Ponda, Kawindana Toi and Hoddo villages. Tourism Sites in and around Dompu, close to Tambora National Park include.

- Mada Prama, located about 4 km from Dompu. There is natural spring water for swimming and bathing set in lush tropical forest surroundings.
- Dermaga Kempo Beach is in Saleh Bay. It is a port that connects Dompu to Nisa Pudu and Nisa Rate Islands.
- Nisa Pudu and Nisa Rate Islands have beautiful panoramas. People can enjoy the sun from either island. They are located 45 km from Dompu and they can be reached by public transportation.
- Hodo Beach is about 45 km from Dompu. Located on Saleh Bay, the beach is an excellent stopover before ascending Mount Tambora. Fresh water is available.
- Doro Bata is about 1 km from Dompu. Here there are the remains of Dompu Palace, which was covered in volcanic ash after the eruption of Mount Tambora in 1815.
- Woja Beach is in the western part of Cempi Bay. The beach has white sand and a beautiful panorama.
- Lapadi, located about 5 km South of Dompu, is an area for traditional horse races, in which the jockeys are children 8 years of age. There is also a livestock "Koteka" (cattle breeding and sheep herding).
- Hu'u Beach is about 40 km from Dompu and about 100 km from Bima airport. It is the largest stretch of beach in West Nusa Tenggara Province. It has the reputation of being one of the best surfing beaches in Indonesia.
- Nangga Doro is about 45 km from Dompu. It has a mountain resort with very hot water springs. The temperature ranges between 80.5 to 81 degrees Celsius

Access and Transportation

Flights are available in small planes daily from Bali to Dompu or Bima. Most people who climb Mount Tambora leave from the village of Pancasila. It is problematic to do this in one day from Sumbawa Besar. To do this, take the Sumbawa Besar/Dompu bus very early in the morning and tell the driver you want to go to Calabai/Pancasila. From Calabai take an ojek (back of a motorcycle) to Pancasila. Pancasila is a village of approximately 1,000 people. There are a couple of shops stocked with only basic food. If you plan to organize the trek yourself, bring food from better-stocked shops in Dompu or Calabai.

It may be wiser and less taxing to go to Dompu, have a good night's rest and make the journey to Calabai/Pancasila the next day. In that way you will arrive in Pancasila reasonably rested. Buy supplies for the climb in Calabai. There are plenty of shops selling noodles, biscuits, fruit, bottled water etc. If you are riding a motorbike be warned that the road is badly damaged and pay it the respect it's due. Allow six hours for the 130 km Dompu-Pancasila trip.

The only other recommended route to the crater starts at the village of Doro Peti close to Dompu. The route goes back on the main road until the savannah starts and then climbs up a poorly maintained jeep track. It is possible to hire transport in Doro Peti. With a 4WD, and in good conditions, it is possible to get as far up the track as 1,830 m (Post 3). It is then only 3 hours on foot to the crater rim. There is no water, food, fuel, villages or accommodation on this route but if you have a 4WD then it might be worth a try. The return follows the same route.

It is possible to climb Tambora but be warned—it is not a stroll in the park. You have to be physically fit and ready for some discomfort and danger. Very few Indonesians, and far fewer foreigners, ever make it to this out of the way place. Since 2004 when records began to be kept by K-PATA (The Tambora Nature Lovers Group), only about 50 people per year have made the trip. For example, in the first six months of 2009 only three groups climbed the mountain -a group of nine Indonesians, one Frenchman and one Australian.

At Pancasila you will be directed to K-PATA headquarters. A fair price for food and lodging is available here. If you are not Indonesian, be warned, they may well try to add several other imaginary charges and ask for double or more of the legitimate fees. The coffee plantation up the road also offers guides and accommodation.

The walk is through virgin jungle. It is incredibly thick in parts and a lot of knife work is required. You will walk for hours along an almost indecipherable trail. At the end of the 2012 climbing season the path was clear all the way up to the summit and work was being done to build shelters at each of the 6 'pos' or stations on the route. For experienced, well-equipped hikers unguided trips would be possible (though great care is required around the crumbling caldera and be aware that help is a long way off).

When it rains, it really rains. Expect to be soaked to the skin if you don't have very good gear. But then again, it's not cold and what gets wet, dries eventually. Leeches are ever present, as are stinging nettles. There is a lot of fallen timber to climb over, under and along. The tree line ends quite abruptly and from there the walking becomes easier. Dawn at the summit is a very special experience. The experience of gazing down into the crater and taking in its immensity will stay with you forever

Facilities

At the mouth of Saleh Bay is a 30,000-hectare islet called Moyo (Indonesian: *Pulau Moyo*) which has a luxurious resort. At Dompu City close to the park, there are many hotels and the park is only an hour drive with a rental car.

Park Office

Balai Taman Nasional Tambora
Gg. Jadi, Dorotangga, Dompu, Nusa Tenggara Barat 84212
Call Center: 081237933233
Instagram: @ btn_tambora

41. MOYO SATONDA NATIONAL PARK

Geographic Location and Size

Moyo Satonda National Park is located between $8^0\ 6"$ - $8^0\ 23"$ S and $117^0.13"$ - $117^0.46"$ E. It consists of most of Moyo and Satonda Islands. Both islands are in Saleh Bay, north of Sumbawa Island. The park consists of both marine and terrestrial areas. The marine part covers 6000 ha and terrestrial part, on Moyo and Satonda islands, covers 25,200.15 ha.

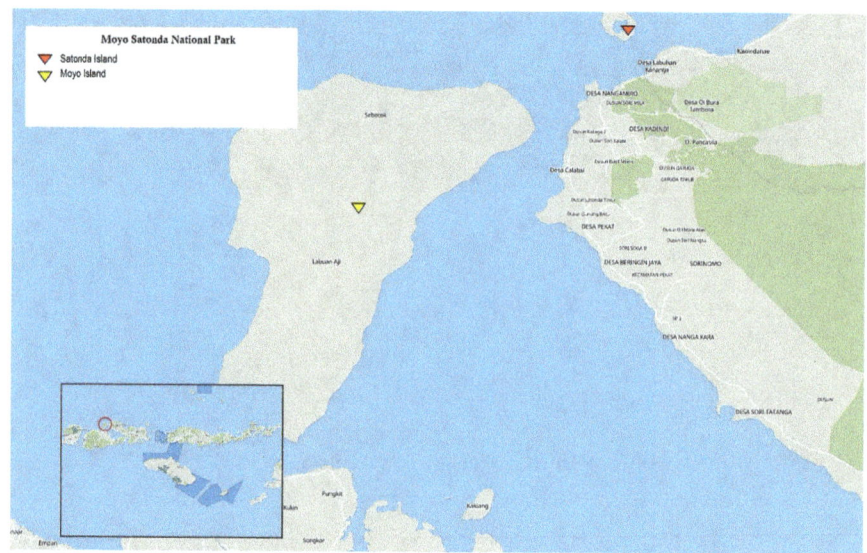

Climate and Topography

Moyo is an island located 2.5 km north of Sumbawa Island. The island is 35,000 ha of undulating terrain with a maximum height of 671 m asl, and a coastline of 88 km. Administratively, this island belongs to the Sumbawa Regency, West Nusa Tenggara Province. Satonda is a small island with a mountain and a lake. Satonda island is 480ha of which 80ha is the lake. The bottom of the lake is rocky. The water tastes salty because the lake follows the tidal pattern of sea water around the island. The phenomenon of the Satonda tidal lake is hailed as a wonder of the world. Sea water flows through underground. If the sea water recedes, the lake recedes.

History

This park is a newly established national park based on Ministry of Environment and Forestry decree of SK 901/MENLHK/SETJEN/PLA.2/8/2022. The park is made up of three previous protected areas, Pulau Moyo Game Reserve (21,200,15 ha), Pulau Moyo Marine Tourism Park (6,000 ha) and Satonda Tourism Park (2,600 ha). This decree was based on a proposal from the Directorat General of KSDAE (Conservation of Natural Resources and Ecosystem) of the Ministry of Environment and Forestry in 2017. The proposal for converting those 3 different protected areas into one was based upon many considerations, including intact primary forest, unique dry forest and the biodiversity within those forests, caves and the salt lake. Following Government Regulation no. 108 of 2015 on management of protected areas, this park matched all criteria in that regulation.

Biodiversity and Ecosystems

The forest on Moyo Island around the Matajitu river and waterfall is evergreen, but in other locations that are not near rivers or springs, the forest is deciduous in the dry season. The species of plants that can be found on Moyo Island include Ketimis, Protium javanicum, Kesambi, Schleichera oleoca, Monggo, Syzigium sp. and Awar-awar, Ficus spp. Wild mammals are limited to long-tailed macaque, Macaca fascicularis, Timor rusa, Cervus timorensis, wild boar, Sus scrofa, and many small mammals including flying foxes and 21 species of other bats, as well as monitor lizards. Birds are diverse and lesser yellow-crested cockatoo, Cacatua sulphurea, is very common here.

There are many diverse coral reefs. Angel Reef is on the west coast of Moyo Island, located south of Panjang Reef and boasts plenty of healthy hard corals in the shallower parts and a vertical wall over 40 meters deep. Once divers approach this impressive wall, they are likely to encounter enormous schools of long-fin bannerfish, Heniochus acuminatus, round batfish, Platax orbicularis, and red tooth triggerfish Odonus niger. The colourful wall is generously covered with soft corals and some hard sheet corals. However, there are still some remnants of dynamite fishing there, with barren blast zones scattered on the wall. Yellowfin tuna, Thunnus albacares, often scour the deep channels of this reef looking for prey, as do blacktip reef sharks, Carcharhinus melanopterus.

Local Communities

Less than 2000 local people live in several villages adjacent to the park. The largest is in Labuhan Aji village. Home stays are available within this village. Most people here are Sumbawan. It is only a few kilometers across the sea from the much larger Sumbawa Island.

Tourism

Moyo island has long been known for the beauty of its long white beaches, forests, caves, waterfalls and reefs. Some interesting places for snorkeling and diving activities include Takat Segele, Labu Aji Reef, Labu Aji Wall, Angle Reef, Berang Sedo, Pantai Satu, Ai Manis, and Tanjung Pasir. All are relatively intact and in good condition. These locations offer the beauty of marine flora and fauna that are amazing. Moyo Island is definitely one of the more beautiful places in Indonesia. It is exceptionally remote and off-grid, even compared to the rest of Nusa Tenggara and Sumbawa's neighbouring islands of Flores and Lombok.

From Labuan Aji Village, Matajitu is about 7 km and can be reached by motorbike taxi (ojek) in about 15 minutes, the only transportation in the village. The road is still a dirt road that has been shored up with broken concrete. During the dry season the road becomes dusty. At the waterfall there is a small cave with stalactites on the roof of the mouth of the cave. Although only about 7 meters high, the waterfall looks beautiful because of the clear bluish green water cascading down four steps and seven ponds. Like a terrace, the flowing water falls step by step, creating a beautiful panorama. These steps or terraces are sedimentary limestone formations. These rocks were formed thousands of years ago. The name Matajitu is from the local community and means that this waterfall comes from a spring that falls directly into the pool below.

Moyo Island is rich in bird diversity. On the path to the waterfall visitors can observe many birds in the morning or evening when they are active. One species is the lesser yellow-crested cockatoo, Cacatua sulphurea, a rare Indonesian bird.

The Amanwana Resort on Moyo Island provides exclusive accommodation that has been used by celebrities including Princess Diana, Mick Jagger, David Bowie, Maria Sharapova and David Beckam.

The best months to visit are June to August during the dry season. The seas start to calm down from April.

Access and Transportation

Flights are available in small planes daily from Bali to Sumbawa. There are two options for a daytrip to Moyo from Sumbawa: fast and slow. There is a speedboat, which takes 30 minutes to reach the Amanwana Resort and 45 minutes to get to Labuan Aji for approximately USD200. Slow boats range from around USD50 to USD100 (depending on the size of boat, your negotiating ability and other random factors) and take 1.5 hours to reach Amanwana and two hours to reach Labuan Aji. This means leaving Sumbawa Besar early to get a reasonable amount of time on the island as it is two hours each way by boat, plus an hour hike each way to the waterfall - so six hours of travel.

From the west coast of Lombok, it is approximately 2 hours to Kayangan harbour on the east coast of Lombok. From here there are fishing boats or regular fast boats to Pototano Port on Sumbawa Island. All crossings can be accessed 24 hours a day. From Pototano Port, there are three docks to go to Moyo island, Labuan Badas, Muara Kali, and Ai Bari. It is recommended to go to Muara Kali because you can ride on a people's boat which is public transportation for the people of Moyo Island. The price offered is around IDR 40,000 for one person. Or if you want to make it even more adventurous, you can rent your own boat, either a fishing boat or a fast boat.

Facilities

There are many guest houses and other inns close to the park at Labuhan Aji Village. There are also seaside hotels on Sumbawa Island directly opposite Moyo Island. Amanwana Resort offers luxury accommodation on Moyo Island itself. There are about 12 glamping tents surrounded by beautiful shady forests and beautiful views of the Flores Sea where animals such as whales, hawksbill turtles, green turtles, morays, and cockatoos can be seen.

Park Office

Balai Besar Konservasi Sumber Daya Alam (BKSDA) Nusa Tenggara Barat
Jl. Majapahit no.54, Mataram, Nusa Tenggara Barat. Indonesia
https://www.instagram.com/bksdantb,
Call Center: 087882030720
Instagram: @ bksda_ntb
Email bksdantb@gmail.com

CHAPTER 5
SULAWESI

Sulawesi is an island of approximately 184,840 km², with a coastline about 6,000 km long. Unlike Indonesia's other large islands, Sumatra, Java, and Borneo, of which Kalimantan is the Indonesian part, Sulawesi has never been connected to a continent. This isolation has resulted in very high levels of endemism; 98% of its mammals are endemic (not including bats), and approximately 89 of the 247 bird species present in Sulawesi cannot be found anywhere else. The best book to read on the island, although now quite old, is The Ecology of Sulawesi (Whitten et al., 1987).

Sulawesi is divided into six provinces, they are:
- North Sulawesi Province. The capital is Manado
- Gorontalo Province. The capital is Gorontalo
- Central Sulawesi Province. The capital is Palu
- South Sulawesi Province. The capital is Makasar
- South East Sulawesi Province. The capital is Kendari
- West Sulawesi Province. The capital is Mamuju.

Sulawesi Island has 9 National Parks. Three of them are Marine National Parks:
- Bunakan Marine National Park, Manado Tua, of 89,065 ha in North Sulawesi Province
- Bogani Nani Wartabone national Park (Dumoga Bone), of 287,115 ha in Gorontalo Province
- Lore Lindu National Park, of 229,000 ha in Central Sulawesi Province
- Taka Bone Rate National Park, of 530,765 ha in South Sulawesi Province
- Rawa Aopa Watumohai National Park, of 105,194 ha in South East Sulawesi Province

- Wakatobi islands Marine National Park, of 1,390,000 ha in South East Sulawesi Province
- Bantimurung Bulusaraung National Park, of 43,750 ha in South Sulawesi Province
- Togean Islands National Park, of 365,241 ha, Central Sulawesi Province
- Gandang Dewata National Park, of 214,201 ha, West Sulawesi Province.

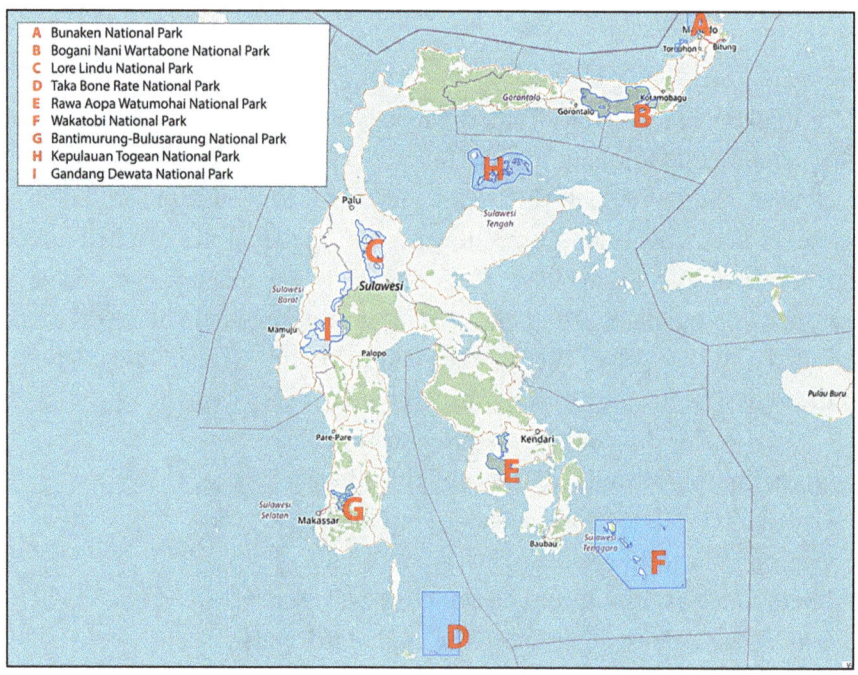

Views of Bantimurung Bulusaraung National Park (Photos by Lili Adidjaja) & Beach view at Wakatobi National Park (Photo by Adhi Nurul Hadi)

Lindu Lake at Lore Lindu National Park (Photo by Nurul Winarni) & Underwater view at Wakatobi National Park (Photo by Adhi Nurul Hadi)

View landscape of Gandang Dewata National Park (Photo by Mahendra) & River at Bogani Nani Wartabone National Park (Photo by KLHK)

Underwater view at Wakatobi National Park (Photo by KLHK) & Beach at Togean National Park (Photo by Adhi Nurul Hadi)

Macaca Nigra in North Sulawesi & *Macaca Maura* in Bantimurung Bulusasaung National Park (Photo by Jatna Supriatna)

Tarsius pelengensis at Sulawesi (Photo by Mochamad Indrawan) & *Tarsius supriatnai* at Gorontalo (Photo by Lynn Clayton)

Macrocephalon maleo (Photo by Sarlito Mamonto) & *Bubalus depressicornis* in Sulawesi (Photo by Ardin Mokodompit)

Babyrousa celebensis in *Sulawesi* (Photo by Jatna Supriatna)

42. BUNAKEN NATIONAL PARK

Geographic Location and Size

Bunaken National Park at Manado Tua is located between 1°37'-1°47' North Latitude and 124°04'-124°48' East Longitude in north-east Sulawesi, and is administered by Wori Regency, Manado Municipality, and Minahasa Regency, North Sulawesi Province. The park is 18 km. from Manado, the capital of North Sulawesi.

The park's area is 89,065 ha, 97% of which consist of water, with the remainder forming the islands of Bunaken, Manado Tua, Mantehage, Nain, and Siladen. There is a core zone for nature preservation and the protection of habitats and ecological processes, and a utilization zone/ Production zone for nature tourism, consisting of an intensive utilization zone and a limited utilization zone.

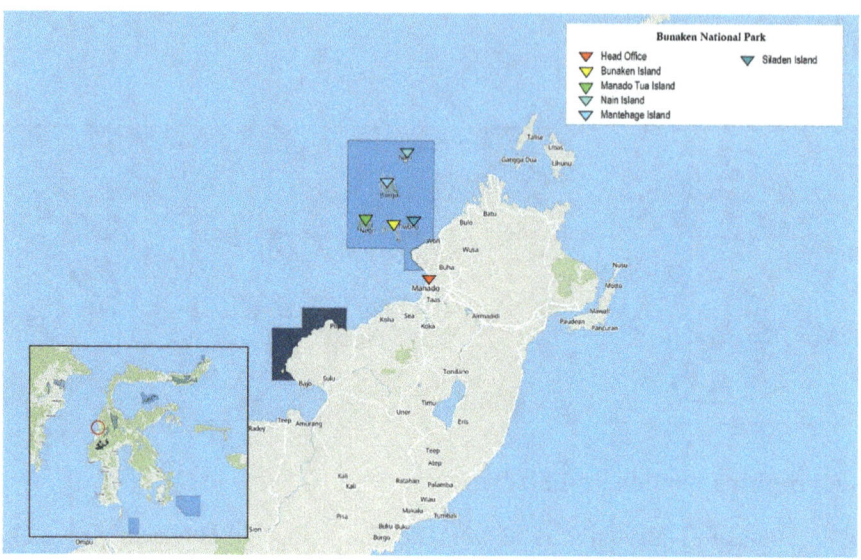

Climate and Topography

From November to March there are strong northern sea winds that bring rain, but it is still considered safe to visit. From April to October there are also strong winds, but the sea is relatively calm and there is no rain. Total rainfall is between 2,000 and 3,000 mm/year and entire days of rain are between 90 and 130. The humidity level is 50% -68%, with a temperature range 26^0-31^0 C.

Most of the park consists of sea with a variable underwater topography from flat to sloping and steep. Most of the corals are on steep reefs, except in the surroundings of Nain Island which has barrier reefs. The current changes constantly and can grow to be very strong, with relatively dangerous whirlpools. Mt. Manado Tua (400m asl), is an extinct volcano with tropical rainforest on its peak.

The waters are generally very deep. Manado Bay, for example, is 1,566m deep with very clear water making it possible to see 35-40m down.

History

In 1980, based on Provincial Governor Decree No. 224, Bunaken Island and its surrounding waters was declared Laut Manado Tourism Development Region.

In 1984, based on Governor Decree No. 201, Arakan-Wawontulap was added to Manado Marine Tourism Region.

In 1986, the Minister of Forestry issued decree No. 328/Kpts-II/1986, declaring the Bunaken Manado Tua Marine Reserve, which includes Bunaken Island, Manado Tua Island, and its surrounding waters and the coastal area in the surroundings of Tanjung Pisok north of Teluk Manado.

On the 1st of April 1989, the Minister of Forestry issued decree No. 444/Menhut-II/1989, to establish the Bunaken Manado Tua Marine National Park, which included Bunaken Island, Manado Tua Island and its surrounding waters and the coastal area in the surroundings of Tanjung Pisok, and the coastal area of Arakan – Wawontulap.

Biodiversity and Ecosystems

Bunaken Marine National Park consists of deep sea, shallow water, coral and mangrove forest ecosystems. The mangroves are dominated by *Rhizophora* spp. and *Sonneratia* spp. and cover approximately 1,800 ha. The seaweeds, *Enhalus sp.* and *Thalassia hemprichii* grow on coral reef flats in shallow waters. Seaweed is the habitat for small fishes and shrimps and is the feeding grounds for herbivorous and carnivorous marine animals such as the dugong, *Dugong dugon*, sea turtles and various species of fish. Terrestrial habitats include the tropical rain forest ecosystem at the peak of Mt. Manado Tua, farm plots and fields, and settlements.

There are approximately 70 coral species, dominated by *Caulerpa racemosa*, *Halodule univervis*, *Pocillopora* sp., *Seriattopora* sp., *Pachyseris* sp., *Porites* sp., *Fungia* sp., *Herpolitha* sp., *Halomitra* sp., *Galaxea* sp., *Pectinia* sp., *Lobophyllia* sp., *Echinopora* sp., *Leptoria* sp., *Tubastrea* sp., *Acropora* sp., *Turbinaria* sp., *Millepora* sp., *Echinopora* sp., *Montipora* sp. and *Thalassodendron ciliatum*.

The coral covers approximately 8,000 ha and consists of fringing, barrier, and patch reefs in relatively good condition. All islands are surrounded by reef flats, except Manado Tua Island. The open seas are inhabited by phytoplankton and zooplankton. They contain aquatic species such as dugong, *Dugong dugon*, giant clam, *Tridacna gigas*, sand clam, *Hippopus hippopus*, green turtle, *Chelonia mydas*, hawksbill turtle, *Eretmochelys imbricata*, leatherback turtle, *Dermochelys coriacea*, Cowries, *Cypraea* spp., sea snails, *Conus* spp. *Trochus* spp., and *Turbo* spp., trigger fish, *Balistoides niger* and other decorative fish including sea horses, *Hippocampus* spp. and schooling bannerfish, *Heniochus diphreutes*. Various coral fish species include parrot fish, *Bolbometopon muricetum*, *Anthias* spp., trumpetfish, *Aulostomus chinensis*, Moorish idol, *Zanclus cornutus* and the rare Indonesian coelacanth, *Latimeria menadoensis*.

The islands are rich in plants such as palms, sago, woka, silar, and coconut. Mango, banana, and other fruit trees are found all over the islands providing food for various birds and bats.

Sulawesi endemic wildlife species found in Bunaken National Park include the Sulawesi black macaque, *Macaca nigra*, Sulawesi bear cuscus, *Phalanger ursinus*, Sulawesi dwarf cuscus, *Strigocuscus celebensis*, tarsier, *Tarsius spectrum*, Sulawesi boar, *Sus celebensis*, Elang-alap, or vinous breasted sparrow hawk, *Accipiter rhodogaster*, white cuckoo shrike, *Coracina leucopygia* and Elang, or Sulawesi serpent eagle, *Spilornis rufipectus*.

Local Communities and Culture

Most of the local population originates from Sangihe. But there are also tribes from Gorontalo, Minahasa, Bajo and Bugis. Interactions between the tribes are familiar with low rates of friction. The primary language used for communicating here is the local lingua franca. The people accept the principle of open access to resources to benefit fishermen from outside the region. Several traditional customs to deal with disasters such as typhoons and volcanic eruptions exist, such as leaving in boats well ahead of such events. The traditional wisdom involved here is being able to predict such events well ahead of time.

Livelihoods are based on the seasons. In general, during the dry season people go out to sea (January – July), and during the wet season (September – December) they farm the land. Most of the people are both farmers and fishers and many of them also cultivate seaweed, *Euchema* sp., especially on Nain Island.

The pattern of land utilization varies depending on livelihoods and social aspects. Land ownership is based on traditional customs and no one holds formal legal land ownership documents. Farming is extensive on Manado Tua Island and Bunaken Island in the form of coconut trees with vegetable crops such as cassava and other root crops underneath. Siladen Island is dominated by coconut plantations, while at Mantehaga Island it is a combination of coconut plantations and farming. Seaweed cultivation plots can be found on Nain Island. The fishery varies from deep sea fishing to fishing activities in the coral reefs using fishing traps, harpoons, simple traditional nets and fishing rods.

The most common religion is protestant Christianity. The culture is dominated by customs of the Sangir Talaud where the Village Head is called Opo Lao. The traditions and customs of the Sangir Talaud are maintained by, amongst other practices, the *Masamper* ceremony, which occurs when blessing a new house. It is done by dancing and stamping feet during a long traditional song. The *Ampa Wayer* is a dance mimicking a sailing boat and usually done while being drinking an alcoholic beverage. The traditional customs of the Minahasa tribe include the *Miramba*, which is a dance to praise the Lord for his kindness and blessings. The tradition of most of the population is influenced by Manado city customs, which follow the Minahasa tradition. This also includes the *Maengket and Cakalele* dance, which is often performed at formal state events to welcome guests and the inauguration of customary Heads, *Tonaas*.

Tourism

Bunaken island, with its long sandy beach and coral reefs filled with various decorative fish and dugongs, is a world-class tourist attraction. The northern part is the best place to sunbathe, dive, swim, and snorkel, especially to a depth of 25m, or sail with a glass-bottomed boat to enjoy the coral reef views. Here there is a shipwreck that is 50-60 years old, and a coral reef which has the form of a cave located 50m below the sea surface. The main tourist sites are listed below.

- Bunaken Island Lekuan Fakui, Mandolin, Tanjung Parigi, Ron's Point, Pangalisang, Muka Kampung, Sachiko Point and East Bunaken.
- Manado Tua Island: Buwalo, Pangalingan Negeri and Tanjung Kopi.
- Mantehage Island: Bango and Tangkasi
- Nain Island: Batu Kapal and Jalur Air
- Siladen Island: Siladen 1 and 2
- Sulawesi Mainland: Tanjung Pisok and Molas
- Siladen Island with its white sandy beaches and breathtaking sea garden.

- Manado Tua island: white sand and fantastic sea garden, hike/ climb Mt. Manado Tua.
- Arakan-Wowontulap – here one can dive and see the seaweed cultivation pads, which by the way is a favorite place of turtles and dolphins.
- The best time to visit Bunakan National Park is during the months of May to October, which is the dry season when the sea is relatively calm.

Access and Transportation

Bunaken Marine National Park and islands located within it can be reached as follows:
- Fly from Jakarta or many other major cities to Manado, then continue by sea transportation
- From Kuala Jengki Harbor (Manado) by motorboat to Bunaken Island takes 30 minutes, to Manado Tua Island takes 45 minutes, to Mantehage Island takes 50 minutes, to Nain Island takes 1hr, and to Siladen Island takes 20 minutes.
- From Manado to Malalayang village by speedboat takes 1 hour.
- From Manado to Arakan and Wowontulap takes 40 min.
- From Manado by land, transportation on an asphalt road to Molas (North Bunakan mainland) takes approximately 30 minutes and to Teling (South Bunakan Mainland) takes approximately 1.5 hours.
- From Molas or Wori beach to the park by speedboat takes 15 to 20 minutes.

Facilities

- In Manado City there are many quality hotels and other kinds of accommodation such as lodgings and inns. There are also diving businesses, which rent out diving equipment, for example the Nusantara Diving Center, Barracuda Dive Resort, Murex Diving Center and Manado Diving Center. There are also professional guides.
- At Teluk Liang, Pangasilang on Bunaken Island and Siladen Island there are many simple but clean guesthouses.
- Other supporting facilities include a helipad, hiking paths, watch posts, observation towers, piers, diving equipment and speedboats.

Park Office

Balai Taman Nasional Bunaken
Jl. Raya Molas Kode Pos 1202 Batusaiki - Manado 95242
Tel and Fax: . (+62-431) 859022
Call Center: 082195399339
Instagram: @ btn_bunaken
e-mail   : info@bunaken.org , tn_bunaken@yahoo.com

43. BOGANI NANI WARTABONE NATIONAL PARK

Geographic Location and Size

Bogani Nani Wartabone National Park is located between $0°25'$-$0°44'$ North Latitude and $16°40'$-$19°29'$ East Longitude. It is in the center of Sulawesi's northern peninsula, between Lembah Dumoga and Gorontalo, North Sulawesi and Gorontalo Provinces. It is flanked by two large rivers the Ongkag Dumoga River and the Bone River. Administratively it is in the jurisdiction of Bolaang Mongondow Regency and Gorontalo Regency.

Bogani Nani Wartabone National Park covers 287,115 ha, and approximately 90% of it consists of primary forest.

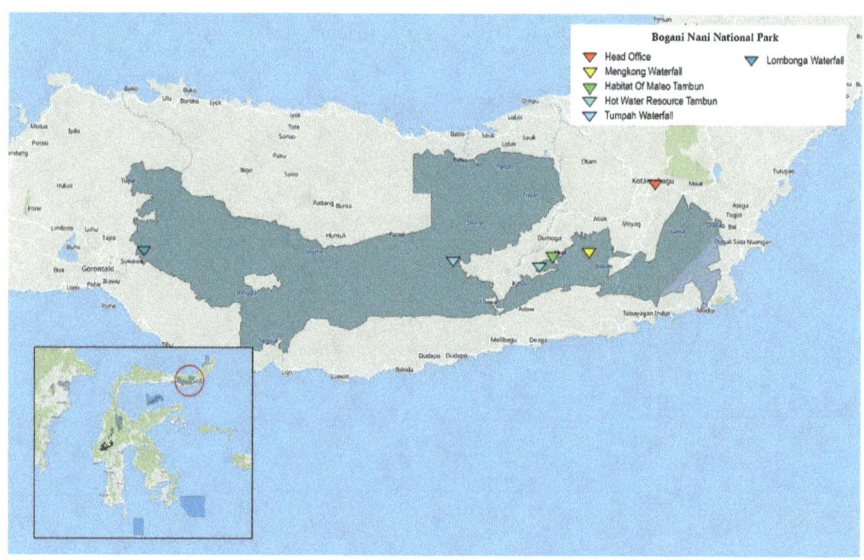

Climate and Topography

The climate in this region is monsoonal with high temperatures between 21.5^0 and 31^0C with high levels of humidity. The annual rainfall varies between 2,000 to 2,700 mm per year, and the average annual rainfall is 2,200 mm/ year. The maximum dry season is 5 months and wet season 4 months.

Most of the park is mountainous with narrow valleys, between 50 -1,970 m asl and slopes of between 15% -75%. There are two rivers, the Dumoga River and the Bone River. The Bulawan highlands include Mt. Sinombayuga (Mt. Bulawa), which is the highest peak at 1,970 m asl. The Perantaraan highlands include Mt. Pau-Pau (1,815 m) and Mt. Gambuta (1,540 m), and the Bone Highlands include Mt. Pinonimposa (1,740 m), Mt. Renga (1,460 m) and Mt. Sulo (1,750 m).

History

In August 1979, based on the Minister of Agriculture Decree No. 476/Kpts/Um/8/1979, the Dumoga Wildlife Reserve of 93,500 ha was created.

In December 1979, based on the Minister of Agriculture Decree No. 764/Kpts/Um/12/1979, the Bone Wildlife Reserve of 110,000 ha was established.

In June 1980, based on the Minister of Agriculture Decree No. 438/Kpts/Um/6/1980, the Bulawa Nature Preserve of 75,200 ha was established.

On the 14th of October 1982, during the III National Park World Congress in Bali, and based on the Minister of Agriculture Decree No. 736/Mentan/X/1982, the three conservation areas mentioned above were merged to become the Dumoga Bone National Park of 278,700 ha.

On the 15th of October 1992 the Minister of Forestry released decree No. 731/Kpts-II/1992, to reinforce and enlarge Dumoga Bone National Park to 287,115ha.

On the 18th of November 1992, based on the Minister of Forestry's Decree No. 1068/Kpts-II/1992, to honour Mr. Bogani Nani Wartabone, a local heroe who resisted Dutch colonisation, the name of the national park was changed from Dumoga Bone National Park to Bogani Nani Wartabone National Park.

Biodiversity and Ecosystems

The ecosystems in this area are unique because they represent the geographical transition from the Indo-Malayan region in the West (Java,

Sumatra and Kalimantan) and the Papua-Australia region in the east. Due to this there is a high level of endemism. The biodiversity rate is also very high, because almost all Sulawesi plant species and wildlife can be found here. The ecosystems present in the park are

Lowland Rainforest located between 300-1,000 m asl. The forest has high biodiversity with very tall trees (30-35 m) typically with small canopies. The species include bintangor, *Palaquium obovatum*, bunut, *Calophyllum soulattri*, kayu cempaka, a magnolia endemic to Indonesia, *Elmerillia ovalis*, *Pterospermum celebensis*, *Diospyros* sp., kauri, *Agathis* sp., various species of fig, *Ficus* spp, kayu Inggris, or rainbow gum, *Eucalyptus deglupta*, matoa, *Pometia pinnata* and benuang, *Octomeles sumatrana*. The plants growing beneath the canopy are mostly rattan palms, especially *Calamus* sp, *Daemonorops* sp and *Korthalsia* sp.

River swamp forest in river valleys located at a low altitude. This can be dominated by large-trunked plant species such as the rainbow gum, *Eucalyptus deglupta*, whose height can reach 70 meters.

Mountain forest located above 1,000 m asl. This forest contains many species of palm trees such as pinang, *Arenga* sp., wanga, *Pigafetta filiaris*, which is a typical Sulawesi species, rattan, *Calamus* sp. and various species of pandan, *Pandanus* spp., as well as orchids. Other species of plants include kayu Bugis, *Koordersiodendron pinnatum*, kayu hitam, *Diospyros pilosanthera*, kayu Inggris, the rainbow gum, *Eucalyptus deglupta*, kayu cempaka, a magnolia, *Elmerillia ovalis*, nibong/palem kipas, a fan palm, *Livistona rotundifolia* and kayu linggua, *Pterocarpus indicus*.

Secondary Vegetation can be found in abandoned fields in the form of shrubs, dominated by the introduced pepper, *Piper aduncum*, *Melastona malabathricum*, bananas, *Musa* sp. and the introduced weed, *Lantana camara*. Areas which are open and dry are dominated mostly by the tall grass, *Imperata cylindrica*.

The moss forest at the peak of the mountains consists of various species of moss, low primary forest, shrub plains, swamp grass and areas of water pools and mud in the surroundings of Kosinggolan.

The wildlife species present in the park are typical of Wallacea, which is the zoogeographical transition zone between south-east Asia and Australia. There are at least 31 mammal species, 51 bird species and 38 butterfly species along with many reptiles and amphibians, and a variety of fish. Typical mammals include flying squirl, *Acerodon celebensis*, Sulawesi bear cuscus, *Ailurops ursinus*, babirusa, *Babyrousa babyrussa*, Sulawesi pig, *Sus celebensis*, tangkasi, *Tarsius dentatus* and dihe, Gorontalo macaque, *Macaca nigrescens* and lowland anoa, *Bubalus depresicornis*.

The bird species include elang-alap Sulawesi, or Sulawesi goshawk, *Accipiter griseiceps* Sulawesi hornbill, *Aceros cassidix*, Oriental warbler, *Acrocephalus orientalis*, pond heron, *Ardeola speciosa*, white woodswallow, *Artamus leucorynchus*, bubut Sulawesi, a coucal, *Centropus celebensis*, Sulawesi kingfisher, *Ceyx fallax*, beilibis jambul, or wandering whistling duck, *Dendrocygna arcuata*, white imperial pigeon, *Ducula luctuosa* and maleo, an endemic megapode, *Macrocephalon maleo*.

Local Communities and Culture

The population on the coastline north of the park has increased rapidly, including in the Dumoga and Bone valley basin, and the eastern part of Gorontalo. Approximately 47,700 people were recorded living in the Dumoga valley in 1980 but by now it has almost doubled. The Regency Capital, Kotamobagu, had a population of 35,000 and Gorontalo approx. 222,000 people but now Kotamibagu is almost 125,000 and with Gorontalo becoming a province, more people have come to the area. Almost 1.1 million lived Gorontalo in 2019.

Tourism

Nature tourism opportunities abound in Bogani Nani Wartabone National Park. There are magnificent views over the landscape, clear crisp fresh air, cross country walking tracks, and camping. The Kosinggolan Dam located on the border of the park is great for fishing. Rare and charismatic wildlife include the anoa, babirusa, monkeys such as the Gorontalo macaque, the tarsier tangkasi and the maleo, an endemic megapode bird. There is also a hot water spring and ancient caves with calcite rock formations. Some locations with nature tourism potential include:

- Tumokang Kosinggolan Recreation Forest. Enjoy the beautiful natural scenery with clean cool air, hot springs and a camping site. Kosinggolan also has natural baths, in the building which is an original example of the local Bolaang Mongondow architecture. Kosinggolan dam is a perfect spot for fishing.
- Toraut Recreation Forest. Here visitors can see the anoa, or dwarf buffalo and the deer pig, babirusa. There are also waterfalls, primary forests, and natural baths.

- Tambun Recreation Forest, which is located on the main highway linking Kotamobagu and Doloduo. East of the village there is nesting habitat of brush turkeys and hot water spring baths. Maleo can be observed from a hide made of sago palm leaves. There are also observation towers and work cabins.
- Lombongo Tourism Forest is in the Bone area at Lombongo (about 17 km to the east of Gorontalo) where a hot water bath has been constructed. From this bath, a walk of about 1 hour will reveal a waterfall near the trail, which is 25 meters high. At Hungayono there is the nesting habitat of the Maleo, limestone caves and hot springs baths, as well as the path to the Pinogu residential area and a guardhouse.

Access and Transportation

From Jakarta and other major cities there are direct flights to Manado or connections from Makassar to Manado.

The park can be reached from Manado via 2 routes:
- By following the coastline Manado-Kapitu -Inobonto-Kotamobagu-Dumoga-Dolodua for approximately 200km, which will take 5 hours by car or motorbike.
- By following the inland route Manado-Kapitu – Motoling – Tompasu Baru – Mt. Ambong – Dolodua for approximately 225 km, which will take 6 hours by car or motorbike through winding roads.

From Makasar there are flights to Gorontalo. It is then a 20 minute drive to Lombongo forest in the western part of the park.

Facilities

- In Mandao, Kotamobagu, Doloduo and Toraut there are hotels and Inns.
- At Toraut there is the Toraut Research Station, the Wisma Cinta Alam guesthouse, walking trails, work cabins and a Guard post.
- At Kosinggolan there is an office, an Information Center, a Guest Inn, walking trails and a camping site.
- At Lombongo there is an office, walking trails, work cabins and Guard Post.

Park Office

Balai Taman Nasional Bogani Nani Wartabone
Jl. AKD Mongkonai Kotak Pos 106
Kotamobagu 95716 - Sulawesi Utara
Tel. (+62-434) 22547
Fax. (+62-434) 2254
Call Center: 081245941865
Instagram: @ btn_boganinaniwartabone
Email: btnbnw@yahoo.co.id

44. LORE LINDU NATIONAL PARK

Geographic Location and Size

Lore Lindu National Park is located between 1004'-1008' south latitude and 119058'-120016' east longitude. Administratively, it is in the Donggala and Poso Regencies, Central Sulawesi Province. Lore Lindu National Park covers 217,991.18 ha.

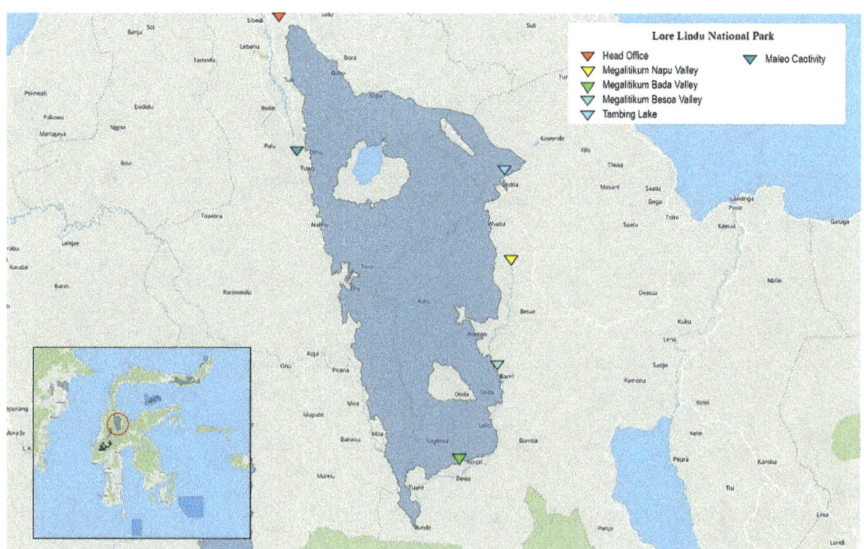

Climate and Topography

Because this park is in the center of the island of Sulawesi rainfall is relatively high and very variable, determined largely by altitude and topography. In the northern part of the park rainfall ranges between 2,000-3,000 mm/year, and in the southern part between 3,000-4,000 mm/year. Average temperatures range from 22°-34°C. The altitude ranges from 500-2,610 m asl, with a topographic terrain that is undulating, hilly and to the north mountainous. The whole park is covered by mountain rainforest vegetation. There are several mountains. The highest peaks are Mt. Rorekatimbu, 2,610 m and Mt. Nokilalaki, 2,355 m.

The mountains are surrounded by valleys that form gaps such as the Fossa Sarasin, Tawelia and Bada. In the past the bottom of the Besoa and Napu Valleys were part of Lindu Lake. The current Lindu Lake is a smaller version of an ancient lake.

History

On the 20th of October 1973, the minister of Agriculture decree No. 552/Kpts/Um/10/1973, announced the declaration of the Lore Kalamanta Wildlife Reserve, covering 131,000 ha.

In 1977 the area was proclaimed as a Biosphere Protected area by UNESCO

On the 25th of January 1978, the Minister of Agriculture decree No. 46/Kpts/Um/1/1978, established the Lindu Lake Tourism Forest, covering 31,000 ha.

On the 10th of November 1981 the Minister of Agriculture decree No. 1021/Kpts/Um/11/1981, established the Sopu River Wildlife Reserve, covering 67,000 ha.

On the 14th of October 1982, the Minister of Agriculture decree No. 736/Mentan/X/1982, merged the three areas into one, which was declared the Lore Lindu National Park, covering 231,000 ha.

On the 26th of February 1992, a decree was issued by the Minister of Forestry, No. 280/Kpts-II/1992, reinforcing the status of the area as Lore Lindu National Park.

On the 10th of May 1993, the Minister of Forestry Decree No. 593/Kpts-II/93, reduced the area Lore Lindu National Park to 229,000 ha.

Lore Lindu National Park has been named as a World Heritage Site by UNESCO.

Biodiversity and Ecosystems

The ecosystems in the park, all rich in biodiversity, can be differentiated by height above sea level

Lowland tropical forests cover the approximately 20% of the park that lies below 1,000m asl. Tree species include *Pterospermum celeb*ensis, ndolia, or ylang ylang, *Cananga odorata*, uru, *Mangelietia* sp., tahiti, a mahogany, *Dysoxylum* sp., durian, *Durio zibethinus*, tea uru, *Artocarpus* sp., benuang, *Duabanga moluccana*, melinjo, *Gnetum gnemon*, and various palm species such as the saguer, *Arenga pinnata*, pinang, *Pinanga* sp., Wanga palm, *Pigafetta filaris* and ratan, *Calamus* spp. There are also secondary forests, which are areas of former cultivation or forest clearings, which are now experiencing ecological succession. These areas are dominated by the tree species wulaya, *Trema orientalis*, kuo, *Alphitonia zizyphoides* and palili, *Lithocarpus* sp. Some more open prairie-like areas are dominated by Sumatran ru, *Casuarina sumatrana* and Wanga palm, *Pigafetta filaris*.

Sub-montane forests occur on the lower slopes between approximately 1,000-1,500 m asl. Plant species here include kaha, *Castanopsis argentea*, palili pance, *Lithocarpus* sp., berbagai jenis, *Eugenia* sp., *Turpinia* sp., *Sterculia* sp., *Vernonia* sp., *Engelhardtia* sp., *Canarium* sp., *Elaeocarpus* sp., *Adinandra* sp. and *Podocarpus* sp., There is also the bamboo, *Dinochloa scandens*. On the upper slopes, species include uru, *Elmeralia* sp. or *Magnolia sp.* on open plains. This is a water catchment area where there is also an abundance of rainbow gum, *Eucalyptus deglupta*, forming pure stands along the Sopu, Dabuki and Lamea Rivers. Areas of secondary vegetation are generally dominated by grasses such as the *Themeda* sp., *Setaria sp.* and *Rotboellia sp*. In the east at altitudes between 1,500-2,000 m asl the forest is dominated by a species of kauri, *Agathis philippinensis*, a celery pine, *Phyllodados hypophyllus*, and *Pandanus* sp.

Sub-alpine forest is located at altitudes above 2,000 m asl, and generally consists of dwarf trees that are overgrown with moss; this area is often covered with fog and mist. It is dominated by species of Leptospermum, Rapanea and Myrsine, which grow to a height of 7-10 m and generally have diameters of approximately 5-15 cm.

There are at least 127 mammals in the park, including Sulawesi bear cuscus, *Ailurops ursinus*, babirusa, *Babyrousa babyrussa*, bajing perut merah, or Sulawesi giant squirrel, *Rubrisciurus rubriventer*, and Sulawesi mountain tarsier, *Tarsius pumilus*. There are at least 328 bird species, including elang-alap Sulawesi, or Sulawesi goshark, *Accipiter griseiceps* kakatua jambul kuning, or yellow-crested cockatoo, *Cacatua sulphurea*, bubut Sulawesi, or bay coucal, *Centropus celebensis*, raja udang Sulawesi, *Ceyx fallax*, red jungle fowl, *Gallus gallus* and burung maleo, *Macrocephalon maleo*. There are at least 117 species of reptiles and amphibians.

Almost half of the wildlife species in the park are endemic to Sulawesi. They include a frog, *Bufo celebensis*, Sulawesi warty pig, *Sus celebensis*, highland anoa or dwarf buffalo, *Bubalus quarlesi*, Sulawesi owl, *Tyto rosenbergii*, Sulawesi starling, *Asilornis celebensis*, Sulawesi cuckoo, *Cuculus crassirostris*, maleo, *Macrocephalon maleo*, Sulawesi dwarf cuscus, *Strigocuscus celebensis*, and the Tonkean monkey, a macaque, *Macaca tonkeana*. *Pterospermum celebensis* is an endemic plant occurring in the park.

Local Communities and Culture

People in the area surrounding Lore Lindu National Park are generally from the Lindu community but there is a small number from the Katu community. They maintain interesting local customs, such as religious ceremonies, dances, clothing, and their traditional rice field farming method (Faruja) which involves plowing the rice fields by using their buffaloes to trample on them. Groups of megalithic stones can also be found here, which are some of the best examples of such monuments in Indonesia.

Since 1937 Lindu Lake has attracted the interest of many scientists, especially health specialists, because of the endemic parasitic worm *Schistosoma japonicum*, which attacks both humans and wildlife. The temporary host of the worm is a type of small snail (*Oncomelanis kupensisi linduensis*) that is commonly found on the banks of Lake Lindu.

Tourism

- The Besoa, Napu and Bada Valleys are home to some of the best megalithic monuments in Indonesia. The area is full of beautiful megalithic rocks, the origins of which are still unknown. The Besoa Valley, swamp and spike plants, and the surroundings of Lindu Lake are the habitat of the endemic maleo.
- Mount Bosu (west Lake Lindu) is the habitat of the anoa (dwarf buffalo).

- The Lakes Lindu, Gimpu, Wuasa and Bada provide panoramic lake views and opportunities for fishing or boating while observing wildlife. Lake Lindu is located at an altitude of 900-1000 m asl and has an area of about 3,000 ha. In the center of the lake there are small islands. Some are sacred to the local Lindu people because they are the tombs of their kings. Lake Lindu is the only location in Indonesia where the parasite causing *Schistosomiasis* is found. Due to this it is recommended not to swim in or drink the lake's water.
- Saluki, Besoa, Bada and Napu Villages have beautiful megalithic rocks such as the *tong* or Kalamba, stone sculptures, dolmen/ sacred or worship rocks, and rocks with a hole/*dakon* rock.
- Mt. Nokilalaki (2,355 m asl) and Mt. Rorekatimbu (2,610 m asl), present panoramic views over the forested landscape, and wildlife can be observed from the walking tracks.
- At Kamarora there is a hot water spring where there are bathing facilities.
- Walking tracks include:
- Bada Valley, Tentena – Bomba; duration 2 days.
- Rimpa Valley-Bada Valley, Masamba-Rimpa Valley-Badangkaia; duration 6 days.
- Bada Valley-Palu valley, Gintu – Tuare – Moa – Pili – Gimpu; duration 3 days.
- Napu Valley – Besoa Valley – Bada Valley, Wuasa – Doa- Gintu; duration 2 days.
- Bada Valley-Besoa Valley – Palu Valley, Gintu - Bomba - Doda - Hanggira – Gimpu; duration 4 days.
- Napu Valley – Lindu Lake, Wuasa - Danau Lindu – Sidonda; duration 3 days.
- Lindu Lake, Sidaunta - Tomado - Bomba – Simpang tiga; duration 3 days
- To climb Mt. Nokilalaki, Lindu Lake – Mt. Nokilalaki – Lindu Lake; duration 2 days.
- There is a Research site for the study of the unique biodiversity and the culture and customs of the people living in the surroundings of the park.

Access and Transportation

Lore Lindu National Park is located near Palu, the capital city of Central Sulawesi Province. There are flights from Jakarta or Makasar. The park has two main entrances, one near Palu in the northwest and the other at Tentena near Poso Lake, in the southeast. The national park can be reached from Palu as follows:

- Palu to Sidoanto on the main road (65 km), duration 2 hours. Continue on foot or horseback to Lake Lindu (20 km), duration 5 hrs., or by motor bike, 2 hrs.
- Palu to Kamarora (50 km) by public transportation, duration 2.5 hrs.
- Palu through Palolo to Wuasa (105) km by public transportation (3-5 hrs), continuing on foot or horseback to Besoa (50 km) through natural forests and grassy plains and meadows (3 days) or by 4-wheeled drive vehicle, duration 4 hrs.
- Palu to Sidoanto to Kulawi to Gimpu (19 km) by public transportation, duration 3-6 hrs., continuing to the Besoa and Bada valleys (10km) on foot or horseback, duration 3 hrs.
- Access can also be arranged using a chartered plane, which can land on the airstrips at Gimpu, Besoa and Bada, duration 3 hrs.

Facilities

Facilities present or provided in the surroundings of the Lore Lindu National Park include:

Quality hotels at Palu and inns at Tentena, Bada Valley, Besoa Valley, Palu Valley and Napu Valley. There is a ranger post at Kamarora and another at Toro.

Other facilities include Watch Posts, Information Center, Work Cabins, Camping Sites, Guesthouses, Watch Towers, Shelters and walking paths.

Park Office

Balai Besar Taman Nasional Lore Lindu
Jl. Prof. Dr. Moh. Yamin No. 21
Palu 94111 - Sulawesi Tengah
Telp. (0451) 457623
Call Center: 08114555989 ; 08114555990
Instagram: @ bbtn_lorelindu
e-mail : tnlorelindu@gmail.com

45. TAKA BONERATE NATIONAL PARK

Geographic Location and Size

Taka Bonerate Marine National Park is located between $6°22'$-$7°04'$ South Latitude and $120°55'$-$121°00'$ East Longitude. It consists of 21 islands and surrounding seas that lie in the Flores Sea, approximately 127-332 kms south of Benteng, the capital city of Selayar Regency, South Sulawesi Province. Administratively, it falls under two istricts of Selayar Regency and Passi Maranu District (7 island clusters) in the south, and Passi Masunggu District (14 island clusters) in the north.

This National Park covers 530,765ha consisting of several small islands, sand, seaweed plains and coral reefs that are rich in marine ecosystem biodiversity.

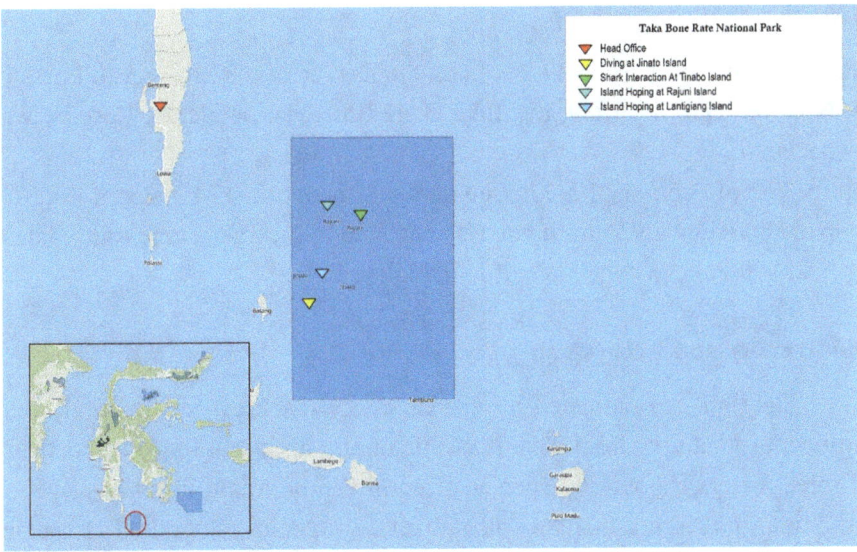

Climate and Topography

The climate is influenced by the western wind, which is the wet season from January-March and the eastern, drier season from July-September and in between are the transition seasons (April-June and October-December). The temperature varies between 28^0-32^0C.

The topography consists of clusters of coral islands and shallow lagoons, grouped into the Macan Islands (14 islands) and the Passi Tallu islands (7 islands). At the lowest ebb, most will show dry land punctuated by small pools. There is also a coral atoll, which is the largest in Indonesia (220,000 ha) and the third largest in the world. At the atoll there are reef edges, swathes of coral, micro atolls and barrier reefs, with a flat coral reef area of 500 km^2. Strong ocean currents are found in the northern part of the Taka Bone Rate Islands.

The topography of the region is unique and interesting. An Atoll that consists of a cluster of reef islands and spacious flat reefs and is underwater, will eventually form a considerable amount of island. In between the islands and reefs are narrow straits that are deep and steep-sided, while on the surface of the reef there are many small deep pools that are surrounded by corals. The outside of the atoll is surrounded by blue water, which is over 1,500 meters deep.

History

On the 20th of February 1989, based on the Decree of the Minister of Forestry No. 100/Kpts—II/1989, the Taka Bone Rate Marine Nature Reserve was declared.

On the 26th of February 1992, based on the Decree of the Minister of Forestry No. 280/Kpts-II/92, the status was changed to the Taka Bone Rate Marine National Park, with an area of 530,765 ha.

Biodiversity and Ecosystems

The park is dominated by marine ecosystems, especially coral reefs, with depths that vary from what is visible on the surface at low tide to a depth of three meters. About 167 species of coral can be found in relatively good condition. In the park there are also about 200 species of corals, 121 species of *Gastropods*, 78 species of *Bivalves* and one kind of tusk shell, *Scaphopoda*. The outer edges of the reef are dominated by *Acropora* spp. of coral. In more sheltered parts, antelope horn corals and rows of leaf corals such as *Montipora foliosa* can dominate the area from the top of the reef to the reef slope. The reef lagoons have the same composition as the outer reefs, but colonies of the of the branched and leafed coral species reach greater proportions. The marine flora includes algae and seaweed, which are often dominated by *Thalassadendron ciliatum*, *Enhalus* sp., *Halimeda* sp. and *Sargasum* sp. On land the flora is dominated by the Coconut, *Cocos nucifera*, sea pandan, *Pandanus* sp., sea pine, *Casuarina equisetifolia*, Ketapang, *Terminalia catappa*, and sea waru, *Hibiscus tiliaceus*.

The marine mammals include hawksbill turtle, *Eretmochelys imbricata*, green turtle, *Chelonia mydas*, leather back turtle, *Dermochelys coriaceae* and loggerhead turtle, *Caretta caretta*.

Local Communities and Culture

Based on stories obtained from the current community leaders, people living in this region from the Bajau and Bugis tribes are descendants of sailing ships' crews and fishermen who fled from pirates as well as the fierce waves of the Banda Sea during the western wind season.

There are about 6,000 local residents spread over six of the 21 islands, Rajuni Island, Tarupa Island, Latondu Island, Jinato Island, Central Passitalu

Island and East Passitalu Island. Most belong to the Bajau (50%) and Bugis (45%) tribes. The rest are from Buton, Muna and Palue in East Nusa Tenggara. In general, they work as fishermen utilizing the coral reef resources, such as sea cucumbers, reef fish, lobsters and shellfish using traditional fishing methods.

Tourism

There are abundant marine tourism opportunities due to the wealth of beautiful coral reefs and the largest atoll in Indonesia. Common marine activities include diving, snorkeling, swimming, sailing, fishing, sun-bathing, and enjoying the unique beach views, or camping on the uninhabited islands.

There is a Research Site, for studying the marine ecosystems and coral reef species, and the way of life and consumption patterns of the people living in the surroundings of the park.

Access and Transportation

Taka Bone Rate National Park can be reached as follows:
- From Makassar (Capital of South Sulawesi) to Bira in Bulukumba Regency (153km) by car – duration 4 hrs.
- From Bira to Pamatata by ferry – duration 2.5 hrs; continue to Benteng Selayar duration 1.5 hr, then continue to the park by motorboat (rental) duration 7hrs. The islands within the park can be reached by motorboat – duration 1-3 hrs.
- The route Makassar – Bulukumba – Benteng Selayar has a routine fixed schedule by bus and boat.
- From Bira straight to the park by rental motorboat – duration 10 hrs.

Facilities

The facilities present in the park's surroundings include motorboat rental, work Cabins and Watch Post. But people who want to visit the park usually stay in the city of Makassar, where hundreds of hotels are available from 5 star down to the bed and breakfast. There are also many hotels in the small cities of Bilukumba and on Selayar island.

Park Office

Balai Taman Nasional Taka Bonerate
Jl. S. Parman No. 40 Benteng, Selayar 92812
Sulawesi Selatan
Telp. (062-414) 22111
Fax. (062-414) 21565
Call Center: 0811418481
Instagram: @ btn_takabonerate

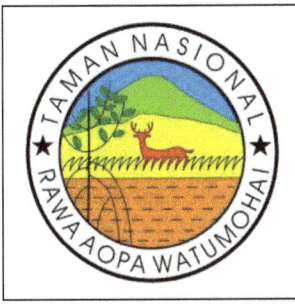

46. RAWA AOPA WATUMOHAI NATIONAL PARK

Geographic Location and Size

Rawa Aopa Watumhoai National Park is located between 4°22'-4°39' South Latitude and 121°44'-122°09' East Longitude. It lies in the most southeastern part of Sulawesi Island in South East Sulawesi Province. It is bordered by the Tirawuta District of Kolaka Regency and the Lambuya and Kendari Regencies in the north, and in the east by the Lambuya Regency and the Tinanggea district of Kendari Regency, in the south by the Tiworo Strait and part of the Rumbia District of Buton Regency and in the west by the Ladongi District of Kolaka Regency. The national park is 70 km south-west of Kendari (Capital of South East Sulawesi Province).

Rawa Aopa Watumohai National Park covers 105,194 ha, divided between Kendari Regency, 46,764 ha, Buton Regency, 45,605 ha and Kolaka Regency, 12,825 ha. Most of the park consists of savannah plains covered with various palm trees and tall grasses.

Climate and Topography

Rainfall is between 1,500-2,000mm/ year and temperatures range between 20^0-33^0C. The rainy season is from December - March and the dry season between July-October. Topography consists of wide expansive beaches, undulating plains to the south, and hilly to mountainous with slopes between 30^0-40^0 in the north. The park has an altitude of between 0-981m asl at the peak of Mount Mendoke.

Most of the plateau region consists of hilly uplands such as the Mokaleleo Mountains in the north and the Matoha highlands (270 m asl) in the East. Mt. Watumohai is 549m above sea level. Swamps cover approximately 16,538ha. There are many rivers including Roraya River, Langko Wala River, Konaweha River, Pohara River, Laea River and Lanowulu River.

History

In 1976 a Decree was released by the Minister of Agriculture, No. 648/Kpts/ Um/1976, regarding the establishment of the Buru Watumohai Tourism Park, with an area of 50,000 ha.

On the 11th of January 1985 a Decree was released by the Minister of Forestry, No. 138/Kpts-II/1985, regarding the establishment of the Rawa Aopa Wildlife Reserve, covering 55,560 ha.

On the 27th of July 1985, based on the Minister of Forestry Decree No. 189/ Kpts-II/1985, the Buru Watumohai Tourism Park was divided into two parts, Buru Dataran Rumbia Park with an area of 55,560 ha and Watumohai Wildlife Reserve, covering 41,244 ha. making the total area 96,804 ha.

On the 1st of April 1989 a decree was released by the Minister of Forestry, No. 444/Kpts-11/1989, at the Pekan Konservasi Alam Ke II in Yogyakarta, that changed the status to National Park with the name Rawa Aopo Watumohai National Park.

On December the 17th 1990, with the completion of the arrangement of the outer boundary of the region in 1986/1987 and through the Minutes of governance, the area boundaries were determined by the Boundary Committee; the Decree of the Minister of Forestry No. 756 / Kpts-II / 90 was released to confirm the status of the Rawa Aopa Watumohai National Park, with an area of 105,194 ha.

Biodiversity and Ecosystems

Rawa Aopa Watumohai National Park contains the following ecosystems.
- Savana, an area of 19,120 ha located in the north on the slopes of Mount Tobonto and Mount Watumohai dominated by weeds, such as the cosmopolitan grass *Imperata cylindrica*, turtle grass, *Androphogon aciculatus*, tiot-tio, *Fimbristylis cylindrica*, rumput teki, *Cyperus rotundus*, Agel, *Arenga pinnata* and bamboo, *Bambussa* spp.
- Lowland tropical rain forest, with a total area of 32,397 ha but scattered in remnants, dominated by the tree species kasumeto, or black ebony, *Diospyros celebica* and Malabar ebony, *Diospyros malabarica*.
- Low mountainous tropical rainforest, covering 32,754 hectares located in the north and west of Rawa Aopa in between Mount Makaleleo and Mount Watumohai. The tree species that are common here include nona, *Metrosideros petiolata*, sisio, *Cratoxylum formosum*, nato, *Palaquium* sp., kayu angin, *Casuarina sp*, palali, *Quercus* sp., bayam, *Intsia bijuga*, saninten, *Castanopsis buruana*, *Ficus* sp., *Macaranga* sp. and damar, or kauri, *Agathis homii*.
- Swamp forest, covers 16,538 ha. It is the only major occurrence of peat bog wetland type in Sulawesi. The plants that are common here include paoti, *Hopea gregaris*, weru, *Planchonia valida*, gelagah, *Saccharum* sp. and Teratai, a water lily, *Nymphaea* sp.
- Mangrove forest, covering 4,385 ha, is located on the south coast and is in a relatively good and intact condition.

In this park there are approximately 9 species of mammals, various kinds of freshwater fish, and several species of reptiles. The mammals include lowland anoa, *Bubalus depressicornis*, babirusa, *Babyrousa babyrussa*, kera hitam, or booted macaque, *Macaca ochreata*, kuskus kecil, or dwarf Sulawesi cuscus, *Strigocuscus celebensis*, tangkasi or spectral tarsier, *Tarsius spectrum*, kerbau, or water buffalo, *Bubalus bubalis*, rusa deer, *Cervus timorensis*, and babi Sulawesi, the Celebes warty pig, *Sus celebensis*. At least 163 bird species have been recorded, including Sulawesi hawk-eagle, Nisaetus lanceolatus, Sulawesi goshark, *Accipiter griseiceps*, dwarf sparrowhawk, *Accipiter nanus*, Sulawesi starling, *Basilornis celebensis*, yellow-crested cockatoo, *Cacatua sulphurea*, sea eagle, *Haliaeetus leucogaster*, Sulawesi scops owl, *Otus manadensis* and wood sandpiper, *Tringa glareola*.

Reptiles include saltwater crocodile, *Crocodylus porosus*, reticulated python, *Python reticulatus*, Asian water monitor, *Varanus salvator*, soa-soa, or sail-finned lizard, *Hydrosaurus amboinensis*, tokek, a gecko, *Gehyra mutilata*, cicak, or common house gecko, *Hemidactylus frenatus* and kadal, a common skink, *Mabuya multifasciata*.

Sulawesi endemic wildlife in the park include Soa-soa, or sail-finned lizard, Sulawesi Eagle, *Spilornis rufipectus*, Anoa, Celebes warty pig, dwarf cuscus, and spectral tarsier.

Endemic plants include Kasumeto, or black ebony, *Diospyros celebica*.

Local Communities and Culture

The Mandumandula river flats is the Bugis community's ceremonial site. There are remainders of the megalithic stone age in the form of a hollow rock shaped like a mortar.

Tourism

In general, the people who visit Rawa Aopa Watumohai National Park come to enjoy beautiful natural views, climb mountains, observe various wildlife species, especially birds, take boats down the rivers and fish in the wide rivers and swamps. The places that can be visited include:

- Lanowulu where there is accommodation, marine tourism sites such as mangrove forest on Lanowulu beach and boating on the river. Saltwater crocodiles can sometimes be seen. There are beautiful sunsets on the savannah plains, and from the watch tower it can be possible to see thousands of deer and wild boar.
- Rawa Aopa where there is a peat swamp with an island located in the middle. There are shelters, and rowing and fishing activities are available on the peat swamp with its clear black water, surrounded by water birds.

- Mount Watumohai is a mountain covered with a low tropical rain forest vegetation with a diverse variety of plant species. There are many species of rattan and liana. There is also beautiful scenery to enjoy. It can be reached along a track of 13 km, which is suitable for two-wheeled drive vehicles.
- The Tinanggea-Kasipute road is a public road constructed across the National Park for easy access. Along the road there are panoramic views of savannahs and in the north mountains and in the south mangroves with various species of animals.

Access and Transportation

Rawa Aopa Watumohai National Park can be reached from Kendari, the Capital of South East Sulawesi Province by vehicle to the entrance at Lanowulu village via several routes:
- Kendari - Lambuya - Rawa Aopa – Lanowulu, a distance of 145km, duration 4 hrs.
- Kendari - Mowila - Rawa Aopa, a distance of 85km, duration 1.5hrs.
- Kendari - Punggaluku - Tinanggea - Lanowulu/Ropaya, a distance of 120km, duration 3 hrs.
- Kendari - Tinanggea by sea, a distance of 200km, using a motorboat, duration 12hrs.

Facilities

The facilities available at the Rawa Aopa Watumohai National Park include
- Lanowulu – Inns and guesthouses, a pier, traditional boats equipped with outboard motors, motorboats and an observation tower.
- Rawa Aopa – shelters and motorboats
- Jalan Poros Tinanggea-Kasipute is a road crossing the savannah (25km), with 4 Observation Towers, and 3 shelters. There is also an office, camping site, and information center.

Park Office

Balai Taman Nasional Rawa Aopa Watumohai
Desa Tatanggai, Kec. Binanggia, Kab. Konawe Selatan
Kendari - Suawesi Tenggara 93721
Tel and Fax: +62-401-3128138
Call Center: 085241916565
Instagram: @ btn_rawaaopawatumohai
Email: btnraw@yahoo.com

47. WAKATOBI ISLANDS NATIONAL PARK

Geographic Location and Size

Wakatobi Islands Marine National Park is located between 5°12'-6°10' South Latitude and 123°20'-123°39' East Longitude. The total size of the park is 1,390,000 ha. It is located in the Tukang Besi Archipelago, Buton Regency, South East Sulawesi Province and is bordered in the North by the Banda Sea (Laut Banda) and Buton Island, in the East by the Banda Sea, in the South by the Flores Sea (Laut Flores) and in the West by Buton Island and the Flores Sea.

The park's name is based on the first 2 letters of 4 islands in the Tukang Besi Archipellago, Wangi-Wangi Island, Kaledupa Island, Tomea Island and Binongko Island. The park is surrounded by coral reefs. Koromaho Island, which is located withn the park, is a major habitat for sea birds.

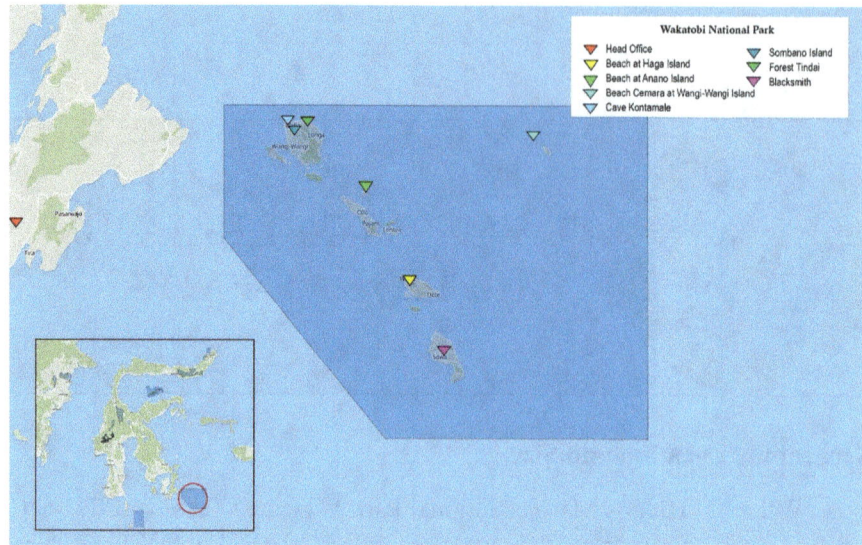

The zoning system of Wakatobi Islands Marine National Park consists of
- Core Zone – 683,500 ha at Karang Tomia and Karang Kapota, Koromaho Island
- Protection Zone – 160,500 ha in the waters of Kaledupa Island
- Utilization Zone – 70,500 ha at Haga and Tolandona Islands
- Traditional Utilization Zone – 300,500 ha
- Rehabilitation Zone – 175,000 ha

Climate and Topography

The temperature ranges between 28.5°-32°C. The best time to visit the park is between April and June because the water is calm or between October and December during dry season.

Coral reef flats cover almost the entire park below high-water mark. The depth varies from what is visible on the surface during low tide to depths of three meters. Various reef types include patch reef, fringing reef, barrier reef and atolls. Most of the Wakatobi Islands Marine National Park consists only of water. The islands are not included in the National Park, which raises serious issues for the management of the park.

History

- In September 1989, the Wakatobi region was proposed as a National Park by the Directorate General of Forest Protection and Nature Conservation (DG PHPA) and the World Wide Fund for Nature (WWF) Indonesia
- On the 7th of March 1994, based on the recommendation of the Governor of South East Sulawesi No. 522.51/2548, the Natural Resources Conservation Sub-Office (BKDSDA), Southeast Sulawesi and the Department of Forestry, Southeast Sulawesi Regional Office, followed up on the recommendation and proposed the Wakatobi region as a Marine Tourism Park.
- On the 4th of September 1995, the Minister of Forestry released Decree No. 462/Kpts-II/1995, which approved the waters of Waikatobi as a Marine Tourism Nature Park of 306,680 ha.
- On the 30th of July 1996, based on the Decree of the Minister of Forestry No. 393/Kpts-VI/1996, the Waikatobi region was declared to be the Waikatobi Islands Marine national Park, covering 1,390,000ha.
- The Waikatobi Islands Marine National Park has been listed as a wetland of international importance, especially for waterbirds, under the Ramsar Convention.

Biodiversity and Ecosystems

More than 167 species of coral have been recorded in the park. The whole park consists of beautiful and unique corals that are still relatively intact, including black coral, *Antipatharia* sp. There are also seagrass meadows, giant clams, *Tridacna* sp., large top shell, *Trochus niloticus*, and pearl oysters, *Pinctada* spp.

The wildlife include dugong, *Dugong dugon*, hawksbill turtle, *Eretmochelys imbricata*, green turtle, *Chelonia mydas*, and many species of fish.

Local Communities and Culture

In general, the people of the Wakatobi Islands are fishermen and farmers, peddlers between islands, merchants or traders and civil servants. Most of the people of these islands are of the Bajo tribe, called sea gypsies, who are nomads of the sea

Tourism

The tourism activities available in the park include snorkeling, diving, swimming, fishing, sun-bathing, sailing and enjoying sea panoramas. Tourists can visit the lighthouse and see a great variety of sea birds on Koromaho Island. Wakatobi is one of the best diving locations in the world, being in the center of the coral triangle, with its rich diversity of corals and fishes.

Access and Transportation

The Waikatobi Marine National Park can be reached by going first to Kendari, the capital city of South East Sulawesi. There are daily flights from Makassar, South Sulawesi. In Kendari, there are so many good 3-to-4-star hotels.
- From Kendari, there are daily commercial flights to Wangi-wangi.
- Tukang Besi within the park can be reached by motorboat from Wangi-wangi, which takes 3-6 hrs.
- Another option is to fly from Kendari by a Twin-engined Otter plane to Baubau on Buton Island, taking 1 hr. Then continue by a 4-wheeled drive vehicle to Pasarwajo, taking 1 hr. From Pasarwajo by motorboat to Wakatobi Islands Marine National Park takes 8 hrs.

Facilities

- On Wangi wangi Island there are many small hotels and resorts.
- On Hoga Island there is a basecamp called "the Wallacea" and Eco Resort Yayasan Zephyr – Sama.
- Other facilities include a Watch Post and Pier.

Park Office

Balai Taman Nasional Kepulauan Wakatobi
Jl. Dayanu Ikhsanuddin No. 71
Bau Bau - Sulawesi Tenggara 93724
Tel and Fax: +62-402-25652
Call Center: 08114057113; 08114002725 ; 081354584541 ; 0813 4380 6005
Instagram: @ btn_wakatobi
Email: tnkw-buton@msn.com
e-mail : tnbabul@tnbabul.org

48. BANTIMURUNG BULUSARAUNG NATIONAL PARK

Geographic Location and Size

This national park is in South Sulawesi Province and is administered by the Maros and Pangkajene Regencies. It is located between 119° 34'-119° 55' East Longitude and between 4° 42'-5° 06' South Latitude. In the north it is bordered by Pangkep, Barru and Bone regencies, in the east by Maros and Bone Regencies, in the South by Maros Regency and in the West by Maros and Pangkep Regencies.

The park covers approximately 43,750 ha. It was created by merging the smaller nature reserves, Bantimurung Nature Reserve, Karaenta Nature Preserve and Bulusaraung Nature Preserve.

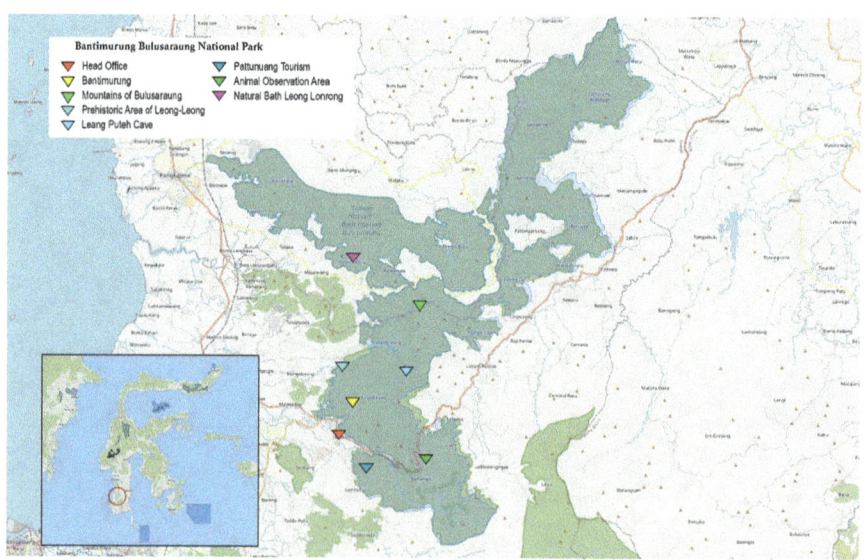

History

On the 18th of October 2004, based on the decree of the Minister of Forestry SK.398/Menhut-II/200, Bantimurung-Bulusaraung National Park was declared. Currently it is managed by the Bantimurung-Bulusaraung National Park Center, located in the Bantimurung Districit, Maros Regency, South Sulawesi.

Biodiversity and Ecosystems

The park is divided into three major ecosystems, forest growing on limestone rocks, better known as karst ecosystems, non-dipterocarp rainforest ecosystems, and lower montane forest ecosystems. The spatial distributions of each are very clear as karst rocks form steep walls with relatively flat peaks. This is very different from the topography of the non-dipterocarp pamah rainforest that has flat to hilly topography, and the montane forest ecosystems, which are characterized by relief that is steep or sometimes undulating.

In the National Park there are two distinct karst formations, one in the area of Maros - Pangkep in the west, and at the northern end in the Mallawa area. Geologists distinguished these two; the first known is the Pangkajene group and the second the Eastern Mountain group.

The geomorphology of the karst in this park is in the shape of karst towers (in some references referred to as The Spectacular Tower Karst). It is the only example found in Indonesia. Other karst areas in general are shaped as cones or are a transitional between karst towers and cones. The karst ecosystems in the Bantimurung Bulusaraung National Park have many caves with stalactites and stalagmites, and other endokarst ornaments.

Six hundred and eighty-three plant species have been recorded in Bantimurung Bulusaraung National Park. Four are protected. They are black ebony, *Diospyros celebica*, the palms, *Livistona chinensis*, and the orchids, *Ascocentrum miniatum* and *Phalaenopsis amboinensis*. Altogether, there are 90 species of orchids. There are also 43 species of fogs in the park.

Species in the low mountain forests include *Litsea* sp., *Agathis philippinensis*, *Ficus* spp., and many others.

The high calcium and magnesium content of the limestone rock which dominates the karst areas limits the species of plants that can survive there. Examples include *Palaquium* sp., *Calophyllum* sp., bandicoot berry, *Leea indica, Polyalthia insignis,* a mangrove, *Pangium edule,* candlenut, *Aleurites moluccana, Cinnamomum* sp., *Litsea ascendens, Eugenia acutangula, Mallotus* spp., *Mangifera* spp., various figs, *Ficus* spp., yellow cheesewood, *Nauclea orientalis,* Kadam, *Anthocephalus cinensis,* and cassia, *Cassia siamea*

The plants in the non-dipterocarpaceae pamah rain forest include black ebony, *Diospyros celebica,* bitti, or Pacific teak, *Vitex cofassus,* nyato, *Palaquium obtusifolium,* cendrana, *Pterocarpus indicus,* beringin, *Ficus* spp., *Sterculia foetida,* dao, *Dracontomelon dao,* aren, a feather palm, *Arenga pinnata, Dillenia serrata,* kemiri, or candlenut, *Alleurites moluccana,* bayur, *Pterospermum celebicum,* mango, *Mangifera* spp., kenanga, *Cananga odoratum, Duabanga moluccana, Eugenia* spp., *Garcinia* spp., Javan plum, *Syzyzigium cumini,* little gooseberry tree, *Buchanania arborescens, Anthocepalus cadamba,* and tamunu tree, *Calophyllum inophyllum.*

Bantimurung Bulusaraung National Park is known worldwide for its butterflies. Alfred Russel Wallace, the co-discoverer with Charles Darwin of the theory of evolution, went so far as to call Bantimurung 'The Kingdom of Butterflies'.

The park has at least 20 species of butterflies that are protected by Government Regulation No. 7/1999. Some unique species found only in South Sulwesi include common birdwing, *Troides helena* Rippon's birdwing, *troides hypolitus,* haliphron birdwing, *Troides haliphron,* swallowtail, *Papilio adamantius,* and violet lacewing, *Cethosia Myrana.*

Unfortunately, butterfly exploitation in the Bantimurung area has been excessive with many species being harvested in the wild and suffering significant population declines. To try to control this situation, in 2005 a recovery project was set up in the form a breeding program in collaboration with the locak community. At least 12 species have been bred in demonstration plots in the Bantimurung Bulusaraung Butterfly Park, which is open to the public.

Bantimurung Bulusaraung National Park is also rich in fauna. There are 43 species that are protected by law and 127 species that are endemic to Sulawesi found in Bantimurung. These include many mammals, birds, reptiles, and frogs.

The largest population of the Dare Sulawesi monkey, or Moor macaque, *Macaca maura,* in South Sulawesi can be found in this national park.

Tourism

Bantimurung-Bulusaraung National Park has many extraordinary attractions, including karst formations, caves with beautiful stalactites and stalagmites, waterfalls and steep gorges, and the best known and most popular of all, the butterflies.

- Bantimurung. Bantimurung has two caves, Gua Batu and Gua Mimpi, both of which are available for tourists to visit. These caves are of great interest of speleologists, due to long tunnels and diverse of stalagmite and stalactite formations. One cave is 500 m long. The Bantimurung Tourism Area is in a strategic location. The Maros-Bone main road goes through it. It is only 42 km from Makassar and 24 km from Sultan Hasanuddin International Airport.
- Leang Leang Prehistoric Tourism Site. Leang-Leang Prehistory Park offers tours of the cultural history of early human occupation. Prehistoric human life can be traced through paintings of human hands and deer on the walls of the cave and a variety of artifacts. The natural panorama here is truly captivating. The cluster of stone cliffs with their distinctive shapes as well as the towering mountains of solid rock are the features of a typical karst landscape. There are various challenging adventure activities that can be undertaken here such as climbing the walls of the steep karst cliffs, caving, both vertically or horizontally, walking along the rocky and clear water rivers, camping to enjoy the outdoors, walking along paths in the forest, hiking through the rocky karst hills, or just enjoying the beautiful scenery. Leang Leang is only 9 km from the Bantimurung tourist area. If starting from Maros, then it is 3 km before the Bantimurung Tourism Complex. Visitor facilities available here include shelters, toilets, walking trails, interpretation boards, and watch posts.
- Karaenta. At Karaenta there is a natural laboratory that offers a wide range of science and knowledge experiences. The Karaenta forest was divided by the road connecting Makassar to the north east cities of South Sulawesi. The abundance of flora and fauna, including moor macaque, *Macaca maura,* as well as the unique karst landscape is interesting to explore. For environmentalists or others with a thirst for the natural sciences, Karaenta is the place to visit. It can be reached by bus or rental car approximately 56km, duration 1-2 hrs, from Makassar.

Access and Transportation

This park is easily accessible from the capital city of South Sulawesi, Makassar. Almost every hour planes land from Jakarta and many other large cities and tourist destinations such Bali. From Makassar, you can rent a car or take a bus from Daya terminal in the north of the city.

Facilities

Makassar is the 4 rth largest city in Indonesia and the largest city port in the eastern Indonesia. Many types of accommodation are available from guest houses to 5 star hotels.

Park Office

Balai Taman Nasional Bantimurung Bulusaraung
Jl. Poros Maros - Bone Km. 42 Bantimurung
Kab. Maros, Sulawesi Selatan
Telp. (+62-411) 3881699, 3880252
Fax. (+62-411) 3880139
Call Center: 085240671664
Instagram: @ btn_bantimurungbulusaraung
e-mail : tnbabul@tnbabul.org

49. TOGEAN ISLAND NATIONAL PARK

Geographic Location and Size

This park is located in between $8^0 54'.33"$ - $8° 32' 35.9"$ S and $119^0.48' 944"$ - $119^0 29' 21"$ E.` Nestled in the middle of the Gulf of Tomini, just south of the Equator, the Togean Islands archipelago is composed of seven principal islands and their satellites on a shallow plateau no deeper than 200 meters. An almost continuous barrier reef protects this plateau. Togean Islands National Park consists of 292,000 ha of sea area, including 132,000 ha of coral reefs, which is the largest in Indonesia, 70,000 ha of land, of which 10,659 hectares are forests and mangroves. It is within the Coral Triangle, an area of extraordinary marine diversity bounded approximately by the Philippines in the north, Indonesia in the west and Papua New Guinea in the east.

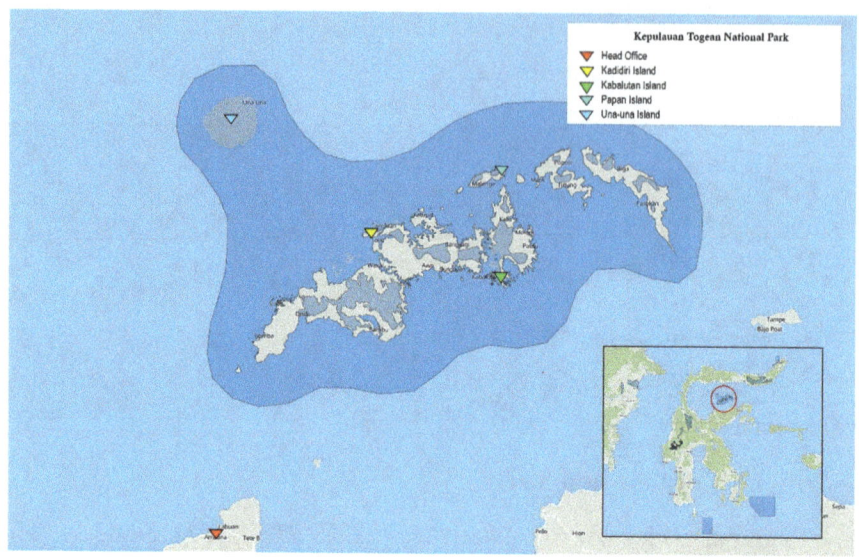

Climate and Topography

Lying 2° south of the equator, the temperature in the Togeans is a more or less constant 30°C all year round. The rainy season runs from December to March. August is the windiest month. All four types of coral formation, patch, fringing, barrier and atoll, can be found here

History

The Minister of Forestry on 19 October 2004, declared Togean Islands a National Park with total area of 362,605 ha. The Togean Islands is recognized by BAPPENAS (National Development Planning Agency) in its Biodiversity Action Plan for Indonesia as one of the priority areas for marine conservation.

Biodiversity and Ecosystems

The islands are covered by rainforest and surrounded by coral reefs, which provide habitat and breeding areas for hawksbill turtle, *Eretmochelys imbricata*, green turtle, *Chelonia mydas,* and dugong, *Dugong dugon*. The Tonkean macaque, *Macaca tonkeana,* is found in the forests. The Togean hawk-owl, *Ninox burhani,* discovered in 1999, is endemic to the islands. The Togian white-eye, *Zosterops somadikartai*, another endemic bird species, was only described in 2008. Many endangered species use the Togean Islands as breeding grounds, including dugongs and hawksbill and green sea turtles. These islands also support one of the last populations of the endangered coconut crab, *Birgus latro*, a giant crab that spends most of its life on land. It is the largest terrestrial arthropod in the world.

Almost 60% of the land area of the Togeans is covered in tropical forest that supports an impressive array of local and Sulawesi endemics. Apart from the Togean macaque, a primate only recently described in 1996 (Froehlich & Supriatna 1996) there is also the Togean lizard, *Varanus salvator togeanus*, babirusa or 'pig deer', *Babyrousa babyrussa togeanus* and the Togean tarsier, *Tarsius togeanus*. A survey by YABSHI (Indonesian Foundation for Advancement of Biological Sciences) in the forest on Malenge island, northern Togeans, found at least 90 bird species.

Conservation International's (CI) Marine RAP survey in 1998, found that biological diversity is high in the Togeans and is comparable to the Calamianes group in the Philippines, Milne bay in PNG, and Komodo Islands in eastern Indonesia. The RAP recorded 262 coral species, 596 reef fish species, 555 species of molluscs belonging to 103 families, 336 gastropods, 211 bivalves, 4 chitons, 2 cephalopods, and 2 scaphopods. Wallace et al. (2001), recorded 91 species of Acropora corals within the Indonesian archipelago, 78 of which are found in Tomini Bay between Togean Island and Central Sulawesi. As well as the endemic *Acropora togianensis*, the survey also recorded 2 species new to science in this park.

Seagrasses are found in channels and passages between the two big islands and in several other areas, often near coral reefs. Dugongs are occasionally reported from these areas, especially in the channels that separate Batudaka and Togean Islands and Talatakoh and Togean Island, although sightings have become rare.

Local Communities and Culture

The Togean islands are inhabited by almost 30,000 people from several ethnic groups: Togeanese, Bajau, Bobongko, Saluanese, Buginese, Gorontalonese, Kaili, Minahasa, Sangir, Chinese, and Javanese. The first four groups are considered to be the natives of these islands. Most of them are farmers (especially of coconut, cloves, and vegetables). The Bajau is the only ethnic group highly dependent on marine resources. Bajau, or Sea Gypsies, adopt a rather secretive, nomadic existence entirely at sea. They live in wooden shacks built on stilts on top of coral reefs. They move from home to home by dugout canoe and exist by subsistence fishing and selling sea cucumbers to the Chinese market. The Bajau hold their breath when diving and use only goggles and spears for hunting.

In general, most community livelihoods show a degree of dependence on natural resources and some ethnic groups still practice traditional natural resources management systems. The Bobongko people, for example, especially in Lembanato village, still adhere to the traditional law of Bayan for managing their sago forest as well as some other rituals connected to the dry farming system (ladang).

Similarly, the Bajau people, known as remarkable fishermen, use only traditional methods for fishing. The adat indigenous institution practiced by the Bajau is called the *bapongka* system

Villagers traditionally use the mangroves for several purposes, including medicine, rituals and firewood.

Tourism

Spread over a 90 km stretch in the middle of Tomini Bay, the winding, hilly coastlines and equatorial waters of the Togean Islands cast a magical spell of green, yellow and blue, in all shades imaginable. Travelers endure a long journey in search of this almost mythical beach paradise and many stay longer than they expected to. They spend lazy days sunbathing, beachcombing, diving and snorkelling, and exploring the dense jungle interiors. The simple lifestyle can be alluring.

Lying in a deep water basin and protected on all sides by the spidery arms of Sulawesi, and miles from anywhere, the calm and clear waters are full of marine life, and the beaches are clean and undisturbed. Diving is possible on many different atolls.

It is recommended to visit from May to December. Accommodation is very limited so book well ahead.

Access and Transportation

For budget-conscious travelers, the best way to the Togeans is via the overnight ferry from Gorontalo, a 10-11 hr ride. The return times from the Togeans are afternoons every Monday, Thursday and Saturday. A daily speed boat service is available but it is a more expensive. Another convenient way to see the Togeans is via a liveaboard boat from Manado.

There are daily flights to Gorontalo from Jakarta. Alternatively, connections to Gorontalo can be made through Manado from several cities in Indonesia and abroad. There are regular flights from Manado to Gorontalo.

An alternative route is to fly to Palu, then from Palu to Luwuk Banggai, a regency in Central Sulawesi. Drive from Luwuk Banggai to Pagimana, which takes 2 hours then take a boat to the Togean islands.

Facilities

On the island of Malenge, there are at least 40 homestays developed by local communities and NGOs. There were also 5 dive resorts of international repute in 2016. The only means of transportation between the islands is by boat. Public boats run between the main islands every day. Aside from public transport, the only option is to charter one of the inexpensive local boats. There are no roads on any of the islands, so to get around on land, it's the old-fashioned mode of transport - walking.

Park Office

Balai Taman Nasional Kepulauan Togean
Jl. Poros Uemalingku, Kel Uentaga Atas
Kec. Ampara Kota 94683, Kab. Tojo Una-Una 94683 - Sulawesi Tengah
Telp./ Fax. : +62-464) 20287
Call Center: 08114500321 ; 081287757169
Instagram: @ btn_kep_togean
Email: togean_tnkt@yahoo.co.id, Website:

50. GANDANG DEWATA NATIONAL PARK

Geographic Location and Size

Gandang Dewata National Park is located in Mamasa, Mamuju Tengah and Mamuju Utara Regencies of West Sulawesi Province. This is a newly established national park declared in 2016. The park consists of one big mountain, Gandang Dewata and several hills surrounding it in the lowlands. The size f this park is 214,201 ha.

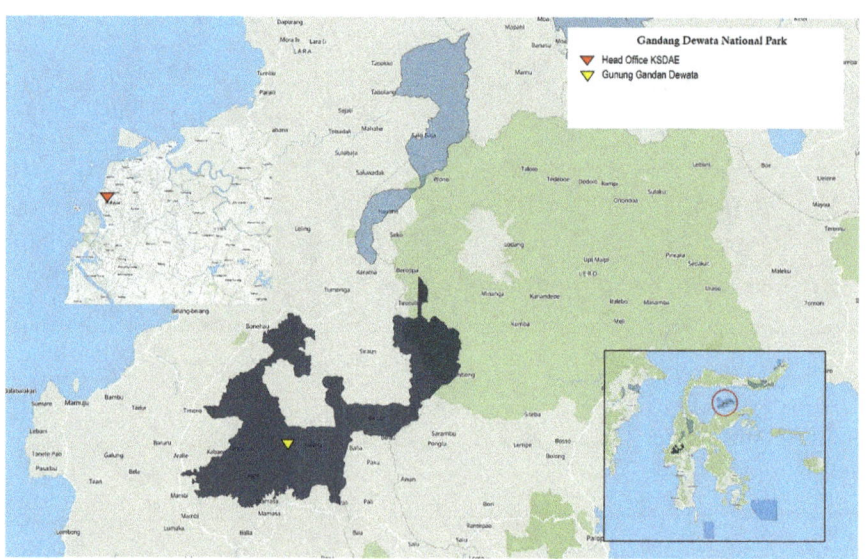

Climate and Topography

Gandang Dewata is in the mountains of the Quarles area. The peak of the mountain is 3,307 m asl. Local relief is so rugged that it can be difficult to hike through the thick forest, deep valleys and steep walls of the karst formations.

History

Gandang Dewata National Park was established in 2016, based on the Ministry of Environment and Forestry decree of October the 3rd 2016, and decree no. SK.773/MENLHK/SETJEN/PLA.2/10/2016. The local government also supported the park and released the following statement.

1. Support the establishment of Gandang Dewata National Park
2. Recognized and honor traditional community rights the area of Pitu Ulluna Salu and Kondosapata Wai Sapalelean and protect the traditional wisdom of people around the park.
3. Support for developing and disseminating knowledge and management practices to communities surrounding the park
4. Support all local governments to legalize traditional community practices by local government regulation.

Biodiversity and Ecosystems

This is a relatively new park, for which there is little information on biodiversity and ecosystems. However, it is certain that it is rich in both. For example, of the 417 species endemic to Sulawesi, 116 were found during a survey of the park by LIPI (Indonesian Institute of Science) in 2016 (LIPI report 2016).. They found many undescribed species of birds, mammals and other taxa. Tri Haryoko of LIPI found 45 species of birds, of which 70% are endemic to Sulawesi in surveys from 500m to 2,000m asl.

New species of mammal were also found including the rat called Tikus akar, *Gracilimus sp*. Tikus ompong, *Ausidentomys vermidas*, Tikus air, *Waiomys mamasae*, and Tikus hidung babi, *Hyorhinomys stuempkei*.

Local People and Culture

The people around the park are mostly from the Torajan tribes, either Mamasa or Mamuju. They practice traditional agriculture and collect forest products for their subsistence. Some communities have converted the forest into ladang, the name for a section of land for planting rice or corn on very steep slopes making them vulnerable to landslides. Many places have experienced landslides. The community of Rante Pongkok, Tobdok Bakaru village of Mamasa Regency, regard the Gandang Dewata Mountain as sacred land.

Tourism

This park is the most remote of all Sulawesi national parks. Visitors coming to this park are mostly adventure tourists come to climb the sacred mountain. There have been few tourists except for those who come to hike and climb. The rivers have great potential for kayaking and white water rafting.

Access and Transportation

From Makassar, there are rental cars and many buses go to Polewali in West Sulawesi Province, Mamasa Regency and Tabulahan in Mamuju Regency. The trip of about 250km will take about 5 hours.

Facilities

There are many small hotels in Mamasa, Mamuju and even Poliwali. There are many hotels of a high standard in Makassar.

Park Office

Balai Besar KSDAE Sulawesi Selatan
Jl. Perintis Kemerdakaan km 13.7
Makasar 90242, Sulawesi Selatan
Tel. +62411-590371, Fax: +62-411-590370
Call Center: 08114600883
Instagram: @ bbksda_sulsel

CHAPTER 6
MALUKU (MOLUCCA)

The Maluku Islands, also called the Moluccas is an archipelago within Indonesia. Tectonically they are located on the Halmahera Plate within the Molucca Sea Collision Zone. Geographically they are located east of Sulawesi, west of New Guinea, and north and east of Timor. The islands are also the historical core of the Spice Islands known to the Chinese and Europeans, but this term usually included other adjacent areas such as Sulawesi. They were known as the Spice Islands due to the nutmeg, mace and cloves that were originally found only there, the presence of which sparked colonial interest from Europe in the 16th century.

Originally occupied mainly by Polynesians, especially the Banda Islands, those people were largely exterminated in the 17th century during the spice wars. A second influx of Austronesian immigrants began in the early twentieth century under the Dutch and continues in the Indonesian era. The Maluku Islands formed a single province from Indonesian independence until 1999, when it was split into two provinces. A new province, North Maluku, incorporates the area between Morotai and Sula, with the arc of islands from Buru and Seram to Wetar remaining within Maluku Province. North Maluku is predominantly Muslim and its capital is Sofifi on Halmahera island. Maluku province has a larger Christian population and its capital is Ambon.

A very good account of the biodiversity and ecology of the islands can be found in The Ecology of Nusa Tenggara and Maluku (Monk et al., 2000). Alfred Russell Wallace (1823-1913), the co-discoverer with Charles Darwin of the Theory of Evolution, lived on Halmahera for some years and travelled extensively throughout eastern Indonesia. His observations on isolation and speciation in these islands led to his ideas on the mechanisms of evolution.

Currently there are two national parks, one in each province:
1. Manusela National Park, of 189,000 ha on Seram island in Maluku province

2. Aketajawe – Lolobata National Park, of 167,300ha on Halmahera Island in North Maluku province.

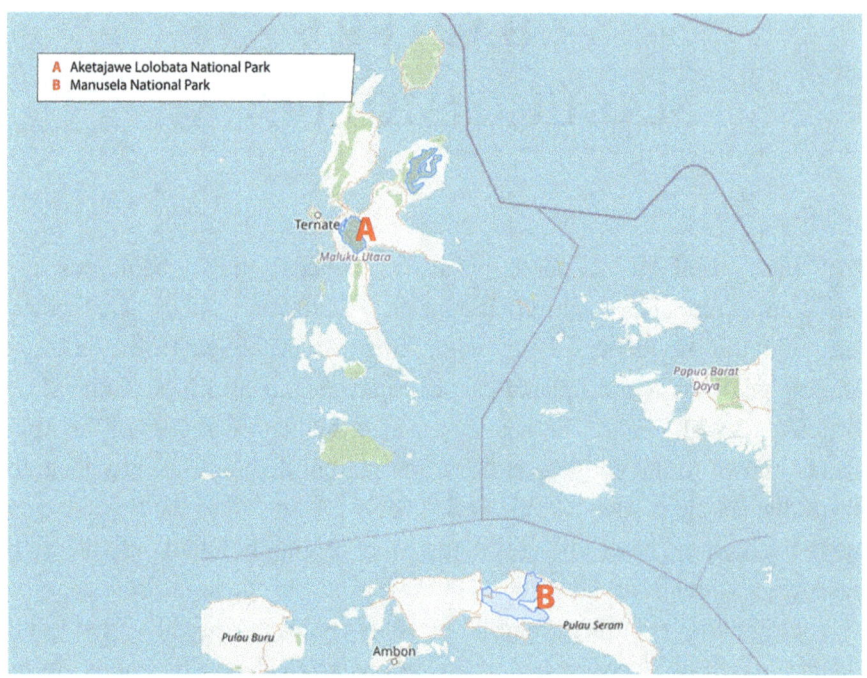

Local people *at* Aketejawe Lolobata National Park (Photo by KLHK) & Manusela National Park (Photo by KLHK)

Pitta maxima ; *Ninox hypogramma* ; *Dicrurus bracteatus* ; *Ceyx lepidus* ; *Hydrosaurus amboinensis* in Aketejawe Lolobata National Park (Photo by Santa Wijaya)

Semioptera wallacii ; *Hydrosaurus amboinensis* in north Molucca & Aketejawe Lolobata National Park (Photo by Santa Wijaya)

Spilocuscus sp. from Halmahera (Photo by Elzabeth Novi Kusumaningrum)

51. MANUSELA NATIONAL PARK

Geographical Location and Size

Manusela National Park is located between $2^0 48'24"-3^0 18'24"$ south latitude and $129^0 09'33"-129^0 46'14"$ East Longitude. It is on the island of Seram, which at 17,500 km² is one of the largest islands in the Malukus. Administratively, it is under the jurisdiction of Wahai District and Tehoru District, Central Maluku Regency, Maluku Province

Manusela National Park covers 189,000 ha, which is 19% of Seram Island

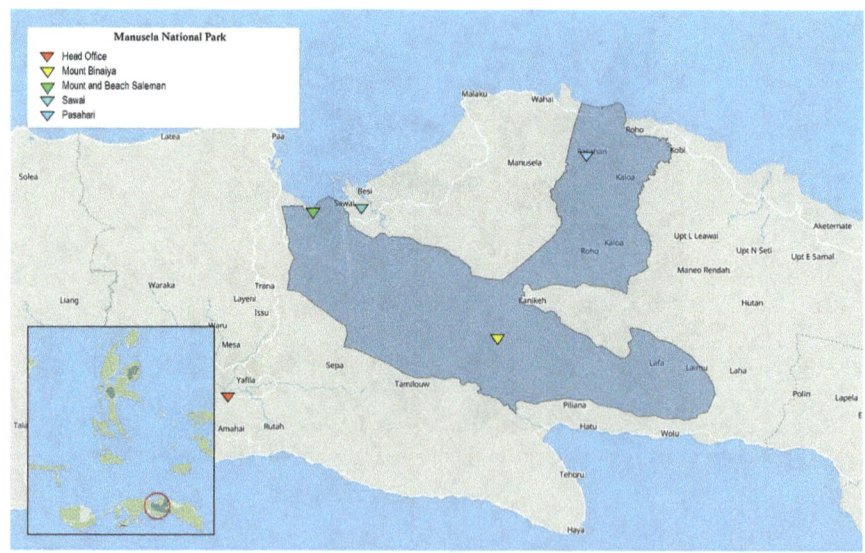

Climate and Topography

The average rainfall ranges between 1,500-2,000 mm/year. In the north, the rainy season is from November to April and in the south from May to October, with a peak in July. The temperature ranges between 25^0-35^0C with a humidity level between 82.9%-93.5%. The distribution of local rainfall is uneven due to the influence of the local topography. This area is affected by the monsoon winds, as follows

- Eastern wind, which occurs from April-September. This wind blows from the southeast and has only a minor influence on the Seram Sea, and therefore does not create high waves.
- Northern wind, which occurs from January-March. This blows from the West and the North, strongly influencing the Seram Sea and creating very high waves.
- The transition season, called Pancaroba, occurs from October-December. The wind during this time comes from the east and southeast, and has no impact on the Seram Sea, which generally remains quiet with no high waves.

Manusela National Park is in a mountainous region with altitude ranging from 0-3,027m above sea level. Slopes of 30-60% are found from Mount Merkele to Mount Pinaia (the highest peak at 3,027 m). There are fast-flowing rivers with steep gorges, such as Wae Nausea, Wae Isal, Wae Kawa, Wae Nua and Wae Talwarang. In the north there is an area consisting of hills and alluvial plains. Along the north coast there are mangroves and coastal forests. On the north coast, there are lowland swamp forests and mangrove swamp forests.

History

On the 7th of December 1972 Decree No. 557/Kpts/U /12/1972 regarding the establishment of the Wae Nua Nature Preserve with an area of 20,000ha, and the Wae Mual Nature Preserve, with an area of 17,500ha, was released by the ministry of Agriculture.

FAO proposed the merger of both these Nature Preserves to form a national park.

In November 1980 decree No. 840/Kpts/Um/11/1980 was issued by the Ministry of Agriculture increasing the size of the Wae Mual Nature Preserve to 35,000ha.

On the 24th of February 1981, based on the Maluku Governor Decree No. 100/G.Mal/1981 and the results of a survey by the Direktorat Bina Program, DirJen Kehutunan, the total area was extended to 189,000ha by including a Protected Forest and a Limited Production Forest, with the recommendation that it should become a National Park.

On the 14th of October 1982, Manusela National Park was declared with an area of 189,000 ha in the Statement of the Minister of Agriculture No. 736/Mentan/X/1982.

On the 23rd of May 1997, decree from the Minister of Forestry No. 281/Kpts-VI/97 confirmed the status of Manusela National Park, and its area of 189,000 ha.

Biodiversity and Ecosystems

The vegetation types in Manusela National Park range from lowland, to low mountain, sub-alpine and alpine. The ecosystems include:

- Mangrove forest, which covers less than 3,000 ha and is found at the back of the slightly higher sandy beach along the north coast. At Tanjung Mual and Wae Isal estuaries, mangroves are well developed. Species include tancang, *Sonneratia alba*, mangrove, *Rhizophora acuminata*, black mangrove, *Bruguiera sexangula*, api-api, *Avicennia* sp. and nipah, *Nypa fruticans*.
- Coastal forest is well developed along the north coast behind the mangroves. Common species include beach morning glory, *Ipomoea pes caprae*, beach spinifex, *Spinifex* sp., Ketapang, or sea almond, *Terminalia catappa*, pandan, *Pandanus tectorius*, cemara laut, or coast she-oak, *Casuarina equisetifolia* and powderpuff mangrove, *Barringtonia racemosa*. On the north coast it is rare to find natural beach vegetation as it has been replaced with cultivated plants such as the coconut, *Cocos nucifera*) and kemiri, or candlenut, *Aleurites moluccana*.
- Swamp forest occurs over small areas, also behind the north coast mangroves. Common species include bangkal, *Nauclea* sp., powderpuff mangrove, *Barringtonia racemosa*, figs, including *Ficus nodosa* and

other species, kayu putih, a paperbark, *Melaleuca leucodendron*, pulai, or blackboard tree, *Alstonia scholaris* and bintangor, *Calophyllum soulattri*.

- River cliff forest is well developed on the steep slopes and cliffs along the Mual and Wae Kawa River valleys, with a dense canopy and undergrowth dominated by ferns and grasses. Common species include pelaka or benuang, *Octomeles sumatrana*, samama, lebau or rainbow gum, *Eucalyptus deglupta*, kasai, *Pometia pinnata*, coast she-oak, *Casuarina equisetifolia*, figs, *Ficus* spp., medang, *Litsea* sp. and laban, *Vitex gofassus*.

- Lowland Tropical Rain Forest is found on most of the plains of the Mual and Wae Kawa valleys at an altitude of 500m asl. The most common species include meranti, *Shorea salanica*, kapur, *Hopea* sp., kayu raja, *Koompassia malaccensis*, kenari, *Canarium indicum*, nyamplung, *Calophyllum inophyllum*, janitri, or quandong, *Elaeocarpus sphaericus*, merbau, *Intsia bijuga* and *Myristica succedanea*. Lowland forest is not dense and the ground is often washed clean of litter by floods during the rainy season.

- Low Mountain Forest is found on Merkele and Kobiponto Mountains between 500 and 1,500 m asl. The most common species include the tropical conifers, *Agathis alba* and *Agathis philippinensis*, cemara gunung, or mountain ru, *Casuarina junghuhniana*, kayu Cina, another conifer, *Podocarpus blumei*, nyamplung, *Calophyllum inophyllum*, Makassar ebony, *Diospyros celebica*, kayu lingua, *Pterocarpus indicus*, leda, *Duabanga moluccana*, lahusa, or manggustan, *Garcinia dulcis* and pinang palm, *Pinanga* sp. Low mountain forests are characterized by a relatively open canopy, with ground vegetation scarce although becoming denser at higher altitudes, and including herbs such as *Impatiens* spp., *Burmannia* spp. and *Dianella* spp.

- Sub-Alpine Mountain Forest is located at 2,400 m asl, and is characterised by short trees covered with moss and orchids, and various species of epiphytic ferns. The ground vegetation includes *Rhododendron* spp. Other plants found here include orchids such as, *Bulbophyllum* spp., *Coloegyne* spp., *Dendrobium* spp. and *Phalaenopsis amboinensis*. Trees include *Dipterocarpus* spp., *Castanopsis* sp., *Melaleuca leucodendron* and *Dacrydium* sp.

Manusela National Park has at least 117 species of birds, 14 of which are endemic. They include Blyth's hornbill, *Aceros plicatus*, salmon-crested cockatoo, *Cacatua moluccensis*, blue-eared lory, *Eos semilarvata*, western crowned pigeon, *Goura cristata*, rainbow bee-eater, *Merops ornatus*, and lazuli kingfisher, *Todiramphus lazuli*.

There is also at least 41 species of reptiles and 6 species of amphibians, including brown tree snake, *Boiga irregularis*, Pacific boa, *Candoia carinata* and sail-finned lizard, *Hydrosaurus amboinensis*. In the adjacent waters and estuaries there is green turtle, *Chelonia mydas*, loggerhead turtle, *Caretta caretta*, leatherback turtle, *Dermochelys coriacea,* hawksbill turtle, *Eretmochelys imbricata*, and saltwater, or estuarine, crocodile, *Crocodylus porosus*. The mammals include fruit bat, *Macroglossus minimus*, Manusela rat, *Melomys fraterculus*, Seram bandicoot, *Rhynchomeles prattorum*, rusa deer, *Cervus timorensis* and spotted cuscus, *Spilocuscus maculatus* (kuskus).

Local Communities and Culture

Parts of Manusela National Park are inhabited by the indigenous people of Manusela and the Nuaulu tribe, who live in small villages in the interior. The survey conducted in 1981 approximately only 19,000 people living in the outer boundaries of the park, with the majority 15,000, living in the narrow coastal strip between the mountains and the Taluti Merkele Gulf. About 4,000 people live in five villages outside the northern boundary, on the coastal plains of Mual. Land use is dominated by subsistence agriculture, with crops such as cloves and coconut plantations.

Tourism

Locations that provide tourism opportunities include
- Many caves found between Marawele and the Manusela.
- Filiana is a good place to see many species of butterfly.
- At Padang Pasahari there are safari tours to see deer and to enjoy the beautiful natural scenery.
- At Wae Isal Plains, there are deer, cassowary birds and various other species of birds.
- Manusela National Park can be explored from north to south by following the path Alakamat/ Wahai - Wasa - Roho –Wasa Mata - Kanikeh - Selumena

The National Parks of Indonesia

- Manusela - Sinahari - Hatumetan – Mosso. This will take 7-8 days. The Kanikeh track to Mount Binaiya takes 2 -3 days, and the track Selumena - Solealama - Wasai takes about 3 days.

The best time to travel to the southern part of the park is December to April. In general, the rivers cannot be accessed by boat, although in certain conditions they can be accessed by using small traditional boats, which can penetrate the interior for a couple of kilometers. The Seram Trans Road is the only land route that connects Masohi Regency with Saleman in the north and Masohi with Tehoru.

Access and Transportation

Manusela National Park can be reached as follows:
- Through the northern route, by ferry, which makes three scheduled trips per week, from Ambon-Sawai-Wahai. This takes 8 hours. Alternatively, Ambon to Masohi by motor boat or ferry takes about 3 hours. Then follow the Trans Seram highway to the north coast then by long boat to Wahai, duration 6 hours. From Wahai walk to the park.
- Through the southern route from Ambon-Tehoru-Saunulu by ferry, which is scheduled 4 times per week. This takes 9 hrs. From Saunulu proceed to Wai Kawa Valley or to Manusela Valley crossing the Merkele Mountains, which is a rough route that will take 2 days on foot.
- By plane to the Pattimura Airport, Ambon. Then continue to the Wahai airstrip.

Facilities

In Ambon city, the Capital of Maluku Province, there are several good hotels and inns. This city has many interesting sights to offer because of the former colonial history.

At Wahai, the Capital of North Seram Regency, there 2 ismall hotels, Losmen Sinar Indah and the Manusela National Park Guesthouse. There is also a landing strip.

At Tehoru there is the Losmen Susi guesthouse.

In Masohi there are several inns and lodgings.

Other facilities provided for visitors include the park office, visitor center, watch post, work cabin, research cabin, observation tower, paths and trails, and diving equipment, snorkeling and other equipments.

Park Office

Balai Taman Nasional Manusela
Jl. Kelang No. 1 Kotak Pos 09
Masohi – Maluku Tengah
Tel: +62-914-22164, Fax: +62-914-21672
Call Center: 08114791000
Instagram: @ btn_manusela
Email: balaitnmanusela@gmail.com

52. AKETAJAWE-LOLOBATA NATIONAL PARK

Geographical Location and Size

Aketajawe Lolobata National Park is located between 128°12'-129°40' East Longitude and 1°27'-00°58' South Latitude. It covers an area of 167,300 ha on Halmahera Island in North Maluku Province, in the East Halmahera Regency.

Climate and Topography

The average annual rainfall is between 1,500-2,000 mm/year. In the north, the rainy season is from November to April. In the south the rainy season is from May to October, with its peak in the month of July. The temperature ranges between 25⁰-35⁰C with a humidity level of between 83% - 93.5%. The rain fall follows the season and is greatly influenced by the central mountains.

History

In 1995 the Aketajawe and Lolobata region was recommended to be given the status of a National Park, and in 2004 as per the Decree of the Minister of Forestry No. 397/Menhut-II/2004 dated 18th October 2004, this national park was created from the protected forest area, limited production forest and permanent production forest located in the Aketajawe Protected Forest Group and Lolobata Forest Group in East Halmahera.

The national park was created based on the recommendation of local government heads namely the Bupati of East Halmahera, Bupati of Central Halmahera, Mayor of the Tidore Islands and the Governor of North Maluku.

Biodiversity and Ecosystems

This area is known to the world because Alfred Wallace, the co-discoverer with Charles Darwin of the theory of evolution, travelled extensively in the islands of eastern Indonesia and lived for some years on Ternate, a small island adjacent to Halmahera. His collection of many endemic birds from these islands is stored in the British Natural History Museum.

The fauna present in the national park is estimated to consist of 51 mammal species, 243 bird species, 42 species of reptiles, and 6 species of amphibians. The endemic mammal found here is the Maluku cuscus, *Phalanger ornatus*. Other species include wild boar, *Sus scrofa*, and deer, *Cervus timorensis*. Bird species endemic to North Maluku include mandar gendang, or Wallace's rail, *Habroptila wallacii*, blue and white kingfisher, *Todiramphus diops*, Halmahera cicadabird, *Coracina parvula* and Halmahera oriole, *Oriolus phaeochromus*. The park is considered by Birdlife International to be very important to the survival of 23 endemic bird species. Reptile species include narrow mouthed toad, *Callulops dubia*, another microhylid endemic to Halmahera, *Cophixalus montanus*, water lizard, *Hydrosaurus warneri*, and a monitor, *Varanus sp.*

Common plant species include kauri, *Agathis sp*, bintangur, *Calophyllum inophyllum*, benuang, *Octomeles sumatrana*, kayu bugis, *Koordersiodendron pinnatum*, matoa, *Pometia pinnata*, merbau, *Intsia bijuga*, kenari, *Canarium mehenbethene gaerta* and nyatoh, *Palaquium obtusifolium*.

The park contains the following ecosystems

- Inland Swamp. The inland swamp consists predominantly of sago palm forest swamp. This ecosystem is very important to the various species of water birds, including the endemic Wallace's rail, or mandar gendang, *Habroptila wallacii*.
- Cultivated Land. Although it is prohibited to cultivate within the national park, in several places there are former plots that were cultivated by communities that used to live there. The presence of these plots has attracted birds usually associated with human settlements, such as finches, *Lonchura spp.*, sunbirds, *Nectarinia spp.*, Willie wagtail, *Rhipidura leucophris* and spotted turtledove, *Streptopelia chinensis* into the national park.
- The river and riparian areas. The river is a water source widely used by both people and wildlife. The Maluku sailfin lizard, *Hydrosaurus amboinensis*, found in the park, depends on the river.
- Grasslands, reeds and ferns. Grasslands or reeds, though small in area, can be found in some locations in the park. Generally, these are lands formerly cultivated. Sometimes they may be overgrown by fern species.
- Forest. Based on altitude, the forests in the national park can be classified, as follows:

Lowland forest (0-750 m asl). Lowland forest is the most common forest in Aketajawe Lolobata National Park and the richest in biodiversity. Burung Indonesia (Birdlife Indonesia) surveys have found that 98% of bird species are found in the lowland forests.

Highland Forest (>750 m asl). The highland forests are generally marked with the presence of carnivorous pitcher plants, *Nepenthes* spp. and moss that covers the tree trunks, especially at altitudes of more than 800m.

Primary forest covers 85.95% of the total park and consists of relatively untouched and undamaged forest. Secondary forest covers 2.66% of the area of the park. It commonly consists of former logging areas or former logging tracks and roads. The secondary forest is usually dominated by tree species such as burflower tree, *Anthocephalus chinensis*, *Duabanga moluccana*, or the invasive *Piper aduncum*. Secondary forest can become the feeding ground for mammals such as deer, *Cervus timorensis* and wild boar, *Sus schrofa*.

Local Communities and Culture

The Tobelo Dalam community consists of 19 groups, spread throughout East Halmahera, Central Halmahera and the Tidore Islands. Of the 19 groups only 2 are directly involved with Akatajawe-Lolobata National Park. The 2 groups are from the Akejira and Woe Sopen, Central Halmahera.

Tourism

The park has a spectacular landscape with an outstanding potential to attract tourists to see its natural scenery and waterfalls, and a variety of endemic birds in their natural habitat. It is the home to the endangered endemic species, the Halmahera Bidadari, or Wallace's standard-wing bird of paradise, *Semioptera wallacei* and the white cockatoo, *Cacatua alba*. The Togutil culture and customs, including their wealth of knowledge in the use of medicinal plants is a potential tourism attraction..

Access and Transportation

The airport is on Ternate Island, which is accessible by air from many cities in Indonesia

To Lolobata: Ternate to Sofifi by speedboat takes 45 minutes or by ferry takes about 2 hours. The National Park office is located in Sofifi, the capital city of North Maluku province. Here, cars to Daru can be rented, which takes 2hrs. From Daru to Poli or Subaim, by sea, takes 1.5 hr.

Aketajawe: Ternate to Bastiong by speedboat, takes 30 minutes. Bastiong to Gita over land, takes 3 hrs. Gita to Akejira/Hijrah river over land takes another 3 hrs.

Facilities

In Ternate, the capital of city of Halmahera Regency, there are many good hotels because this city is the oldest city in Indonesia established by Sultan Ternate 600 years ago. Smaller hotels and Bed and Breakfast types of accommodation are also available.

Park Office

Balai Taman Nasional Aketajawe Lolobata
Jl. Empat puluh, Sofifi Kota, Tidore Kepulauan
Maluku Utara
Call Center: 085342258008
Instagram: @ btn_aketajawelolobata
Email: aketajawe@gmail.com, Website: www.aketajawelolobata.org

CHAPTER 7
PAPUA

Papua and West Papua together cover more or less half of the island of New Guinea. The original province of Irian Jaya was changed to Papua Province in 1999. In 2003 it was split into two new provinces. Papua is still the name of the eastern part, while the new province of West Papua, Papua Barat in Indonesian, occupies the Bird's Head Peninsula, the Bomberai Peninsula and off shore islands in the west. The divide is clear to see in the landscape but not in the people or the distribution of tribes within the landscape. West Papua is smaller than Papua, but it is equally rich in natural resources, such as timber, oil, gas, minerals and biodiversity, both terrestrial and marine.

New Guinea is a tropical island that is the second largest island in the world after Greenland and the highest island in the world. It covers an area of more than 785,750 km^2. Papua and West Papua together cover 416,000 km2. The equator passes along the north coast of the smaller island of Waigeo off the western end of New Guinea, yet there is a glacier at 4,600m asl near the summit of Puncak Jaya. Unfortunately, as a result of global warming, this glacier is retreating and may disappear in the future. The island contains the largest remaining tract of tropical jungle forest left in the world, as well as the highest level of coral reef biodiversity on earth. More than 250 tribes inhabit Papua and West Papua. Each tribe is rich in culture and customs related to its surrounding natural environment. Like the neighbouring country of Papua New Guinea, each tribe also has a distinct separate language.

Papua is home to many species of animals and plants that are mostly not found anywhere else in the world. It contains an abundance of various species of beautifully feathered birds, butterflies, tree kangaroos, possums, bandicoots, orchids, trees of the *Araucaria* and *Rhododendron* genera and others. It is very rich in biodiversity and tropical ecosystems. Because the province also has the largest intact tracts of rain forest in Indonesia, Papua's challenge now is to reconcile the interests of development with conservation.

Papua and West Papua together have half of the total biodiversity of Indonesia with 164 species of mammals, approximately 330 reptiles and amphibians, 650 birds, 250 species of fish and 1,200 freshwater fish, and at least 150,000 species of insects. As a result, Papua makes a major contribution to the very high level of biodiversity in Indonesia, which is one of the highest in the world.

Until recently, Papua was one of the most isolated places in the world. But it has begun to change, attracting the attention of developers and extractive industries. The major threats to the terrestrial and aquatic ecosystems come from forest conversion for transmigration settlements, plantations, logging, construction of dams and roads, and mining, especially oil and gas exploration. Marine ecosystems are threatened by overfishing through the use of bombs and poisons.

Useful books and reports on the biodiversity of Papua include (Supriatna et al., 2000, The Irian Jaya Biodiversity conservation Priority Areas, Ecology of Papua 1 and 2 (Marshall et al., 2007) and Birds of New Guinea (Gregory, 2017).

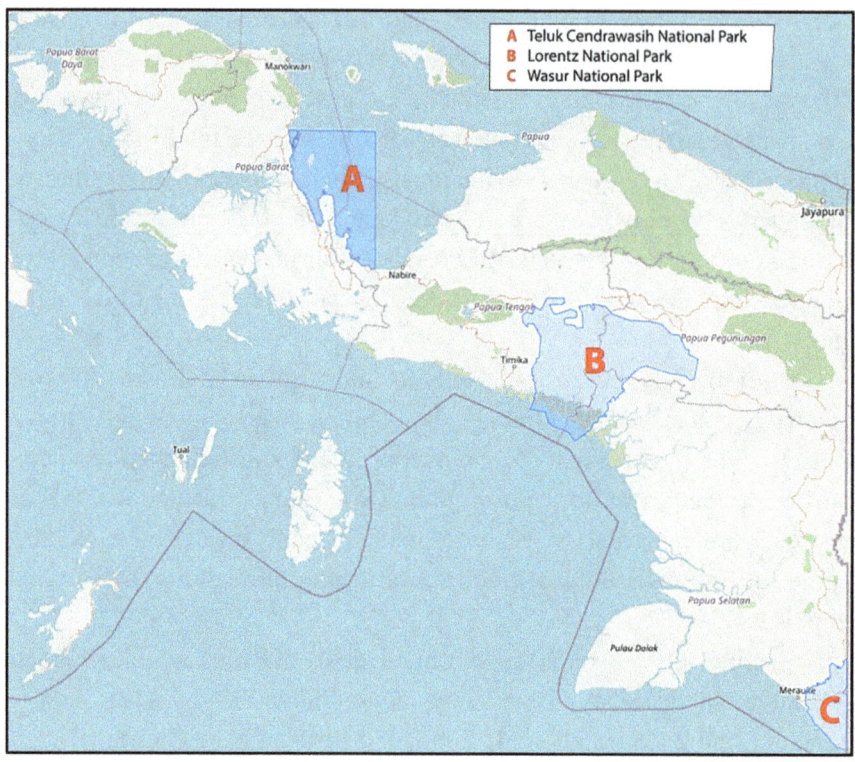

Papua has three National Parks.
- Cendrawasih Bay Marine National Park covering 1,453,500 ha,
- Lorentz National Park covering 2,505,600 ha, and
- Wasur National Park covering 413,810 ha.

Savannah at Wasur National Park & Landscape of Lorentz National Park (Photo by Maikel Simbiak & KLHK)

The National Parks of Indonesia

Casuarius casuarius & *Goura victoria* at Papua (Photo by Bruce Behleer - CI)

Cicinnurus regius; Lophorina niedda at Papua (Photo by Tim Laman – CI)

Paradisaea apoda & *Parotia sefilata* in Papua (Photo By Tim Laman – CI)

Dendrolagus pulcherrimus in Papua & *Spilocuscus rufoniger* in Lorentz National Park (Photos by CI & KLHK) & *Seleucidis melanoleucus* in Papua (Photo By Tim Laman – CI)

Rhincodon typus & View of underwater life at Teluk Cendrawasih National Park (Photos by KLHK)

53. CENDERAWASIH BAY MARINE NATIONAL PARK

Geographical Location and Size

Cenderawasih Bay Marine National Park is located between 0°43'-3°22' south latitude and 134°06'-135°10' east longitude. It is within the large Cendrawasih Bay on the north coast of Papua and includes parts of the Regencies of Manokwari and Paniai, Papua Barat Province.

The 1,453,500 ha includes 500 km of the coastline of the mountainous mainland as well as ocean waters, islands and coral reefs. The park has 18 islands comprising the Auri Islands ark that stretches from the east to Kwatisore Cape in the south and on to Rumberpon Island in the north.

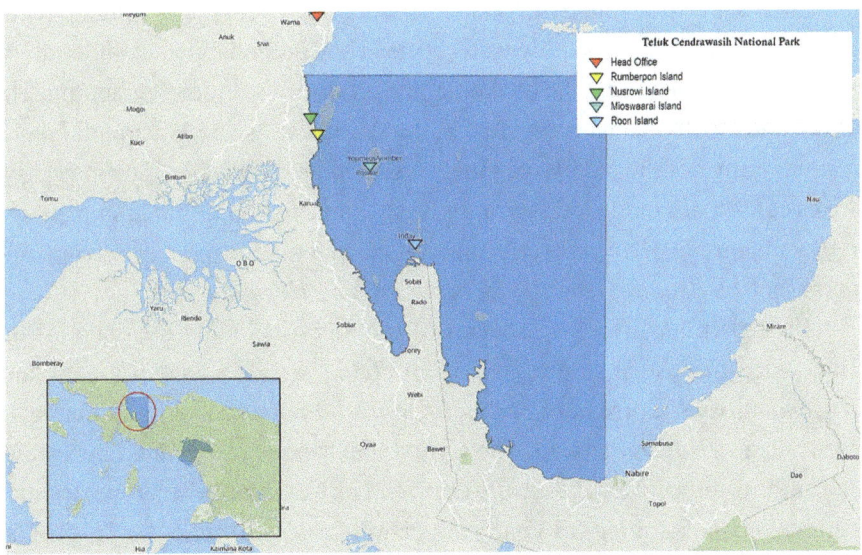

Climate and Topography

The average rainfall is between 1,300-3,700 mm/year. The average temperature is between 21°-33°C with a humidity levels of between 82%-83%.

The topography of the region is varied, ranging from coastal plains to gentle slopes, hills and mountains. The altitudinal range is from 0-1000 m asl. It includes approximately 80,000 ha of coral reefs.

History

On the 3rd of February 1990, based on the Decree of the Minister of Forestry No. 58/Kpts-II/1990, the Cenderawasih Bay Marine Reserve was established, with an area of 1,453,500 ha.

On the 24th of March 1990, based on the Minister of Forestry Statement No. 448/Menhut-VI/1990, the status was changed to the Cenderawasih Bay National Marine Park.

On the 9th of February 1993 the Minister of Forestry issued decree No. 472/Kpts-II/93 to confirm Cenderawasih Bay Marine National Park, covering 1,453,500 ha.

Biodiversity and Ecosystems

Both coastal and marine ecosystems are found in the park. Marine ecosystems include coral reefs, seagrass beds and shallow waters of less than 20 m. There is a wide range of coral reefs including patch reefs, fringing reefs, barrier reefs, atolls and shallow water reefs. They are inhabited by at least 145 species of coral including the hard or stony corals of the order Sceractinia. The coastal and small islands ecosystem at the Auri Islands and Warirundi Island, are commonly dominated by sea pine, *Casuarina equisetifolia*. The mangrove forests consist of mangrove, *Rhizophora* sp., tancang, *Sonneratia* sp., api-api, or grey mangrove, *Avicennia* sp., tingi, or spurred mangrove, *Ceriops* sp., and others such as *Bruquiera* sp., *Xylocarpus* sp. and *Heritiera* sp.

The park has over 195 species of molluscs including members of the Tridacnidae tribe such as the giant clam, *Tridacna gigas*, small giant clam, *Tridacna maxima*, horse hoof clam, *Hippopus hippopus* and Kima Lubang, or crocus clam, *Tridacna crocea*. The Cymatidae tribe is represented by species such as the trumpet triton, *Charonia tritonis*. The Cassidae tribe by species such as kima kepala kambing, or goat head clam, *Cassis cornuta,* the Trocchidae tribe by species such as Lola clam, *Trochus niloticus* and the Trubunidae tribe by species such as lagu rock, *Turbo marmoratus*. There are shrimp, thorny creatures, *hydrozoa*, sponge and sea cucumber. There are approximately 355 species of fish, 209 of which are found in the coral reefs.

Mammals such as the dugong, *Dugong dugon*, dolphins, *Sousa chinensis* and *Delphinus delphinus*, and several species of whales are typical and frequently encountered. Turtles such as hawksbill, *Eretmochelys imbricata*, green turtle, *Chelonia mydas*, olive ridley sea turtle, *Lepidochelys olivacea* and leatherback turtle, *Dermochelys coriacea* are found in the waters of the islands of the Auri Archipelago and Wairundi Island. The estuarine crocodile, *Crocodylus porosus*, is found around Nusambier island. On the land there are echidnas, *Tachyglossus aculeatus,* flying squirrels, lizards and many species of birds including the Junai mas, or Nicobar pigeon, *Caloenas nicobarica*, lesser bird of paradise, *Paradisaea minor*, sea eagle, *Haliaeetus leucogaster*, pied imperial pigeon, *Ducula bicolor,* little egret, *Egretta garzetta*, royal spoonbill, *Platalea regia*, as well as cockatoos, parrots, parakeets and various species of water birds.

Local Communities and Culture

There are more than 70,000 people on the islands and the mainland in and around the national park. They come from the Yaur, Wandamen and Wamesa tribes. In general, they cultivate sago, fish in the coastal areas and hunt animals for their livelihoods.

They still uphold ancestral beliefs, which have been inherited from one generation to the next. In several villages there are rules prohibiting the eating certain kinds of animals, which if not followed will be fatal for the person breaking the rule. These traditions are found in the following villages:

- Aisandami Village (Waropen and Auri tribes) where it is forbidden to eat the hawksbill turtle and the hornbill bird.
- Goni Village, Yaur District (Sadi tribe) where it is forbidden to eat hawksbill turtle and dugongs.
- Yeretnar Village, where it is forbidden to eat crocodile, clams and the bubara fish.

Tourism

- Rumberpon Island has large reed plains populated with deer, *Cervus timorensis* and wild boar, *Sus* sp.
- Mioswar Island has a hot spring approximately 300 m from the beach. It does not contain sulfur but has a high salinity level.
- Kumbur and Nutabari Islands have extensive coral reefs, and burung Junai Mas, or Nicobar pigeon.
- On Pepaya and Anggrameos Islands there are fruit bats, also called flying foxes *Pteropus* sp.
- Numfor Island has historic caves and remnants of settlements, with human skulls, antique plates and carved chests

Access and Transportation

Cenderawasih Marine National Park can be reached through
- Manokwari, the capital of Papua Barat Province, 95 km from the northern boundary of the park by motor boat, schedule 3 times a week and takes 6-10 hours, or by PELNI boat, which takes 18-20 hrs.
- Manokwari to Wasior or Ransiki District by small plane. From Wasior/Ransiki to the islands of the Auri Archipelago by motorboat, which takes 3-4 hrs.
- Nabire, the capital of Paniai Regency, 38 km from the southern boundary of the park, by motorboat takes 2-3 hrs.

Facilities

- The Forest Office, which can provide information about the park, with several facilities including work cabin, checkpoints or Watch Posts, motor boat, and tour guide. Accommodation can be obtained in settlements along the coast by contacting the local village head.
- In Wasior there is a work cabin, checkpoint or Watch Post, motorboat.
- At Windesi, a Watchpost
- At Ransiki, a Work Cabin
- On Rumberpon Island, a Watch Post and Pier.
- On Roon Island, a Watch Post
- At Aisandami Village, a Watch Post.

- At Nabire, a Watch Post and motorboat
- At Kwatisore, a Watch Post

Cottages in the style of the indigenous people's traditional houses on Papaya Island, Anggrameos Island, Roreba Island, Roon Island, Mioswar Island and Rumberpon Island, because they have sources of fresh water and are accessible.
- In Manokwari, the capital city of West Papua Province, there are many good hotels.

Park Office

Kantor Balai Besar Taman Nasional Teluk Cendrawasih
Jalan Esau Sesa Sowi Gunung, Manokwari, Papua Barat,
Po Box: 229, Telp: 062-986-212303, Fax: 062-986-214719
Call Center: 08114860098
Instagram: @ bbtn_telukcenderawasih
email: telukcendrawasih@gmail.com

54. LORENTZ NATIONAL PARK

Geographical Location and Size

Lorentz National Park is located at 4⁰ 75'- 4⁰ 45' S and 137⁰ 83'33"- 137⁰ 49' 60' E. Administratively the region falls under four regencies of Papua Province,
- Fak-Fak Regency in the west.
- Jayawijaya Regency in the east.
- Paniai Regency in the north.
- Merauke Regency in the south-east.

Lorentz National Park covers 2,505,600 ha, and is the largest conservation area in Indonesia and in South East Asia.

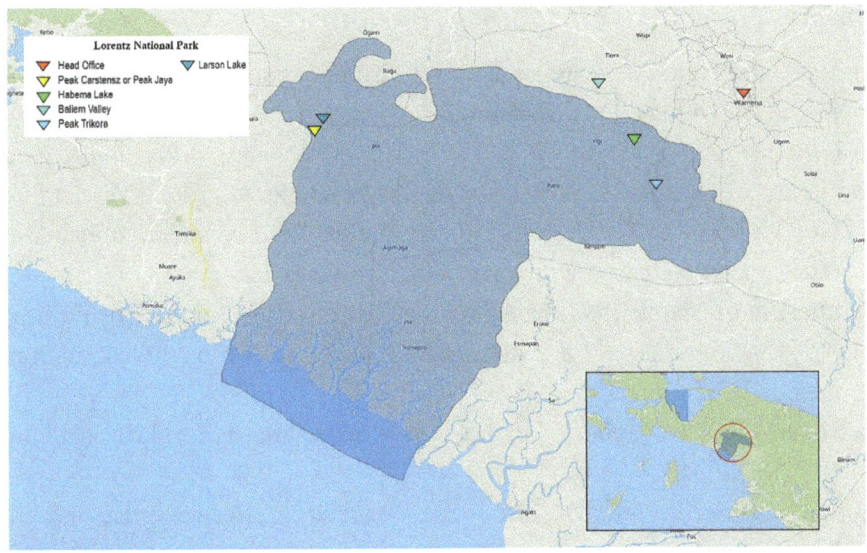

Climate and Topography

The park has variable average rainfall, from less than 2,000 mm to above 6,000 mm. The higher the altitude, the lower the temperature becomes. Between the height of 1,500-1,700 m asl the maximum temperature is 30^0C, while the daily average temperature on the lowlands is 27^0C.

The landscapes of the park vary from the snow cover mountain peaks of Jayawijaya Mountain, the highest in Indonesia, to the lowlands and coastal areas of the Arafura Sea. There are at least 34 different ecosystems over an altitudinal range of 0-5,030 meters above sea level. The topography varies from flat, to undulating to high mountain peaks with steep valleys. Most of the park is covered with tropical forest that has a natural breathtaking beauty, but vegetation types range from beaches with mangroves and nipa palm to sub-alpine and alpine.

In the highest parts of the park there are around 6,900 ha of snow fields, representing one of only three places in the world that have glaciers so close to the equator. The mountains and foothills of the undulating topography are crossed by a large number of streams and rivers that flow towards the southern coastal plain, to then form a muddy meandering river, punctuated by a number of permanent and seasonal lakes and a vast marshland. There are three large lakes, Lake Larson, Discovery Lake and Lake Hoguyugu.

History

In 1909, the Dutch Researcher Dr. H.A. Lorentz was the first foreigner to interacted with the Dani Tribe in his expedition to Mt. Trikora.

In 1916, the region was established as the Lorentz Nature Monument.

On the 25th of January 1978, based on the Decree of the Minister of Agriculture No. 44/Kpts/Um/1/1978, the Lorentz Nature Reserve was established in the Merauke Regency, with an area of 2,150,000 ha.

On the 19th of March 1997, based on the Decree of the Minister of Forestry No. 154/Kpts-II/1997, the status was changed to National Park, with an area of 2,505,600 ha.

In November 1997, Lorentz National Park was declared a World Heritage Site by UNESCO.

Lorentz National Park has also been declared a Biosphere Reserve by UNESCO.

Lorentz National Park has also been inscribed on the ASEAN Heritage Site list.

Biodiversity and Ecosystems

Information on the biodiversity and ecosystems of this park can be found in the book Ecology of Papua (Marshall and Behler, 2007). The ecosystems that occur in the park range from alpine and sub-alpine grasslands and shrublands, high mountain forest, lower mountain forest, lowland tropical forests to mangrove forests and peat swamps and the coastal zone along the Arafura Sea.

Lowland forests are found up to 1,000 m above sea level and can be divided into alluvial lowland rainforest and rainforest hills. The alluvial rainforest is characterized by a high diversity of species, a multi-layered canopy and trees that can reach up to 50 meters, from the genera *Pometia, Ficus, Alstonia* and *Terminalia*. The sub-canopy consists of ferns, vines, epiphytes and buttressed root trees dominated by the genera *Garcinia, Diospyros, Myristica, Maniltoa* and *Microcos*, as well as a large number of palms such as rattan, and orchids.

The hill rainforest is lower and denser than the alluvial forest, although the composition is similar. The taller trees on the hilltops include conifers such as *Araucauria* sp. and *Agathis* sp., while genera of the family Dipterocarpaceae such as *Anisoptera, Vatica*. and *Hopea*. are found in the more sheltered areas.

There are extensive swamps in the lowlands, which include freshwater swamps, tidal swamps, riverine forests, grassy plains, *Pandanus* and sago palm stands and peat swamp forest. These swamps are flooded year-round. The mangrove formation is extensive and includes species of the following genera: *Avicennia, Sonneratia, Rhizophora, Bruguiera, Ceriops, Nypa* and *Xylocarpus*. The low mountain forests are located between 1,000-3,000m asl, and contain a diversity of tree species. *Castanopsis* is a common genus. Clusters of the moss-covered southern beech, *Nothofagus* sp. occur here as do dense coniferous forests dominated by species of the genera *Podocarpus, Dacrycarpus*. and *Papuacedrus*.

Over 3,000 m asl, there is a dramatic change in the vegetation. Ferns such as *Cyathea* spp., savannah and heath formations, and grasslands are commonly found. Heath vegetation includes species of the *Rhododendron, Vaccinium, Coprosma, Rapanea*. and *Saurauia* genera. Above an altitude of 4,000 m asl, the community of low shrubs and *Deschampsia* sp. grasslands form a transition over 200 meters to the alpine zone characterized by short grass, heath and tundra vegetation. The plants that grow on the surface here include the herbs *Ranunculus* sp., *Potentilla* sp., *Gentiana* sp. and *Epilobium* sp., the grasses *Poa* sp. and *Deschampsia* sp., and bryophytes. The highest altitudes are covered by ice and permanent snow.

Characteristic plants of Lorentz National Park include nipah, *Nypa fruticans, Myristica* sp., *Pometia* sp., *Rhizophora apiculata, Terminalia canaliculata, Terminalia catappa, Celthis latifolia, Paraserianthes falcataria, Hopea papuana, Casuarina papuana, Casuarina equisetifolia, Calophyllum inophyllum, Nauclea orientalis, Symplocos cochinchinensis, Rhododendron* spp.,, *Poa nicicola, Ficus* spp., *Podocarpus* spp., *Pinus* spp., and a variety of wild orchids such as *Dendrobium* spp., *Bulbophyllum* spp., *Calanthe tripica*, and *Gramathophylum* spp.

There are at least 350 species of birds, 20 of which are endemic. This represents 63% of all birds found in Papua and Papua Barat. They include the king bird-of-paradise, *Cicinnurus regius*, black bird-of-paradise, *Epimachus fastuosus*, greater bird-of-paradise, *Paradisaea apoda* and Irian eagle, *Harpyopsis novaeguineae*

Approximately 123 of 154 Papuan mammals occur in the park, including the two monotremes (egg laying mammals), short-beaked echidna, *Tachyglossus aculeatus* and long-beaked echidna, *Zaglossus bruijni*. Others include long-nosed antechinus, *Antechinus naso*, spotted cuscus, *Spilocuscus maculatus*, dingiso, a tree-kangaroo, *Dendrolagus mbaiso*, Doria's tree-kangaroo, *Dendrolagus dorianus*, Rummier's mouse, *Coccymys ruemmleri*, long-tailed pygmy possum, *Cercatus cudatus*, brown forest wallaby, *Dorcopsis muelleri*, striped possum, *Dactylopsila trivergata*, freshwater turtle, *Carettochelys insculpta,* New Guinea snake-necked turtle, *Chelodina novaeguineae*, Morea python, *Morelia boeleni*, Papuan crocodile, *Crocodylus novaeguineae*, estuarine crocodile, *Crocodylus porosus* and the Saratoga fish, *Scleropages jardinii*. The coral reef invertebrates include at least 120 gastropods, 78 bivalves and one Scaphopoda. On Santigiang Island, the rocky beach species include oysters, *Crassostrea* spp., rock snails *Drupa* spp., cone snails, *Conus* spp., cowries, *Cypraea* spp. and sea snails, *Turbo* spp., *Trochus* spp. and *Tectus* spp. There are five species of giant clam, Kima lubang, *Tridacna crocea*, small giant clam, *Tridacna maxima*, scaly clam, *Tridacna squamosa*, sand clam, *Hippopus hippopus*, and giant clam, *Tridacna gigas*..

New Guinea endemic species in Lorentz National Park include the tree kangoroo, *Dendrolagus mbasio*, Doria's tree kangoroo, *Dendrolagus dorianus,* dwarf cassowary, *Casuarius benneti*, mountain quail, *Anurophasis monorthonyx* and orange-cheeked honeyeater, *Oreornis chasogenys*.. Endemic plants include *Poa nicicola, Hopea papuana*, and *Casuarina papuana*.

Local Communities and Culture

The high diversity of different cultures contributed to the classification of Lorentz National Park as a World Heritage Area, an ASEAN Heritage Site and a Biosphere Reserve. Approximately 11,000 people lived in and around the park in 2006, consisting of various tribes. The Nduga, Amunge (Damal), Ekagi, Comoros and West Dani tribes live in the mountainous areas, hunting, farming crops and raising pigs. They maintain several cultural activities such as cooking by burying the food under hot rocks and carry nokens, knotted or woven bags handmade from wood fibre soft stems or leaves. The Nakai, Sempan and Asmat tribes live in the southern part of the park on the coastal plains, rivers and swamps. They are semi-nomadic hunter gatherers but carry out simple forms of agriculture. These tribes embrace a traditional leadership

The National Parks of Indonesia

system, which means that every tribe and sub-tribe has a tribal chief, and every village has a chief of war who is respected by society. They practice carving and spend nights on river banks in temporary camps with their rowboats. The mining town of Freeport is located near the northwestern boundary at the foot of the Carstenz Mountains.

Tourism

Lorentz National Park is one of only three places in the world that has tropical glaciers. Most of the park can be reached on foot, by plane or motorboat. Tourism activities include

- The Cartensz area has covered by snow year round. Appropriate activities include mountain climbing, rock climbing and and hiking.
- The Puncak Jaya region can be reached through Ilaga District on foot, taking 6 days, as follows:
  - Day 1, through Pinapa , Bubunana, Liminggame and stay over night in Kama 1 Village
  - Day 2, to Desa Kama 2 on foot for 1 hr, continue to the Bogowigaluk area of sub-alpine forest, on to the Nggoklome area, where there is a lake then follow the hills downwards across grassy plains and stay the night in Kunundo cave.
  - Day 3, cross the Kunundo cave area where there is a natural rock formation, which is an icon of this area. There is also a large cave with a waterfall. stay the night at Mapa Pirom on Bukit Mapa.
  - Day 4, cross the Wenggenawi area of muddy grass plains, and Hugayugu Lake to stay the night in Kumbanikeme.
  - Day 5, continue to Larson Lake across grassy plains and beside Discovery Lake.
  - Day 6, from Larson Lake continue to Dugundugu Lake and Rochiman Lake, then climb the steep New Zealand Pass, down to the lake valley, and stay the night there, near Cartensz Pyramid and Puncak Jaya.
- Mapnduma Village can be reached by plane or on foot for 5 days from Wamena.
- Agimuga District can be reached by plane from Timika or by boat from Mapurajaya, which takes 14 hrs.
- Jila Village can be reached by plane from Wamena or Ilaga, or by motor boat from Mapurajaya and Sawaerma.

Access and Transportation

Lorentz National Park can be reached from Cordillera Mountain to the north of the Arafura Sea (150 km), or from Nabire by plane to Kago in the Illage Valley. Puncak Jayawijaya can also be reached through the PT Freeport mining village.

- Flight Routes: Nabire to Illage, 50 minutes, Nabiri to Beoga, 50 minutes, Nabire to Sugapa, 55 minutes, Nabire to Homeyo 65 minutes, Nabire to Jila 55 minutes, Wamena to Jila 105 minutes, Wamena to Agimura, 110 minutes.
- Outboard motorboat routes: Timika to Agimuga 18 hrs., Jila to Agimuga 8hrs., Agats to Sawaerma 3.5 hrs.
- Rowing boat routes: times from village to village vary, but the average is 1 day depending on the tides.
- On foot: Durations vary, usually more than 1 day depending on the weather and difficulty of the topography.

Facilities

Timika and Nabire are good bases for access to the park. In Timika, there are several good hotels and in Nabire, there are many hotels.

Park Office

Balai Taman Nasional Lorentz
Jl. SD Percobaan Potikelek, Kotak Pos 176,
Wamena 99511, Jayawijaya, Papua
Tel: and Fax: +62-969-34098
Call Center: 082193133616
Instagram: @ btn_lorentz
Email: lorentz.btn@gmail.com

55. WASUR NATIONAL PARK

Geographic Location and Size

Wasur National Park is located between 08°05'-09°07' South Latitude and 140°27'-141°02' East Longitude. It is 2 km the east of Merauke city and is under the jurisdiction of Merauke Regency, Papua Province. In the southeast, the park borders Papua New Guinea. Wasur National Park covers 413,810 ha.

Climate and Topography

The cooler dry season occurs from June to November/December, while the rainy season is from December to May. Average annual rainfall is 2,883 mm with 278 days of rain. The temperature is between 22.2º and 33ºC.

The majority of the park consists of lowlands, and the remainder of undulating plains to an altitude of 90m above sea level.

History

On the 2nd of May 1978, based on the Minister of Agriculture Decree No. 253/Kpts/Um/1/1978, Wasur Wildlife Reserve was established, with an area of 210,000 ha.

Also on the 2nd of May 1978, based on the Minister of Agriculture Decree No. 252/Kpts/Um/1/1978, Rawa Biru Swamp Reserve was established, with an area of 4,000 ha.

On the 24th of March 1990, based on the Minister of Forestry Decree No. 448/Menhut-VI/1990, the status of the region was changed to The Wasur National Park, covering 413,810 ha.

On the 23rd of May 1997, the Minister of Forestry Decree No. 276/Kpts-Vi/1997 was released to confirm the status of the Wasur National Park, covering 413,810.

Biodiversity and Ecosystems

Based on the Ecology of Papua (Marshall and Behler, 2007), information on biodiversity and ecosystem can be summarized as follows. Two-thirds of the park is dominated by savanna vegetation, which has scattered trees over a grassy understory. Other ecosystems found in the park include swamp forest, mangrove forest, monsoon forest, coastal forest, lowland forest, bamboo forest, Melaleuca swamp forest, and fairly extensive sago swamp forest. The hilly areas are dominated by *Eucalyptus* spp. trees. This is the largest and least disturbed wetland habitat in the Asia-Pacific region. Wasur National Park is a stopover site for migratory birds and is included on the Ramsar list, wetland areas designated to be of international importance under the RTamsar convention.

Melaleuca swamp forest, dominated by *Melaleauca leucodendron*, occurs in low lying areas and is reasonably evenly distributed. In the dry season the bases of the trees are overgrown with grass and covered in piles of litter, which shelters wildlife during the day. During the rainy season it is flooded with water approximately 50-100 cm deep. The mangrove forests can be more than 3 kms wide and are dominated by trees and shrubs of *Avicennia* sp., *Sonneratia* sp., *Rhizophora* sp., *Bruguiera* sp., and *Acanthus* sp. plus several ferns such as *Acrostichum spp*. The grass swamp forest is dominated by *Phragmites karka* and teki-tekian, *Cyperus rotundus*. The open water vegetation consists of water lilies, *Nymphaea* sp., lotus, *Lotus* sp., bladderworts, *Utricularia* sp., duckweed, *Lemna* sp. and water fern, *Azolla* sp.

In Wasur National Park there are at least 81 species of mammals, 27 of which are endemic. Prominent mammals include New Guinean quoll, *Dasyurus albopunctatus*, diadem leaf-nosed bat, *Hipposideros diadema*, little bent-wing bat, *Miniopterus australis,* Cape York rat, *Rattus leucopus*, cuscus, *Phalanger intercastellanus*, short-beaked echidna, *Tachyglossus aculeatus* and agile wallaby, *Macropus agilis* and many others.

There are at least 419 bird species, 100 of which are waterbirds, 74 are endemic to Papua and many are seasonal migrants. They include white-necked heron, *Ardea pacifica*, pied egret, *Egretta picata*, Australian white ibis, *Threskiornis molucca*, pied oystercatcher, *Haematopus longirostris*, sandpiper, *Actitis hypoleucos*, little eagle, *Hieraaetus morphnoides*, Papuan eagle, *Harpyopsis novaeguineae*, southern cassowary, *Casuarius casuarius*, brolga, *Antigone rubicunda*, magpie goose, *Anseranas semipalmata*, Australian bustard, Ardeotis australis, noisy pitta, *Pitta versicolor*, spangled kookaburra, *Dacelo tyro*, crimson finch, *Neochmia phaeton*, grey-crowned mannikin, *Lonchura nevermanni* and Fly River grassbird, *Poodytes albolimbatus*. In the dry season, there are thousands of waterbirds, many of them migrants from Australia.

The herpetofauna includes Papuan freshwater crocodile, *Crocodylus novaeguineae*, estuarine crocodile, *Crocodylus porosus*, pig-nosed turtle, *Carettochelys insculpta,* New Guinea snake-necked turtle, *Chelodina novaeguineae*, mangrove monitor, *Varanus indicus*, spotted tree monitor, *Varanus timorensis*, Papuan monitor, *Varanus salvadorii*, frilled lizard, *Chlamydosaurus kingii*, death adder, *Acanthophis antarcticus* and coastal taipan, *Oxyuranus scutellatus*.

Local Communities and Culture

Surrounding the park there are approximately 2,000 people, consisting of the Marind, Kanum, Marori and Yei tribes. The Kanum Tribe is the largest, occupying 5 villages. The tribes are allowed to hunt and use the resources of the park, but only in accordance with their traditions and customs.

Wasur National Park is the source of drinking water for Merauke city and its surroundings.

Tourism

Birdwatching is the major tourism attraction, for example at Rawa Biru Lake and close to the park entrance. Towards the center of the park there are deer, wild boar and wallabies. The region is sometimes known as the Serengeti of Papua, after the east African plains that are so rich in wildlife.

Access and Transportation

Wasur National Park can be reached by regular plane from Jakarta to Jayapura and from Jayapura to Merauke. The airport is about 10 km west of thepark. From Merauke proceed to the villages of Ndalir and Wasur by four wheel drive vehicle, motorcycle, bicycle or even on horseback. To continue visiting Yanggandur, Rawa Biru and Sota, go by motorbike or four wheel drive via the Trans Papua highway. The park can also be reached from Merauke by boat to the beach. There are national shipping company boats, which sail regularly. There are also rowboats or canoes during the rainy season or at high tide.

Facilities

The facilities in Wasur National Park are accommodation, camping sites, hiking routes and watch posts. Accommodation is also available at
- Merauke, where there are several good hotels including close to the airport.
- Yanggandur, where there is a WWF Guesthouse
- Onggaya and Soa, where there are guesthouses.

Park Office

Balai Taman Nasional Wasur
Jl. Garuda Lepro Seri No. 3 Kotak Pos 109
Merauke - Papua/ Irian Jaya 99611
Telp. (+62-971) 322495, 325406, 325408
Fax. (+62-971) 32540
Call Center: 081386804730
Instagram: @ btn_wasur
e-mail   : info_tnwasur@yahoo.com

REFERENCES

Allen, G.R. & M. Adrim 2003. "Coral reef fishes of Indonesia". *Zoological Studies* 42 (1): 1-72.

Ario, A. 2010. *Panduan Lapangan: Kucing kucing Liar Indonesia.* Yayasan Obor, Jakarta, 110 p.

Balai KSDA II Sumatra Utara 2006. *Zonasi Taman Nasional Batang Gadis*, Direktorat Jendral Perlindungan Hutan dan Konservasi Alam, Departemen Kehutanan, Medan.

Bappenas 2003. *Indonesian Biodiversity* Strategy and Action Plan for 2003-2020. Jakarta.

Behleer, B, T.K. Pratt, D.A. Zimmerman 2001. *Burung-burung di Kawasan Papua: Papua, Papua Niugini dan Pulau-pulau Satelitnya.* LIPI-Birldlife International Indonesia Program, Jakarta., 497p.

Bonaccorso ,F.J. "Bats of Papua New Guinea". *Conservation International Tropical Field Guide Series.* CI, Washington,DC., 489 p.

Chandra, Y. 2022. *Biodiversity of Perum Perhutani: Enam Dekade Pengelolaan Keankeragaman Hayati Perum Perhutani di Pulau Jawa.* Perhutani, Jakarta, 172 p.

Chesner, C.A., Westgate, J.A., Rose, W.I., Drake, R., Deino, A. (March 1991). Eruptive History of Earth's Largest Quaternary caldera (Toba, Indonesia) Clarified". Geology 19: 200–203. Bibcode:1991Geo....19..200C. doi:10.1130/0091-7613(1991)019<0200:EHOESL>2.3.CO;2

Coates, B.J. & K.D. Bishop. 2000. *Panduan Lapangan: Burung-burung di Kawasan Wallacea: Sulawesi, Maluku dan Nusa Tenggara.* Birdlife International Indonesia Program and Dove Publications, Bogor, 246 p.

Conservation International Indonesia. 2004. Keanekaragaman jenis Mamalia dan Burung di Kawasan Taman Nasional Batang Gadis. Laporan Tehnik, Northern Sumatra Corridor, CI Jakarta.

Departemen Kehutanan 2004. Naskah Surat Keputusan Menteri Kehutanan no.126/ Mehut-II/2012 tentang Penunjukan Taman Nasional Batang Gadis, di Kabupaten Mandailing Natal, tanggal 29 April 2004, Departemen Kehutanan, Jakarta.

Departemen Kehutanan. 2009. Buku *Statistik Taman Nasional Aketajawe Lolobata. Direktorat Jenderal Perlindungan Hutan dan Konservasi Alam*. Balai Taman Nasional Aketajawe Lolobata. Ternate.

Diamond, J. 1986. "The Design of a Nature Reserve System for Indonesian New Guinea". Dalam: *Conservation Biology: The Science of Scarcity and Diversity* (Soule, M.E., ed.). Sunderland, Massachusetts: Sinauer and Associates, Inc.

Direktorat Bina Kawasan Pelestarian Alam DitJen PHPA Departemen Kehutanan 1994. *12 National Parks: A Field Guide*. Jakarta.

Direktorat Pemanfaatan Jsaa Lingkungan Hutan Konservasi, KSDAE-KLHK 2017. *Pariwisata Alam 54 Taman Nasional Indonesia*. Jakarta.

Dudley, N. & A. Phillips. 1999 *Conservation of Paper Parks to Effective Management Developing a Target*. Washington, D.C.: IUCN/WWF Forest Innovation Project.

Eagles, Paul F.J. 2002. "Trends in Park Tourism: Economics, Finance and Management". In: Journal of Sustainable Tourism Volume 10, Issue 2, p. 133. Doi:10.1080/09669580208667158.

Fennel, D.A. 1999. *Ecotourism: An introduction*. London: Routledge.

Fillon, F.L. A. Jacqueamot, & R. Reid. 1995. *The important of Wildlife to Canadians*. Canadian Wildlife Service, Ottawa, Canada.

Flannery, T. 1995. *Mammals of the South-west Pacific and Moluccan Islands*. REED Books, Australia, 464 p.

FWI/GFW. 2002. *The State of the Forest: Indonesia*. Bogor, Indonesia: Forest Watch Indonesia, and Washington DC: Global Forest Watch.

Gillison, A.N. 2000. Rapid vegetation survey. In: A.N. Gillison (Coord.) Above-ground Biodiversity Assessment Working Group Summary Report 1996-99 Impact of Different Land Uses on Biodiversity. pp. 25-38. Alternatives to Slash and Burn project. ICRAF, Nairobi

Gillison, A.N. 2001. *Vegetation Survey and Habitat Assesment of The Tesso Nilo Forest Complex*.

Gregory, P. 2017. *The Birds of New Guinea: including Bismark Archipelago and Bougainville*. Lync edicions, 464 p.

Harada, K. , A. Muzakir, M. Rahayu & Widada. 2001. *Traditional people and Biodiversity Conservationin Gunung Halimun National Park*. Report on Research and Conservation of Biodiversity in Indonesia. Volume VII, Biodiversity Conservation Project, Pusat Konservasi Alam Departemen Kehutanan, Jica and LIPI, Bogor.

Holmes, D.A. & S. Ash. 1990. *The Birds of Sumatra and Kalimantan*. Oxford University Press, Singapore.

Holmes, D.A. and K. Phillips. 1996. *The Birds of Sulawesi*. Oxford Unipversity Press (Image of Asian series), Singapore.

Honey, M. 1999. *Ecotourism and Sustainable Development: Who own paradise?* Washington, D.C.: Island Press.

Hulme, D. & M.W. Murphree. 2001. *African wildlife and livelihood.* Heinemann, London.

Inger, R.F. and R. Stuebing. 1999. *Panduan Lapangan Katak Katak Borneo.* Natural History Publication, Kota Kinabalu, Sabah, 225 p.

Iskandar, D.T. 1998. *Amfibi Jawa dan Bali.* Puslitbang Biologi-LIPI, Bogor, 117, 26 pictures.

Iskandar, D.T. 2000. *Turtles and Crocodiles of Insular Southeast Asia and New Guinea.* Pal Media Citra, Bandung, 189p.

Jepson, P. & R. Ounstead (eds). 1997. *Birding Indonesia: A Bird-watchers Guide to the World's largest Archipelago.* Periplus Edition (HK) Ltd318 p.

Kottelat, M., & T. Whitten 1996. *Freshwater biodiversity in Asia, with special reference to Fish.* World Bank Technical Paper No. 343. The World Bank, Washington, D.C.

Kottelat, M., Whitten, T., Kartikasari, S.J. & Wiryoatmodjo, S. 1996. *Freshwater fishes of western Indonesia and Sulawesi.* Periplus, Singapore.

Laumonier, Y. 1990. "Search for Phytogeographic Provinces in Sumatra". In: P. Baas, K.Kalkman and R. Geesink (Eds.), The Plant Diversity of Malesia, pp 193-211. Kluwer Academic Publishers, Dordrecht, Netherlands.

Laumonier, Y. 1997. The Vegetation and Physiography of Sumatra. Geobotany 22. Kluwer Academic Publishers, Dordrecht, Netherlands. 222 pp.

LIPI, 2003. Keanekaragaman Hayati Di Tesso Nilo Provinsi Riau.179p.

Leopold, A. 1949. *A Sand county almanac, and Sketch Here and There.* Oxford University Press, New York.

MacKinnon, J. & K. Philipps. 1993. *Field guide to the birds of Sumatra, Borneo, Java and Bali.* Oxford University Press, Oxford.

MacKinnon, J. & M.B. Artha. 1982. *National Conservation Plan for Indonesia.* FAO, Bogor.

MacKinnon, J., K. MacKinnon, G. Child, & J. Thorsell. 1986. *Managing Protected Area in the Tropics.* IUCN, Gland

MacKinnon, J. & K. MacKinnon 1986. *Review of the protected areas system in the Indo-Malayan Realm.* IUCN, Gland.

MacKinnon, J., K. Phillips & B. van Ballen. 1994. *Burung-burung di Sumatra, Jawa, Bali dan Kalimantan.* LIPI-Birdlife International Indonesia, Jakarta, 509 p.

MacKinnon, K, G. Hatta, H. Halim & A. Mangalik. 1996. *The Ecology of Kalimantan.* Periplus Editions, Singapore, 802 p.

Manembu, N.A., 1991. *Suku Sempan, Nakai, Nduga dan Amungme di Kawasan Lorentz.* Laporan PHPA/WWF Project 4521, Jayapura.

McNeely, J.A. 1994. "Protected areas for the 21st century: Working to provide benefits to society". *Biodiversity and Conservation* 3: 390-405.

Mahyar, U.W. and A. Sadili. 2003. *Jenis-jenis anggrek Taman Nasional Gunung Halimun*. Biodiversity Conservation Project LIPI_JICA-PHKA, Bogor, 208 p.

Marshall, A.J. & B. M. Behleer. 2007. *The Ecology of Papua*. Periplus, Singapore, 1467 p. (vol 1 &2).

McQueen, J.B. 1996. *The complete Guide to America's National Parks 1996-1997 edition*. National Park Foundation, Washington DC, 576 p.

Mittermeier, R., Gil, P.R. & Mittermeier, C.G. (eds). 1998. *Megadiversity: Earth's Biologically Wealthiest Nations*. Canada: Cemex Corp.

Monk, K.A., Y. DeFretes, G. Reksodihardjo-Lilley. 2000. *Ekologi Nusa Tenggara dan Maluku*. Prehallindo, Jakarta, 966 p.

Muda, Y.T.D. 2005. "Faktor faktor yang Berpengaruh Terhadap Keputusan Petano dalam Memilih Pola Agroforest "Napu" (Studi di Daerah Penyangga TN Kelimutu, Kabupaten Ende, NTT)". Tesis, Sekolah Pasca Sarjana, Institut Pertanian Bogor, Bogor.

Mulyani, S. 1997. "Pendekatan Sistem Kawasan Konservasi Alam Terpadu untuk Pengembangan daerah Penyangga (Studi Kasus di Taman Nasional Siberut)". Tesis, Sekolah Pasca Sarjana, Institut Pertanian Bogor, Bogor.

O'Brien, B.R. 1999. *Our National Parks and the Search for Sustainability*. Texas Press, Austin, 246 p.

Oppenheimer, Clive. 2002. "Limited global change due to largest known Quaternary eruption, Toba ≈74 kyr BP?", Quaternary Science Reviews 21: 1593–1609, Bibcode:2002QSRv...21.1593O, doi:10.1016/S0277-3791(01)00154-8.

Payne, J. C.M. Francis, K. Phillipps, S.N. Kartikasari .2000. *Panduan lapangan Mamalia di Kalimantan, Sabah, Sarawak dan Brunei Darussalam*. WCS-Sabah Society, Bogor, 386.

Poulsen, Michael K., Frank R. L. & Yusup C. 1999. *Evaluasi Terhadap Usulan Taman Nasional Lolobata dan Ake Tajawe*. BirdLife. Bogor.

Prabandari, F. 2001. "Perancangan Program Pemberdayaan Masyarakat Desa sekitar Taman Nasional Bromo Tengger Berdasarkan kharakteristik Pemanfaatan Hasil Hutan dan Lahan Hutan(Studi Kasus di kawasan penyangga Taman Nasionak Bromo Tengger, Jawa Timur)", Tesis, Sekolah Pasca Sarjana, Institut Pertanian Bogor.

Purnama, S.I.S. 2005. "Penyusunan Zonasi Taman Nasional Manupeu Tanahdaru Sumba berdasarkan Kerentanan Kawasan dan Aktifitas Masyarakat". Tesis, Sekolah Pascasarjana, Institut Pertanian Bogor, Bogor.

Putra, A.D.M., Muda, A.L. & Mahroji. 2021. *Burung-Burung indah Maluku Utara*. Balai Taman Nasional Aketajawe Lolobata, Sofifi, 205 p.

Rampino, Michael R., Self, Stephen. 1992. "Volcanic Winter and Accelerated Glaciation following the Toba Super-eruption" (PDF). Nature 359 (6390): 50–52. Bibcode:1992Natur.359...50R. doi:10.1038/359050a0.

Rampino, M.R & S. Self. 1993. "Climate–Volcanism Feedback and the Toba Eruption of ~74,000 Years ago". Quaternary Research 40: 269–280. Bibcode:1993QuRes..40..269R. doi:10.1006/qres.1993.1081.

Rampino, Michael R., Self, Stephen, 1993. "Bottleneck in the Human Evolution and the Toba Eruption". Science 262 (5142): 1955. Bibcode:1993Sci...262.1955R. doi:10.1126/science.8266085. PMID 8266085

Robson, C. 2011. *Field Guides to the birds of Southeast Asia*. New Holland, Australia, 544 p

Savino, J. & M.D. Jones. 2007. *Supervolcano: The Catastrophic Event That Changed the Course of Human History: Could Yellowstone Be Next*. Career Press. p. 140. ISBN 978-1-56414-953-4.

Sellars, R.W. 1997. *Preserving Nature in the National Parks: A History*. Yale University Press, new Heaven, 379 p.

Siregar, R.E. and N. Winarni. 2014. "Penguatan Efektivitas Bentang Alam Taman nasional Batang gadis Melalui Pendekatan Kolaboratif dan Partisipatif". Laporan Lokakarya Penyusunan Rencana Kerja Kolaborasi Penyelesaian Konflik Tata Batas. SRI dan RCCC Universitas Indonesia, 480 p.

Sofian, D. & V. Hartinie. 2011. *The exoctic journey to Danau Sentarum National Park*. Balai Taman Nasional Danau Sentarum, Sindatng, Kalimantan Barat.

Sriyanto, A., S.Wellesley, D. Suganda, E. Widjanarti and D. Sutaryono. (eds). *Guidebook of 41 National Parks in Indonesia*. Dirjen PHKA, UNESCO and CIFOR, Jakarta, 168 p.

Stattersfield, A.J., M.J. crosby, A.J. Long and D.C. Wege. 1998. *Endemic Bird Areas of the World: Priorities for Biodiversity Conservation*. BirdLife International, Cambridge.

Suhandi, A. & I. Setiawati.1999. *Draft Pertimbangan Pengelolaan Kawasan Pengembangan Ekowisata*. Jakarta: Yayasan Indecon.

Supriatna, J. 1999. The Irian Jaya Biodiversity Conservation Priority-Setting Workshop, Final Report. Washington D.C.: Conservation International.

Supriatna, J. 1996. "Community based Ecotourism: A Lesson learned from Togean Island, Central Sulawesi". Paper presented at the Earthwatch Annual Meeting, Watertown, Massachusett.

Supriatna, J. 1997. "Keanekaragaman Hayati Gunung Salak, Jawa Barat: Permasalahan, Pemanfaatan dan Pelestarian". Dalam: Prosiding Managemen Bioregional Gunung Salak, Gunung Gede-Pangrango dan G. Halimun, Program Studi Biologi, Program Pasacasarjana, Universitas Indonesia, Depok, hal. 27-34.

Supriatna, J. 1998. "Revisited Usaha Pelestarian Keanekaragaman Hayati di Jawa Barat". Makalah dipresentasikan padaSeminar Pengembangan kawasan Terpadu SIJALU (Simpang-Jayanti-Tilu), Bandung, Jawa Barat.

Supriatna, J. 2000. "Konsep Bioregion dalam Pengembangan Kawasan Konservasi Berbasis Keanekaragaman Hayati". Makalah dipresentasikan pada seminar

Pengembangan Wilayah Berbasis Keanekaragaman hayati, UPT Balai Kebun Raya, LIPI, Cibodas, 12 April.

Supriatna, J. 2001. "All sectors of society must work together to save biodiversity". *Nature* 401: 14.

Supriatna, J. 2004. "Keterwakilan Ragam Hayati di Kawasan Lindung, Menurut Unit Biogeografi". Bahan Penataran Pegawai PHKA, Diklat Kehutanan, Serpong.

Supriatna, J. 2005. "Pengelolaan Kawasan Konservasi dengan Pendekatan Ekosistem". Bahan Penataran Petugas PHKA, Pusdik Kehutanan, Serpong.

Supriatna, J. 2006 *Kekayaan Hayati Laut Indonesia Berada di Pusat Segitiga Terumbu Karang Dunia.* Sinar Harapan.

Supriatna, J. dan Sunjaya 2002. "Mengelola Sumber Daya Alam Berbasis Masyarakat (Community Based Natural Resource Management-CBNRM)". Makalah dipresentasikan pada Penataran Wartawan Lingkungan, Yayasan Kehati, Serpong 13 Maret 2000.

Supriatna, J., I.H. Wijayanto, B.O. Manullang, D. M. Anggraeni, Wiratno & S. Ellis. 2002. *The State of Siege for Sumatra's Forest and Protected Areas: Stakeholders View during Devolution, and Political plus Economic Crises in Indonesia.* Proc. IUCN/WCPA-East Asia, pp. 439-456, Taipei, Taiwan.

Supriatna, J., A. Sanjaya, I. Setiawati, M.R. Syachrizal. 2000. "Ekowisata Sebagai Usaha Pemnafaatan yang Berkelanjutan. di Kawasan Lindung". Makalah dalam Workshop Komisi Koordinasi Pemanfaatan Obyek Wisata Alam, Balaikpapan, 6-8 Maret 2000.

Supriatna, J., Y. de Fretes & I.H. Wijayanto. 2000. "The Irian Jaya Biodiversity conservation Priority Areas: Lessons Learned". Dalam: *2nd Southeast Asia Regional Forum, The World Commission on Protected Areas* (Sganty ed), IUCN, Pakse, Laos, Vol. 2: 365-375.

Supriatna, J. 2010. *Melestarikan alam Indonesia.* Yayasan Obor Indonesia

Supriatna, J. 2022. *Field guide to the Primates of Indonesia.* Springer Nature, Switzerland, 233 p.

Suyanto, A. 2001. *Kelelawar di Indonesia.* Pusat Penelitian dan Pengembangan Biologi Lembaga Ilmu Pengetahuan Indonesia (LIPI). Bogor.

Undang-Undang No 5 Tahun 1999. Tentang Konservasi Sumberdaya Alam dan Ekosistemnya. Jakarta.

van Schaik, C.P. & J. Supriatna. 1996. *Leuser: A Sumatran Sanctuary.* Yayasan Bina Sains Hayati Indonesia, Jakarta. 348 p.

Verstappen, H.T. 1973. *A geomorphological Reconnaissance of Sumatra and Adjacent Islands.* (Indonesia). I.T.C. Entschede, Netherlands.

Wells, M., A. Khan, W. Wardojo,, & P. Jepson, P. 1997. "Investing in Biodiversity: A Review of Indonesia's Integrated Conservation and Development Projects". Jakarta: The World Bank – Indonesia Country Program.

Whitten, A.J., S.J. Damanik, J. Anwar & N. Hisyam 1987. *The Ecology of Sumatra.* Gadjah Mada University Press, Jogjakarta, 583 p.

Whitten, A.J. M. Mustafa, G,S. Henderson 1987. *Ekologi Sulawesi.* Gadjah Mada University Press, Jogjakarta, 844 hal.

Whitten, T., J. Whitten, C.G. Mittermeier, J. Supriatna & R. Mittermeier. 1998. *Indonesia.* Dalam: Mittermeier, R.A., Gil, P.R. & Mittermeier, C.G. (eds). Megadiversity: Earth's Biologically Wealthiest Nations. Pp. 75-107. Canada: Cemex Corp

Whitten, T., J. Whitten, R. A. Mittermeier, C.G. Mittermeier, J. Supriatna, and P.P. van Dijk 1999. "Wallacea". Dalam: *Hotspots: Earth Biologically Richest and most endangered Terrestrial Ecoregions.* R.A. Mittermeier, N. Myers and C.G. Mittermeier (eds), CEMEX , Mexico City,. Pp. 297-304.

Whitten, T., J. Whitten, P.P. van Dijk , J. Supriatna, R.A. Mittermeier, C.G. Mittermeier and 1999. "Sundaland". Dalam: *Hotspots: Earth Biologically Richest and most endangered Terrestrial Ecoregions.* R.A. Mittermeier, N. Myers & C.G. Mittermeier (eds), CEMEX, Mexico City, Pp. 279-290.

Whitten, T., R.E. Soeriatmadja, S.A. Affif 1996. *The Ecology of Java and Bali.* Periplus Editions, Singapore.

Whitten, T., M. Mustafa, and G.S. Henderson. 2002. The Ecology of Sulawesi. Persiplus Editions, Singapore. 754p.

Whitten, T., Damanik, S.J. and Anwar, J. 1987. *The ecology of Sumatra.* 2nd edn. Gadjah Mada University Press, Yogyakarta.

Whitten, T., Soeriaatmadja, R.E., and Afiff, S.A. 1996. *The ecology of Java and Bali.* Singapore: Periplus Editions.

Whitten, T. and Whitten, J. 1992. *Wild Indonesia.* New Holland, London.

Whitmore, T.C. (ed). 1984. *Tropical rain forests of the Far East.* 2nd edn. Oxford: Oxford University Press.

Wibisono, H.T., Figel J,J. Arif S.M. Ario,A. And Lubis, A.H. 2009. "Assesing the Sumatran tiger, Panthera tigris sumatrae population in Batang Gadis National Park, a new protected area in Indonesia". *Oryx* 43: 634.

Widianti.D. 2001. *Data base taman nasional di Indonesia.* Konpalindo, Jakarta.

Wiratno. 2019. *The ten (new) ways managing conservation areas in Indonesia.* Direktorat Jendral Konservasi Sumberdaya Alam dan Ekosistem, Kementerian Lingkungan Hidup dan Kehutanan, Jakarta.

Wikramanayake, E.D., E. Dinerstein, C. Loucks, D. Olson, J. Morrison, J. Lamoreux, M. McKnight, and P. Hedao. 2002. *Terrestrial ecoregions of the Indo-Pacific: a conservation assessment.* Island Press: Washington, D.C

World Bank. 2001. *Indonesia: Environment and Natural Resource Management in a time of transition.* Washington, DC.

Wulffaart, S., Tatengkeng, P. and Salo, A. 2006. *The Ecology of Tropical Rainforest in Kayan Mentarang National Park in the Heart of Borneo.* Ministry of

Forestry of the Government of Indonesia and The World Wide Fund for Nature Indonesia, Jakarta.
www.dephut.go.id/ph/kembang_hph/wil.htm.
www.ksdae.menlhk.go.id/publikasi-buku.html

INDEX

A

Aceh, 12, 16, 18
Agile gibbon (*Hylobates agilis*), 24, 38, 44, 49, 54, 62, 74, 82, 211
Agricultural plantations, 49
Air Hitam Laut Village, 54, 55
Aketajawe Lolobata National Park, 329, 331
Alas, 17
Alas Purwo National Park, 152, 154, 156, 158
Alas River, 17
Alas Valley, 13
Anak Dalam, 10, 27, 44, 59
Anak Dalam tribe (Orang Rimba), 59
Angel Reef, 266
Arowana (*Scleropages fomosus*), 82

B

Badui, 121
Bagas Godang, 40
Bajau, 294, 313
Bali, 87, 88, 148, 227
Bali Barat National Park, 171, 172, 174
Bali Tiger (*Panthera tigris balica*), 4, 172
Baluran, 165
Baluran National Park, 89, 164, 165, 167, 169
Bamboo forest, 26, 154, 351
Bantimurung Bulusaraung National Park, 270, 306–308

Batak Karo, 17
Batang Gadis National Park (BGNP), 36
Batang Gadis River, 36, 37, 39
Batanghari, 20, 52, 57
Batanghari River, 21, 27, 43, 55, 58
Bata wood (*Irvingia malayana*), 48
Batin, 49
Bear cat (*Arctictis binturong*), 54
Benu River, 52, 54
Berastagi, 17, 19
Berbak National Park, 52–55, 60
Betang, 190, 201
Betung Kerihun National Park, 178, 180, 198, 201
Binturong (*Arctitis binturong*), 38, 54, 82, 184, 224
Biodiversity, ix, xi, xiii, xix, xxi, 1, 6, 14, 32, 38, 48, 67, 79, 81, 102, 127, 135, 160, 183, 216, 223, 225, 240, 250
Biogeography, xxiii
Bird migration site, 61
Birds migrating, 61, 62, 74
Birdwatching, 353
Black Sumatran langur (*Presbytis sumatrana*), 38
Blekok cina (*Ardeola bacchus*), 82
Bogani Nani Wartabone National Park, 269, 280, 283
Bohorok, 17, 18
Bondol eagle (*Haliastur indus*), 96
Borobudur, 138

Brazil, xx
Bromo Tengger Semeru National Park, 89, 145, 147
Bugis, 54, 104, 206, 277
Bukit Baka Bukit Raya National Park, 178, 188, 189, 191
Bukit Barisan Mountains, 1
Bukit Barisan Selatan National Park, 66, 67
Bukit Duabelas National Park, 57–59
Bukit Hulu Supin, 43
Bukit Tapan Wildllife Reserve, 26
Bukit Tiga Puluh National Park, 42, 43
Bunaken Island, 275, 277, 278
Bunaken Marine National Park, 275
Bunaken National Park, 274, 276
Burung Indonesia, 38
Butterflies, 43, 251, 254, 308

C

Caltex Pacific Indonesia (CPI), 79
Camp Leakey, 206, 207
Cenderawasih Bay Marine National Park, 338
Cibodas Botanical Garden, 111, 114, 116
Clouded leopard (*Neofelis diardi*), 4, 15, 24, 44, 68, 74, 82
Coastal forest, 153, 166, 183, 216, 324, 351
Coconut (*Cocos nucifera*), 96, 127
Columbia, xx, xxi
Conservation International Indonesia (CII), 37
Convention of Biological Diversity (CBD), xiv
Convention on International Trade in Endangered Species (CITES), xiv, 4
Coral reef, 2, 94, 96, 102, 105, 276
Coral Triangle, 304, 311

Costa Rica, xxvi

D

Damar (*Agathis* sp.), 58, 67
Danau Sentarum National Park, 178, 193, 196
Dara-laut kumis (*Chlidonias hybrida*), 82
Dayak, 185, 190, 201, 206, 212, 224
Debt for Nature Swap, xxvii
Deer (*Cervus unicolor*), 38, 62, 184, 211
Dipterocarpaceae family, 8, 14
Dugong (*Dugong dugon*), 174, 241, 276, 304, 312
Durian (*Durio carinatus*), 15, 48, 54, 185, 190, 205, 217, 223, 288

E

Ecosystems, xii, xx, xxii, 2, 14, 32, 38, 43, 53, 67, 81, 94, 102, 127, 140, 153, 166, 183, 199, 205, 216, 232, 240, 259, 281, 307, 317
Ecotourism, 141, 216, 219, 258
Elephant (*Elephas maximus sumatranus*), 5, 15, 24, 49
Elephant population, 49
Emil Salim, 79
Endemic, xxi, 6, 7, 33, 38, 85, 120, 141, 178, 195, 212, 227
Endemic mammals, 3, 4
Ethnic Kluet, 17

F

False gharial (*Tomistoma schlegelii*), 195, 217
Financing, xxvi
Fishing, 55, 64, 94, 97, 104, 128, 155, 191, 196, 206, 242

Fishing cat (*Prionailurus viverrinus*), 4, 15, 24, 68, 103, 167
Flat-headed cat (*Prionailurus planiceps*), 4, 16, 24, 44, 49, 68, 74, 82, 184, 195, 211, 224
Freshwater swamp forest, 2, 15, 32, 53, 64, 73, 102, 183, 205
Funding, xxvi, 216, 258

G

Gandang Dewata National Park, 316, 317
Gayo, 17
Genting Island, 126, 128
Giant clam (*Tridacna gigas*), 96, 128, 174, 276, 339
Golden cat (*Catopuma temminckii*), 38, 62
Golden cat (*Pardofelis temincki*), 4, 15
Green sea turtle (*Chelonia mydas*), 68
Grizzled langur (*Presbytis melalophos*), 24, 38
Gunung Ciremai National Park (TNGC), 130
Gunung Gede, 110
Gunung Gede Pangrango, xvi, 111, 115
Gunung Gede Pangrango National Park (TNGGP), 89, 109, 116
Gunung Halimun Salak National Park, 89, 118
Gunung Halimun Salak National Park (Taman Nasional Gunung Halimun Salak/ TNGHS), 89, 118
Gunung Leuser National Park, 5, 12, 14, 16, 17, 19
Gunung Maras National Park, 84
Gunung Merapi National Park, 89, 134, 140
Gunung Merbabu National Park, 89, 139, 141

Gunung Palung National Park, 178, 182, 184
Gunung Pangrango, 110, 114
Gunung Rinjani National Park, 231, 236
Gurah Tourism Forest, 17, 18

H

Handeuleum Island, 104, 106
Harim (meaning zone), 39
High Mountain Forest, 67, 135, 223, 345
Hiking trails, 137, 220
Honey bear (*Helarctos malayansu*), 38, 43, 58, 68, 190, 200, 217
Human settlements, 39, 49, 331
Hunting, xxvi, 4, 33, 104, 196, 200, 234

I

Ichtyopis glutinosa, 38
Illegal logging, 49, 210
Illegal poaching, 6
Indragiri River, 43
International Union for Conservation of Nature (IUCN), xiv, 240
Islamic boarding schools (pesantren), 40

J

Jalak Bali (*Leucopsar rothchildi*), 172, 174
Jamu, ix, xx, 17
Java, xi, xix, 17, 24, 54, 75, 87, 100, 112, 134, 139, 145, 158, 164
Javan leopard (*Panthera pardus melas*), 103, 113, 132, 136, 141, 161, 167
Javan owl (*Otus angelinae*), 132
Javan rhino (*Rhinoceros sondaicus*), 120
Javan tiger (*Panthera tigris sondaica*), 120, 161
Joko Widodo, 258

K

Kalimantan, xi, xix, xxiii, 6, 178, 184, 188
Kaliurang, 137
Kanchil/mouse deer (*Tragulus javanicus kanchil*), 58, 120, 167, 174
Kapuas River, 186, 194, 198, 201
Karimun Jawa Islands National Park, 125
Kasepuhan communities, 121
Kayan Mentarang National Park, 178, 222
Kayong Malay, 185
Kayu Aro, 26
Kelimutu National Park, 245, 246, 248
Kelompok Hutan Kahayan (Sebangau peat swamp forest), 210
Kemujan Island, 126, 128
Kenduri Seko, 25
Kepulauan Seribu, 95
Kepulauan Seribu Marine National Park, 89, 93, 95, 96
Kerinci Regency, 23, 25
Kerinci Seblat National Park, 20, 22, 25, 27, 28
Kerora village, 242
Kluet Preserve Sanctuary, 18
Kluet River, 17
Komodo Island, 239, 240
Komodo National Park, 227, 238, 239, 241, 242
Kuantan Indragiri River, 43
Kubu (Jungle Men), 44
Kumai Tanjung Harapan, 206
Kutai National Park (KNP), 215

L

Lake Gunung Tujuh, 25, 28
Lake Ranau, 67, 69
Lake Toba, 2, 17
Laut Pasir Tengger, 146, 148
Layar Cape, 105
Leopard cat (*Prionailurus bengalensis sumatrans*), 4, 15, 24, 43
Limestone montane forest, 2
Lio tribe, 247
Local communities, xiv, xv, xxv, 17, 24, 27, 34, 39, 44, 49, 59, 68, 85, 97, 104, 113, 121, 128, 141, 148, 185, 190, 247, 251
Long-tailed macaque (*Macaca fascicularis*), 16, 24, 38, 44, 49, 62, 68, 74, 128, 141, 167, 184, 200
Lore Lindu National Park, 269
Lorentz National Park, 336, 343, 344
Low mountain forest, 23, 73, 135, 140, 223, 307
Lubuk larangan, 37, 39

M

Maluku Islands, 319
Mammalian diversity, xx
Manado Tua Island, 275, 277, 278
Mandailing, 17
Mandailing Natal, 36, 39
Mangrove ecosystems, xix, 94
Mangrove forest, 2, 32, 33, 61, 67, 102, 104, 154, 166, 175, 183, 216, 275, 339
Manupeu-Tanah Daru National Park, 227, 253
Manusela National Park, 319, 322–324, 326, 327
Marbled cat (*Pardofelis marmorota*), 4, 24, 44, 49, 68, 74, 82, 103, 121, 174, 184, 195, 211, 224
Marine fish diversity, xxi
Marine National Park, xix, 89
Mega Rice Project (MRP), 210
Megawati Sukarnoputri, 38
Melayu Cuo, 82

Mentawai gibbon/bilou (*Hylobates klossii*), 6
Mentawai Islands, 2, 6, 30
Mentawai leaf monkey/joja, 6
Mentawai pig-tailed macaque/bokkoi (*Macaca pagensis pagensis*), 6
Merbabu Community Rembug Forum, 141
Meru Betiri, 159, 161
Meru Betiri National Park, 89, 158, 160
Meupeuk pare berkah, 121
Minangkabau, 17, 24
Mitra Kutai, 216
Moist montane forest, 2
Mount Indrapura, 22, 26
Mount Merapi, 136
Mount Palung, 67
Mount Rinjani, 232
Mount Seblat, 26
Mount Sorik Marapi, 40
Mount Tambora National Park, xix
Mount Tompok Tunggal, 67
Moyo Island, 265, 266
Moyo Satonda National Park, 264
Mt. Kerinci, 2
Mt. Satalibo, 239
Muslim Sasak Tribe, 233

N

Naborgo-naborgo, 37
Nandur, 121
Napu, 68, 74
New Guinea, xix, xxi
Nganyaaran, 121
Ngaruwah, 121
Nipah (*Nypa fruticans*), 54, 61
North Sumatra, xvi, 6, 12, 16, 36
Nusa Tenggara, 21, 87, 227
Nyamuk Island, 128

O

Oil palm plantations, 16, 33, 49
"Orang pendek", 25
Orangutans (*Pongo abelii*), 3, 6, 16, 18, 184, 190
Otus mentawi, 7, 34

P

Padang Satwa Ladeh Panjang, 26
Painted terrapin (*Callagur borneoensis*), 7
Pamekahan Village, 68
Panaitan Island, 101, 102, 105
Pancaroba, 126, 323
Papua, 7, 334
Parang Island, 126, 128
Pasak Bumi (*Eurycoma longifolia*), 48, 190
Peacock (*Pavo muticus*), 154, 161
Peat swamp forest, 2, 52, 53, 78, 82, 205, 209, 215
Pere Konde, 247
Pig-tailed langur/simakobu, 6
Pig-tailed macaque (*Macaca nemestrina*), 16, 38
Plant diversity, xxi, 8
Porcupine (*Hystrix brachyura*), 38, 128
Presbytis potenziani potenziani, 6
Presbytis potenziani siberu, 6
Protected area (PA), xxii
Protection forest (hutan lindung), 85
Pua, 196
Pulp wood plantations, 49

R

Rafflesia arnoldi, 8, 14, 23, 24, 68
Rambut Island ("birds heaven"), 96, 97
Rana blythii, 7
Ranidae (true frogs), 7

Rawa Aopa Watumhoai National Park, 296
Rawa Bento, 23
Rawa Ladeh, 23
Reduce Emission from Deforestation and Degradation (REDD), xxvii

S

Sahul, xxiii
Sandy beach forest, 2
Satonda Island, 264, 265
Savanna grasslands, 154
Seagrass beds, xx, 339
Seagrass ecosystems, xx
Sebangau National Park, 209, 212
Seberida Nature Preserve, 43
Segara Anak Lake, 232, 234, 236
Selective logging, xxv
Sembilang National Park, 9, 60, 61
Senyulong crocodile/Sunda gharial (*Tomistoma schlegelii*), 74
Seren Tahun, 121
Siamang (*Sympalangus syndatylus*), 82
Sibayak volcano, 17
Siberia, 61, 62, 74
Siberut Island, 30, 33
Siberut National Park, 30, 32, 33
Sikerei, 34
Silvery leaf monkey (*Trachypithecus cristatus*), 16, 38
Southeast Asia Heritage Site, 22
Spice Islands, 319
Subalpine forest, 2
Sulawesi, 54, 218, 269, 280
Sumatra, xix, xxiii, 1, 3–5, 7, 21, 24, 34, 38, 45, 53, 61, 74
Sumatran coucang (*Nycticebus coucang*), 38, 49
Sumatran ground cuckoo (*Carpococcyx radiceus*), 38

Sumatran rhino (*Dicerorhinus sumatrensis*), 68, 74
Sumatran rhinoceros (*Dicerorhinus sumatrensis*), 5, 10, 15, 24
Sumatran tiger (*Pathera tigris sumatrae*), 4
Sumba, 227
Sumba hornbill (*Aceros everetti*), 250
Sumbawa, 227, 228, 238
Sumbawan, 234, 266
Sun bear (*Helarctos malayanus*), 5, 6
Swamp forest, 2, 15, 16, 24, 32, 43, 52, 53, 61, 73
Syzygium spp., 250

T

Tahija, J., 79
Taka Bonerate Marine National Park, 292
Talang Mamak, 44
Tambora mountain, 258
Tanjung Puting National Park, 178, 203, 205, 212
Tapir (*Tapirus indicus*), 5, 24, 38, 43, 62, 68, 74
Tarsier (*Tarsius bancanus*), 6, 24, 85
Tengger tribe, 148
Tesso Nilo National Park, 47, 50
Tetua Adat, 200
The Bohorok Orangutan Rehabilitation Center, 18
The Buffer Zone, 113, 126
The Core Zone, 126, 225
The Land Zone, 95
The Lau Pengurukan, 18
The Littoral Zone, 94
The Program of Works for Protected Areas (PoWPA), xiv
The Protected Zone, 94
Three horns toad (*Megophrys nasuta*), 38

Tiwu Ata Bupu, 247
Tiwu Ata Polo, 247
Tiwu Nuwa Muri Koofai, 247
Togean Islands National Park, 270, 311
Tongtong heron (*Leptotilos javanicus*), 68
Traditional communities, 26, 40, 49
Traditional medicinal, xx
Tree kangoroo (*Dendrolagus mbasio*), 347

U
Ujung Kulon National Park, 89, 100, 102, 103
Uma, 34, 201
Umbrella leaved palm (*Johannesteijsmannia altifrons*), 54

V
Van Aken, A.Ph., 13

Van Heurn, F.C., 13

W
Wakatobi Islands, 270
Wakatobi Islands Marine National Park, 302, 303
Wallace, A.R., 308, 319
Wasur National Park, 350, 351
Way Kambas National Park (WKNP), 5, 72, 73, 75
White-winged wood duck (*Asarcornis scutulata*), 54
Wijayakusuma (*Pisonia grandis*), 127
Wild goat (*Naemorhedus sumatraensis*), 38
Worldwide Fund for Nature (WWF), 223

Z
Zamrud National Park, 78, 80
Zollinger, H., 259

GPSR Compliance

The European Union's (EU) General Product Safety Regulation (GPSR) is a set of rules that requires consumer products to be safe and our obligations to ensure this.

If you have any concerns about our products, you can contact us on

ProductSafety@springernature.com

In case Publisher is established outside the EU, the EU authorized representative is:

Springer Nature Customer Service Center GmbH
Europaplatz 3
69115 Heidelberg, Germany